A·N·N·U·A·L EDITION·S

Global Issues

Nineteenth Edition

03/04

EDITOR

Robert M. Jackson

California State University, Chico

Robert M. Jackson is a professor of political science and dean of the School of Graduate, International, and Sponsored Programs at California State University, Chico. In addition to teaching, he has published articles on the international political economy, international relations simulations, and political behavior. His special research interest is in the way northern California is becoming increasingly linked to the Pacific Basin. His travels include China, Japan, Hong Kong, Taiwan, Singapore, Malaysia, Portugal, Spain, Morocco, Costa Rica, El Salvador, Honduras, Guatemala, Mexico, Germany, Belgium, the Netherlands, Russia, and Czechoslovakia.

McGraw-Hill/Dushkin

530 Old Whitfield Street, Guilford, Connecticut 06437

Visit us on the Internet
http://www.dushkin.com

UNIT 3
The Global Environment and Natural Resources Utilization

Three articles in this section discuss natural resources and their effects on the world's environment.

UNIT 4
Political Economy

Eight articles present various views on economic and social development in the non-industrial and industrial nations.

The concepts in bold italics are developed in the article. For further expansion, please refer to the Topic Guide and the Index.

UNIT 5
Conflict

Seven articles in this section discuss the basis for world conflict and the current state of order and disorder in the international community.

The concepts in bold italics are developed in the article. For further expansion, please refer to the Topic Guide and the Index.

UNIT 6
Cooperation

Six selections in this section examine patterns of international cooperation and the social structures that support this cooperation.

Unit Overview   156

The concepts in bold italics are developed in the article. For further expansion, please refer to the Topic Guide and the Index.

World Wide Web Sites

The following World Wide Web sites have been carefully researched and selected to support the articles found in this reader. The easiest way to access these selected sites is to go to our DUSHKIN ONLINE support site at *http://www.dushkin.com/online/*.

AE: Global Issues 03/04

The following sites were available at the time of publication. Visit our Web site—we update DUSHKIN ONLINE regularly to reflect any changes.

General Sources

U.S. Information Agency (USIA)
http://usinfo.state.gov
  USIA's home page provides definitions, related documentation, and discussions of topics of concern to students of global issues. The site addresses today's Hot Topics as well as ongoing issues that form the foundation of the field.

World Wide Web Virtual Library: International Affairs Resources
http://www.etown.edu/vl/
  Surf this site and its extensive links to learn about specific countries and regions, to research various think tanks and international organizations, and to study such vital topics as international law, development, the international economy, human rights, and peacekeeping.

UNIT 1: Global Issues in the Twenty-First Century: An Overview

The Henry L. Stimson Center
http://www.stimson.org
  The Stimson Center, a nonpartisan organization, focuses on issues where policy, technology, and politics intersect. Use this site to find varying assessments of U.S. foreign policy in the post–cold war world and to research other topics.

The Heritage Foundation
http://www.heritage.org
  This page offers discussion about and links to many sites having to do with foreign policy and foreign affairs, including news and commentary, policy review, events, and a resource bank.

IISDnet
http://www.iisd.org/default.asp
  The International Institute for Sustainable Development presents information through links to business, sustainable development, and developing ideas. "Linkages" is its multimedia resource for policymakers.

The North-South Institute
http://www.nsi-ins.ca/ensi/index.html
  Searching this site of the North-South Institute, which works to strengthen international development cooperation and enhance gender and social equity, will help you find information and debates on a variety of global issues.

UNIT 2: Population and Food Production

The Hunger Project
http://www.thp.org
  Browse through this nonprofit organization's site, whose goal is the sustainable end to global hunger through leadership at all levels of society. The Hunger Project contends that the persistence of hunger is at the heart of the major security issues threatening our planet.

Penn Library: Resources by Subject
http://www.library.upenn.edu/cgi-bin/res/sr.cgi
  This vast site is rich in links to information about subjects of interest to students of global issues. Its extensive population and demography resources address such concerns as migration, family planning, and health and nutrition in various world regions.

World Health Organization
http://www.who.int
  This home page of the World Health Organization will provide you with links to a wealth of statistical and analytical information about health and the environment in the developing world.

WWW Virtual Library: Demography & Population Studies
http://demography.anu.edu.au/VirtualLibrary/
  A definitive guide to demography and population studies can be found at this site. It contains a multitude of important links to information about global poverty and hunger.

UNIT 3: The Global Environment and Natural Resources Utilization

Friends of the Earth
http://www.foe.co.uk/index.html
  This nonprofit organization pursues a number of campaigns to protect Earth and its living creatures. This site has links to many important environmental sites, covering such broad topics as ozone depletion, soil erosion, and biodiversity.

National Geographic Society
http://www.nationalgeographic.com
  This site provides links to material related to the atmosphere, the oceans, and other environmental topics.

National Oceanic and Atmospheric Administration (NOAA)
http://www.noaa.gov
  Through this home page of NOAA, part of the U.S. Department of Commerce, you can find information about coastal issues, fisheries, climate, and more. The site provides many links to research materials and to other Web resources.

Public Utilities Commission of Ohio (PUCO)
http://www.puc.state.oh.us/consumer/gcc/index.html
  PUCO's site serves as a clearinghouse of information about global climate change. Its links explain the science and chronology of global climate change.

SocioSite: Sociological Subject Areas
http://www.pscw.uva.nl/sociosite/TOPICS/
  This huge site provides many references of interest to those interested in global issues, such as links to information on ecology and the impact of consumerism.

United Nations Environment Programme (UNEP)
http://www.unep.ch
  Consult this home page of UNEP for links to critical topics of concern to students of global issues, including desertification, migratory species, and the impact of trade on the environment.

www.dushkin.com/online/

UNIT 4: Political Economy

Belfer Center for Science and International Affairs (BCSIA)
http://ksgwww.harvard.edu/csia/

BCSIA is the hub of Harvard University's John F. Kennedy School of Government's research, teaching, and training in international affairs related to security, environment, and technology.

Communications for a Sustainable Future
http://csf.colorado.edu

Information on topics in international environmental sustainability is available on this Gopher site. It pays particular attention to the political economics of protecting the environment.

U.S. Agency for International Development
http://www.info.usaid.gov

Broad and overlapping issues such as democracy, population and health, economic growth, and development are covered on this Web site. It provides specific information about different regions and countries.

Virtual Seminar in Global Political Economy/Global Cities & Social Movements
http://csf.colorado.edu/gpe/gpe95b/resources.html

This site of Internet resources is rich in links to subjects of interest in regional environmental studies, covering topics such as sustainable cities, megacities, and urban planning. Links to many international nongovernmental organizations are included.

World Bank
http://www.worldbank.org

News, press releases, summaries of new projects, speeches, publications, and coverage of numerous topics regarding development, countries, and regions are provided at this World Bank site. It also contains links to other important global financial organizations.

UNIT 5: Conflict

DefenseLINK
http://www.defenselink.mil

Learn about security news and research-related publications at this U.S. Department of Defense site. Links to related sites of interest are provided. The information systems BosniaLINK and GulfLINK can also be found here. Use the search function to investigate such issues as land mines.

Federation of American Scientists (FAS)
http://www.fas.org

FAS, a nonprofit policy organization, maintains this site to provide coverage of and links to such topics as global security, peace, and governance in the post–cold war world. It notes a variety of resources of value to students of global issues.

ISN International Relations and Security Network
http://www.isn.ethz.ch

This site, maintained by the Center for Security Studies and Conflict Research, is a clearinghouse for information on international relations and security policy. Topics are listed by category (Traditional Dimensions of Security, New Dimensions of Security, and Related Fields) and by major world region.

The NATO Integrated Data Service (NIDS)
http://www.nato.int/structur/nids/nids.htm

NIDS was created to bring information on security-related matters to within easy reach of the widest possible audience. Check out this Web site to review North Atlantic Treaty Organization documentation of all kinds, to read *NATO Review,* and to explore key issues in the field of European security and transatlantic cooperation.

UNIT 6: Cooperation

American Foreign Service Association
http://www.afsa.org/related.html

The AFSA offers this page of related sites as part of its Web presence. Useful sites include DiploNet, Public Diplomacy, and InterAction. Aso click on Diplomacy and Diplomats and other sites on the sidebar.

Carnegie Endowment for International Peace
http://www.ceip.org

An important goal of this organization is to stimulate discussion and learning among both experts and the public at large on a wide range of international issues. The site provides links to *Foreign Policy,* to the Moscow Center, to descriptions of various programs, and much more.

Commission on Global Governance
http://www.sovereignty.net/p/gov/gganalysis.htm

This site provides access to *The Report of the Commission on Global Governance,* produced by an international group of leaders who want to find ways in which the global community can better manage its affairs.

OECD/FDI Statistics
http://www.oecd.org/statistics/

Explore world trade and investment trends and statistics on this site from the Organization for Economic Cooperation and Development. It provides links to many related topics and addresses the issues on a country-by-country basis.

U.S. Institute of Peace
http://www.usip.org

USIP, which was created by the U.S. Congress to promote peaceful resolution of international conflicts, seeks to educate people and to disseminate information on how to achieve peace. Click on Highlights, Publications, Events, Research Areas, and Library and Links.

UNIT 7: Values and Visions

Human Rights Web
http://www.hrweb.org

The history of the human rights movement, text on seminal figures, landmark legal and political documents, and ideas on how individuals can get involved in helping to protect human rights around the world can be found in this valuable site.

InterAction
http://www.interaction.org

InterAction encourages grassroots action and engages government policymakers on advocacy issues. The organization's Advocacy Committee provides this site to inform people on its initiatives to expand international humanitarian relief, refugee, and development-assistance programs.

We highly recommend that you review our Web site for expanded information and our other product lines. We are continually updating and adding links to our Web site in order to offer you the most usable and useful information that will support and expand the value of your Annual Editions. You can reach us at: *http://www.dushkin.com/annualeditions/.*

World Map

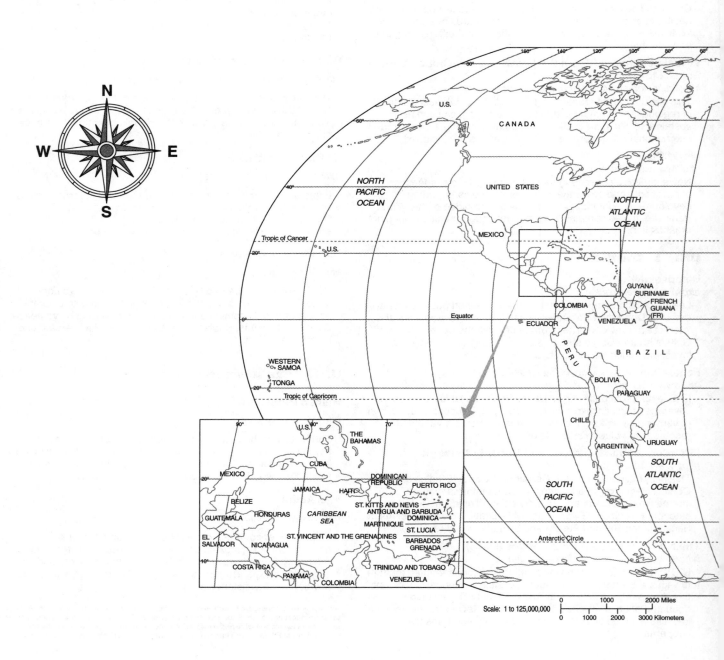

Scale: 1 to 125,000,000

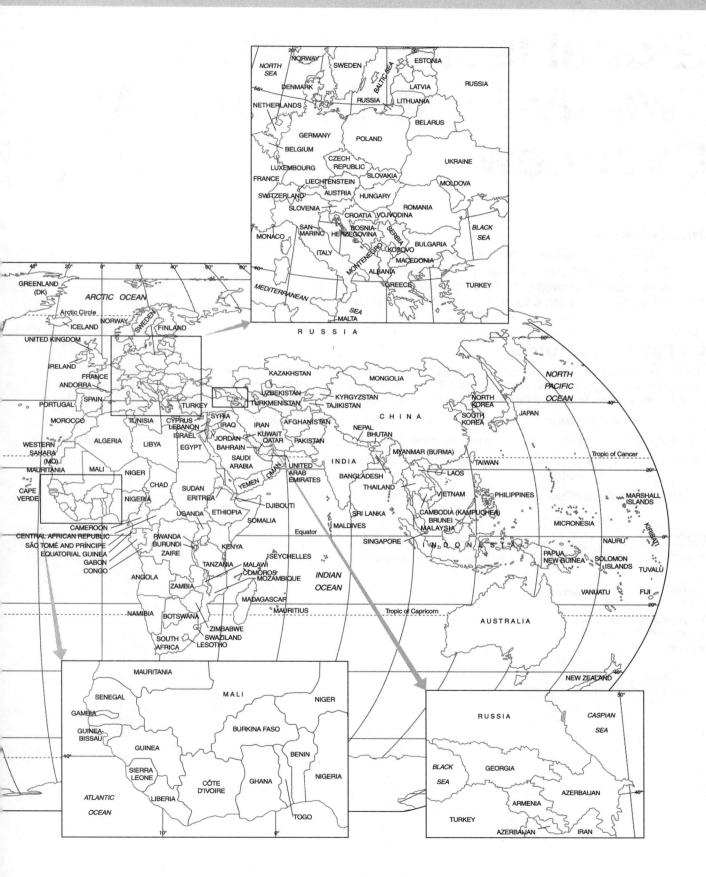

UNIT 1

Global Issues in the Twenty-First Century: An Overview

Unit Selections

1. **A Special Moment in History**, Bill McKibben
2. **Clash of Globalizations**, Stanley Hoffmann
3. **A New Grand Strategy**, Benjamin Schwarz and Christopher Layne
4. **Mr. Order Meets Mr. Chaos**, Robert Wright and Robert Kaplan

Key Points to Consider

- Do the analyses of any of the authors in this section employ the assumptions implicit in the allegory of the balloon? If so, how? If not, how are the assumptions of the authors different? All the authors point to interactions among different factors. What are some of the relationships that they cite? How do the authors differ in terms of the relationships they emphasize?

- What assets that did not exist 100 years ago do people now have to solve problems?

- What events during the twentieth century had the greatest impact on shaping the realities of contemporary international affairs?

- What do you consider to be the five most pressing global problems of today? How do your answers compare to those of your family, friends, and classmates?

- Describe international affairs in the year 2050. How are they different from today and why?

 **Links: www.dushkin.com/online/**
These sites are annotated in the World Wide Web pages.

The Henry L. Stimson Center
http://www.stimson.org
The Heritage Foundation
http://www.heritage.org
IISDnet
http://www.iisd.org/default.asp
The North-South Institute
http://www.nsi-ins.ca/ensi/index.html

Imagine yellow paint being brushed onto an inflated, clear balloon. The yellow color, for purposes of this allegory, represents *people*. In many ways the study of global issues is first and foremost the study of people. Today, there are more human beings occupying Earth than ever before. In addition, we are in the midst of a period of unprecedented population growth. Not only are there many countries where the majority of people are under the age of 16, but also due to improved health care, there are more older people alive than ever before. The effect of a growing global population, however, goes beyond sheer numbers, for this trend has unprecedented impacts on natural resources and social services. An examination of population trends and the related topic of food production is a good place to begin an in-depth study of global issues.

Imagine that our fictional artist next dips the brush into a container of blue paint to represent *nature*. The natural world plays an important role in setting the international agenda. Shortages of raw materials, drought, and pollution of waterways are just a few examples of how natural resources can have global implications.

Adding blue paint to the balloon reveals one of the most important underlying concepts found in this book. Although the balloon originally was covered by both yellow and blue paint (people and nature as separate conceptual entities), the two combined produce an entirely different color: green. Talking about nature as a separate entity, or people as though they were somehow removed from the forces of the natural world, is a serious intellectual error. The people-nature relationship is one of the keys to understanding many of today's most important global issues.

The third color to be added to the balloon is red. This color represents *social structures*. Factors falling into this category include whether a society is urban or rural, industrial or agrarian, and consumer-oriented or dedicated to the needs of the state. The relationship between this component and the others is extremely important. The impact of political decisions on the environment, for example, is one of the most significant features of the contemporary world. Will the whales or bald eagles survive? Historically, the forces of nature determined which species survived or perished. Today, survival depends on political decisions—or indecision. Understanding the complex relationship between social structure and nature (known as "ecopolitics") is central to the study of global issues.

Added to the three primary colors is the fourth and final color of white. It represents the *meta* component (i.e., those qualities that make human beings different from other life forms). These include new ideas and inventions, culture and values, religion and spirituality, and art and literature. The addition of the white paint immediately changes the intensity and shade of the mixture of colors, again emphasizing the relationship among all four factors.

If the painter continues to ply the paintbrush over the miniature globe, a marbling effect becomes evident. From one area to the next, the shading varies because one element is more dominant than another. Further, the miniature system appears dynamic. Nothing is static; relationships are continually changing. This leads to a number of important insights: (1) there are no such things as separate elements, only connections or relationships; (2) changes in one area (such as the weather) will result in changes in all other areas; and (3) complex and dynamic relationships make it difficult to predict events accurately, so observers and policy makers are often surprised by unexpected events.

This book is organized along the basic lines of the balloon allegory. The first unit provides a broad overview of a variety of perspectives on the major forces that are shaping the world of the twenty-first century. From this "big picture" perspective, more in-depth analyses follow. Unit 2, for example, focuses on population and food production. Unit 3 examines the environment and related natural resource issues. The next three units look at different aspects of the world's social structures. They explore issues of economics, national security, conflict, and international cooperation. In the final unit, a number of "meta" factors are presented.

The reader should keep in mind that, just as it was impossible to keep the individual colors from blending into new colors on the balloon, it is also impossible to separate global issues into discrete chapters in a book. Any discussion of agriculture, for example, must take into account the impact of a growing population on soil and water resources, as well as new scientific breakthroughs in food production. Therefore, the organization of this book focuses attention on issue areas; it does not mean to imply that these factors are somehow separate.

With the collapse of the Soviet empire and the end of the cold war, the outlines of a new global agenda have begun to emerge. Rather than being based on the ideology and interests of the two superpowers, new political, economic, environmental, cultural, and security issues are interacting in an unprecedented fashion. Rapid population growth, environmental decline, uneven economic progress, and global terrorist networks are all parts of a complex state of affairs for which there is no historic parallel. As we begin the twenty-first century, signs abound that we are entering a new era. In the words of Abraham Lincoln, "As our case is new, so we must think anew." Compounding this situation, however, is a whole series of old problems such as ethnic and religious rivalries.

The authors in this first unit provide a variety of perspectives on the trends that they believe are the most important to understanding the historic changes at work at the global level. This discussion is then pursued in greater detail in the following units.

It is important for the reader to note that although the authors look at the same world they often come to different conclusions. This raises an important issue of values and beliefs, for it can be argued that there really is no objective reality, only differing perspectives. In short, the study of global issues will challenge each thoughtful reader to examine her or his own values and beliefs.

A Special Moment in History

Bill McKibben

We may live in the strangest, most thoroughly different moment since human beings took up farming, 10,000 years ago, and time more or less commenced. Since then time has flowed in one direction—toward *more*, which we have taken to be progress. At first the momentum was gradual, almost imperceptible, checked by wars and the Dark Ages and plagues and taboos; but in recent centuries it has accelerated, the curve of every graph steepening like the Himalayas rising from the Asian steppe....

But now—now may be the special time. So special that in the Western world we might each of us consider, among many other things, having only one child—that is, reproducing at a rate as low as that at which human beings have ever voluntarily reproduced. Is this really necessary? Are we finally running up against some limits?

To try to answer this question, we need to ask another: *How many of us will there be in the near future?* Here is a piece of news that may alter the way we see the planet—an indication that we live at a special moment. At least at first blush the news is hopeful. *New demographic evidence shows that it is at least possible that a child born today will live long enough to see the peak of human population.*

Around the world people are choosing to have fewer and fewer children—not just in China, where the government forces it on them, but in almost every nation outside the poorest parts of Africa.... If this keeps up, the population of the world will not quite double again; United Nations analysts offer as their mid-range projection that it will top out at 10 to 11 billion, up from just under six billion at the moment....

The good news is that we won't grow forever. The bad news is that there are six billion of us already, a number the world strains to support. One more near-doubling—four or five billion more people—will nearly double that strain. Will these be the five billion straws that break the camel's back?...

LOOKING AT LIMITS

The case that the next doubling, the one we're now experiencing, might be the difficult one can begin as readily with the Stanford biologist Peter Vitousek as with anyone else. In 1986 Vitousek decided to calculate how much of the earth's "primary productivity" went to support human beings. He added together the grain we ate, the corn we fed our cows, and the forests we cut for timber and paper; he added the losses in food as we overgrazed grassland and turned it into desert. And when he was finished adding, the number he came up with was 38.8 percent. We use 38.8 percent of everything the world's plants don't need to keep themselves alive; directly or indirectly, we consume 38.8 percent of what it is possible to eat. "That's a relatively large number," Vitousek says. "It should give pause to people who think we are far from any limits." Though he never drops the measured tone of an academic, Vitousek speaks with considerable emphasis: "There's a sense among some economists that we're *so* far from any biophysical limits. I think that's not supported by the evidence."

For another antidote to the good cheer of someone like Julian Simon, sit down with the Cornell biologist David Pimentel. He believes that we're in big trouble. Odd facts stud his conversation—for example, a nice head of iceberg lettuce is 95 percent water and contains just fifty calories of energy, but it takes 400 calories of energy to grow that head of lettuce in California's Central Valley, and another 1,800 to ship it east. ("There's practically no nutrition in the damn stuff anyway," Pimentel says. "Cabbage is a lot better, and we can grow it in upstate New York.") Pimentel has devoted the past three decades to tracking the planet's capacity, and he believes that we're already too crowded—that the earth can support only two billion people over the long run at a middle-class standard of living, and that trying to support more is doing damage. He has spent considerable time studying soil erosion, for instance. Every raindrop that hits exposed ground is like a small explosion, launching soil particles into the air. On a slope, more than half of the soil contained in those splashes is carried downhill. If crop residue—cornstalks, say—is left in the field after harvest, it helps to shield the soil: the raindrop doesn't hit hard. But in the developing world, where firewood is scarce, peasants burn those cornstalks for cooking fuel. About 60 percent of crop residues in China and 90 percent in Bangladesh are removed and burned, Pimentel says. When planting season comes, dry soils simply blow away. "Our measuring stations pick up African soils in the wind when they start to plough."

The very things that made the Green Revolution so stunning—that made the last doubling possible—now cause trouble. Irrigation ditches, for instance, water 27 percent of all arable land and help to produce a third of all crops. But when flooded soils are baked by the sun, the water evaporates and the minerals in the irrigation water are deposited on the land. A hectare (2.47 acres) can accumulate two to five tons of salt annually, and eventually plants won't grow there. Maybe 10 percent of all irrigated land is affected.

… [F]ood production grew even faster than population after the Second World War. Year after year the yield of wheat and corn and rice rocketed up about three percent annually. It's a favorite statistic of the eternal optimists. In Julian Simon's book *The Ultimate Resource* (1981) charts show just how fast the growth was, and how it continually cut the cost of food. Simon wrote, "The obvious implication of this historical trend toward cheaper food—a trend that probably extends back to the beginning of agriculture—is that real prices for food will continue to drop…. It is a fact that portends more drops in price and even less scarcity in the future."

A few years after Simon's book was published, however, the data curve began to change. That rocketing growth in grain production ceased; now the gains were coming in tiny increments, too small to keep pace with population growth. The world reaped its largest harvest of grain per capita in 1984; since then the amount of corn and wheat and rice per person has fallen by six percent. Grain stockpiles have shrunk to less than two months' supply.

No one knows quite why. The collapse of the Soviet Union contributed to the trend—cooperative farms suddenly found the fertilizer supply shut off and spare parts for the tractor hard to come by. But there were other causes, too, all around the world—the salinization of irrigated fields, the erosion of topsoil, and all the other things that environmentalists had been warning about for years. It's possible that we'll still turn production around and start it rocketing again. Charles C. Mann, writing in *Science*, quotes experts who believe that in the future a "gigantic, multi-year, multi-billion-dollar scientific effort, a

kind of agricultural 'person-on the-moon project,'" might do the trick. The next great hope of the optimists is genetic engineering, and scientists have indeed managed to induce resistance to pests and disease in some plants. To get more yield, though, a cornstalk must be made to put out another ear, and conventional breeding may have exhausted the possibilities. There's a sense that we're running into walls.

… What we are running out of is what the scientists call "sinks"—places to put the by-products of our large appetites. Not garbage dumps (we could go on using Pampers till the end of time and still have empty space left to toss them away) but the atmospheric equivalent of garbage dumps.

It wasn't hard to figure out that there were limits on how much coal smoke we could pour into the air of a single city. It took a while longer to figure out that building ever higher smokestacks merely lofted the haze farther afield, raining down acid on whatever mountain range lay to the east. Even that, however, we are slowly fixing, with scrubbers and different mixtures of fuel. We can't so easily repair the new kinds of pollution. These do not come from something going wrong—some engine without a catalytic converter, some waste-water pipe without a filter, some smokestack without a scrubber. New kinds of pollution come instead from things going as they're supposed to go—but at such a high volume that they overwhelm the planet. They come from normal human life—but there are so many of us living those normal lives that something abnormal is happening. And that something is different from the old forms of pollution that it confuses the issue even to use the word.

Consider nitrogen, for instance. But before plants can absorb it, it must become "fixed"—bonded with carbon, hydrogen, or oxygen. Nature does this trick with certain kinds of algae and soil bacteria, and with lightning. Before human beings began to alter the nitrogen cycle, these mechanisms provided 90–150 million metric tons of nitrogen a year. Now human activity adds 130–150 million more tons. Nitrogen isn't pollution—it's essential. And we are using more of it all the time. Half the industrial nitrogen fertilizer used in human history has been applied since 1984. As a result, coastal waters and estuaries bloom with toxic algae while oxygen concentrations dwindle, killing fish; as a result, nitrous oxide traps solar heat. And once the gas is in the air, it stays there for a century or more.

Or consider methane, which comes out of the back of a cow or the top of a termite mound or the bottom of a rice paddy. As a result of our determination to raise more cattle, cut down more tropical forest (thereby causing termite populations to explode), and grow more rice, methane concentrations in the atmosphere are more than twice as high as they have been for most of the past 160,000 years. And methane traps heat—very efficiently.

Or consider carbon dioxide. In fact, concentrate on carbon dioxide. If we had to pick one problem to obsess about over the next fifty years, we'd do well to make it CO_2—which is not pollution either. Carbon *mon*oxide is pollution: it kills you if you breathe enough of it. But carbon *di*oxide, carbon with two oxygen atoms, can't do a blessed thing to you. If you're reading this indoors, you're breathing more CO_2 than you'll ever get outside. For generations, in fact, engineers said that an engine burned clean if it produced only water vapor and carbon dioxide.

Here's the catch: that engine produces a *lot* of CO_2. A gallon of gas weighs about eight pounds. When it's burned in a car, about five and a half pounds of carbon, in the form of carbon dioxide, come spewing out the back. It doesn't matter if the car is a 1958 Chevy or a 1998 Saab. And no filter can reduce that flow—it's an inevitable by-product of fossil-fuel combustion, which is why CO_2 has been piling up in the atmosphere ever since the Industrial Revolution. Before we started burning oil and coal and gas, the atmosphere contained about 280 parts CO_2 per million. Now the figure is about 360. Unless we do everything we can think of to eliminate fossil fuels from our diet, the air will test out at more than 500 parts per million fifty or sixty years from now, whether it's sampled in the South Bronx or at the South Pole.

This matters because, as we all know by now, the molecular structure of this clean, natural, common element that we are adding to every cubic foot of the atmosphere surrounding us traps heat that would otherwise radiate back out to space. Far more than even methane and nitrous oxide, CO_2 causes global warming—the greenhouse effect—and climate change. Far more than any other single factor, it is turning the earth we were born on into a new planet.

… For ten years, with heavy funding from governments around the world, scientists launched satellites, monitored weather balloons, studied clouds. Their work culminated in a long-awaited report from the UN's Intergovernmental Panel on Climate Change, released in the fall of 1995. The panel's 2,000 scientists, from every corner of the globe, summed up their findings in this dry but historic bit of understatement: "The balance of evidence suggests that there is a discernible human influence on global climate." That is to say, we are heating up the planet—substantially. If we don't reduce emissions of carbon dioxide and other gases, the panel warned, temperatures will probably rise 3.6° Fahrenheit by 2100, and perhaps as much as 6.3°.

You may think you've already heard a lot about global warming. But most of our sense of the problem is behind the curve. Here's the current news: the changes are already well under way. When politicians and businessmen talk about "future risks," their rhetoric is outdated. This is not a problem for the distant future, or even for the near future. The planet has already heated up by a degree or more. We are perhaps a quarter of the way into the greenhouse era, and the effects are already being felt. From a new heaven, filled with nitrogen, methane, and carbon, a new earth is being born. If some alien astronomer is watching us, she's doubtless puzzled. This is the most obvious effect of our numbers and our appetites, and the key to understanding why the size of our population suddenly poses such a risk.

STORMY AND WARM

What does this new world feel like? For one thing, it's stormier than the old one. Data analyzed last year by Thomas Karl, of the National Oceanic and Atmospheric Administration, showed that total winter precipitation in the United States has increased by 10 percent since 1900 and that "extreme precipitation events"—rainstorms that dumped more than two inches of water in twenty-four hours and blizzards—had increased by 20 percent. That's because warmer air holds more water vapor than the colder atmosphere of the old earth; more water evaporates from the ocean, meaning more clouds, more rain, more snow. Engineers designing storm sewers, bridges, and culverts used to plan for what they called the "hundred-year storm." That is, they built to withstand the worst flooding or wind that history led them to expect in the course of a century. Since that history no longer applies, Karl says, "there isn't really a hundred-year event anymore… we seem to be getting these storms of the century every couple of years." When Grand Forks, North Dakota, disappeared beneath the Red River in the spring of last year, some meteorologists referred to it as "a 500-year flood"—meaning, essentially, that all bets are off. Meaning that these aren't acts of God. "If you look out your window, part of what you see in terms of weather is produced by ourselves," Karl says. "If you look out the window fifty years from now, we're going to be responsible for more of it."

Twenty percent more bad storms, 10 percent more winter precipitation—these are enormous numbers. It's like opening the newspaper to read that the average American is smarter by 30 IQ points. And the same data showed increases in drought, too. With more water in the atmosphere, there's less in the soil, according to Kevin Trenberth, of the National Center for Atmospheric Research. Those parts of the continent that are normally dry—the eastern sides of mountains, the plains and deserts—are even drier, as the higher average temperatures evaporate more of what rain does fall. "You get wilting plants and eventually drought faster than you would otherwise," Trenberth says. And when the rain does come, it's often so intense that much of it runs off before it can soak into the soil.

So—wetter and drier. *Different*.…

The effects of… warming can be found in the largest phenomena. The oceans that cover most of the planet's surface are clearly rising, both because of melting glaciers and because water expands as it warms. As a result, low-lying Pacific islands already report surges of water washing across the atolls. "It's nice weather and all of a sudden water is pouring into your living room," one Marshall Islands resident told a newspaper reporter. "It's very clear that something is happening in the Pacific, and these islands are feeling it." Global warming will be like a much more powerful version of El Niño that covers the entire globe and lasts forever, or at least until the next big asteroid strikes.

If you want to scare yourself with guesses about what might happen in the near future, there's no shortage of possibilities. Scientists have already observed large-scale shifts in the duration of the El Niño ocean warming, for instance. The Arctic tundra has warmed so much that in some places it now gives off more carbon dioxide than it absorbs—a switch that could trigger a potent feedback loop, making warming ever worse. And researchers studying glacial cores from the Greenland Ice Sheet recently concluded that local climate shifts have occurred with incredible rapidity in the past—18° in one three-year stretch. Other scientists worry that such a shift might be enough to flood the oceans with fresh water and reroute or shut off currents like the Gulf Stream and the North Atlantic, which keep Europe far warmer than it would otherwise be. (See "The Great Climate Flip-flop," by William H. Calvin, January *Atlantic*.) In the words of Wallace Broecker, of Columbia University, a pioneer in the field, "Climate is an angry beast, and we are poking it with sticks."

But we don't need worst-case scenarios: best-case scenarios make the point. The population of the earth is going to nearly double one more time. That will bring it to a level that even the reliable old earth we were born on would be hard-pressed to support. Just at the moment when we need everything to be working as smoothly as possible, we find ourselves inhabiting a new planet, whose carrying capacity we cannot conceivably estimate. We have no idea how much wheat this planet can grow. We don't know what its politics will be like: not if there are going to be heat waves like the one that killed more than 700 Chicagoans in 1995; not if rising sea levels and other effects of climate change create tens of millions of environmental refugees; not if a 1.5° jump in India's temperature could reduce the country's wheat crop by 10 percent or divert its monsoons....

We have gotten very large and very powerful, and for the foreseeable future we're stuck with the results. The glaciers won't grow back again anytime soon; the oceans won't drop. We've already done deep and systemic damage. To use a human analogy, we've already said the angry and unforgivable words that will haunt our marriage till its end. And yet we can't simply walk out the door. There's no place to go. We have to salvage what we can of our relationship with the earth, to keep things from getting any worse than they have to be.

If we can bring our various emissions quickly and sharply under control, we *can* limit the damage, reduce dramatically the chance of horrible surprises, preserve more of the biology we were born into. But do not under-

estimate the task. The UN's Intergovernmental Panel on Climate Change projects that an immediate 60 percent reduction in fossil-fuel use is necessary just to stabilize climate at the current level of disruption. Nature may still meet us halfway, but halfway is a long way from where we are now. What's more, we can't delay. If we wait a few decades to get started, we may as well not even begin. It's not like poverty, a concern that's always there for civilizations to address. This is a timed test, like the SAT: two or three decades, and we lay our pencils down. It's *the* test for our generations, and population is a part of the answer....

The numbers are so daunting that they're almost unimaginable. Say, just for argument's sake, that we decided to cut world fossil-fuel use by 60 percent—the amount that the UN panel says would stabilize world climate. And then say that we shared the remaining fossil fuel equally. Each human being would get to produce 1.69 metric tons of carbon dioxide annually—which would allow you to drive an average American car nine miles a day. By the time the population increased to 8.5 billion, in about 2025, you'd be down to six miles a day. If you carpooled, you'd have about three pounds of CO_2 left in your daily ration—enough to run a highly efficient refrigerator. Forget your computer, your TV, your stereo, your stove, your dishwasher, your water heater, your microwave, your water pump, your clock. Forget your light bulbs, compact fluorescent or not.

I'm not trying to say that conservation, efficiency, and new technology won't help. They will—but the help will be slow and expensive. The tremendous momentum of growth will work against it. Say that someone invented a new furnace tomorrow that used half as much oil as old furnaces. How many years would it be before a substantial number of American homes had the new device? And what if it cost more? And if oil stays cheaper per gallon than bottled water? Changing basic fuels—to hydrogen, say—would be even more expensive. It's not like running out of white wine and switching to red. Yes, we'll get new technologies. One day last fall *The New York Times* ran a special section on energy, featuring many up-and-coming improvements: solar shingles, basement fuel cells. But the same day, on the front page, William K. Stevens reported that international negotiators had all but given up on preventing a doubling of the atmospheric concentration of CO_2. The momentum of growth was so great, the negotiators said, that making the changes required to slow global warming significantly would be like "trying to turn a supertanker in a sea of syrup."

There are no silver bullets to take care of a problem like this. Electric cars won't by themselves save us, though they would help. We simply won't live efficiently enough soon enough to solve the problem. Vegetarianism won't cure our ills, though it would help. We simply won't live simply enough soon enough to solve the problem.

Reducing the birth rate won't end all our troubles either. That, too, is no silver bullet. But it would help.

There's no more practical decision than how many children to have. (And no more mystical decision, either.)

The bottom-line argument goes like this: The next fifty years are a special time. They will decide how strong and healthy the planet will be for centuries to come. Between now and 2050 we'll see the zenith, or very nearly, of human population. With luck we'll never see any greater production of carbon dioxide or toxic chemicals. We'll never see more species extinction or soil erosion. Greenpeace recently announced a campaign to phase out fossil fuels entirely by mid-century, which sounds utterly quixotic but could—if everything went just right—happen.

So it's the task of those of us alive right now to deal with this special phase, to squeeze us through these next fifty years. That's not fair—any more than it was fair that earlier generations had to deal with the Second World War or the Civil War or the Revolution or the Depression or slavery. It's just reality. We need in these fifty years to be working simultaneously on all parts of the equation—on our ways of life, on our technologies, and on our population.

As Gregg Easterbrook pointed out in his book *A Moment on the Earth* (1995), if the planet does manage to reduce its fertility, "the period in which human numbers threaten the biosphere on a general scale will turn out to have been much, much more brief" than periods of natural threats like the Ice Ages. True enough. But the period in question happens to be our time. That's what makes this moment special, and what makes this moment hard.

Bill McKibben is the author of several books about the environment, including *The End of Nature* (1989) and *Hope, Human and Wild* (1995). His article in this issue will appear in somewhat different form in his book *Maybe One: A Personal and Environmental Argument for Single-Child Families*, published in 1998 by Simon & Schuster.

Clash of Globalizations

Stanley Hoffmann

A NEW PARADIGM?

WHAT IS THE STATE OF international relations today? In the 1990s, specialists concentrated on the partial disintegration of the global order's traditional foundations: states. During that decade, many countries, often those born of decolonization, revealed themselves to be no more than pseudostates, without solid institutions, internal cohesion, or national consciousness. The end of communist coercion in the former Soviet Union and in the former Yugoslavia also revealed long-hidden ethnic tensions. Minorities that were or considered themselves oppressed demanded independence. In Iraq, Sudan, Afghanistan, and Haiti, rulers waged open warfare against their subjects. These wars increased the importance of humanitarian interventions, which came at the expense of the hallowed principles of national sovereignty and nonintervention. Thus the dominant tension of the decade was the clash between the fragmentation of states (and the state system) and the progress of economic, cultural, and political integration—in other words, globalization.

Everybody has understood the events of September 11 as the beginning of a new era. But what does this break mean? In the conventional approach to international relations, war took place among states. But in September, poorly armed individuals suddenly challenged, surprised, and wounded the world's dominant superpower. The attacks also showed that, for all its accomplishments, globalization makes an awful form of violence easily accessible to hopeless fanatics. Terrorism is the bloody link between interstate relations and global society. As countless individuals and groups are becoming global actors along with states, insecurity and vulnerability are rising. To assess today's bleak state of affairs, therefore, several questions are necessary. What concepts help explain the new global order? What is the condition of the interstate part of international relations? And what does the emerging global civil society contribute to world order?

SOUND AND FURY

TWO MODELS made a great deal of noise in the 1990s. The first one—Francis Fukuyama's "End of History" thesis—was not vindicated by events. To be sure, his argument predicted the end of ideological conflicts, not history itself, and the triumph of political and economic liberalism. That point is correct in a narrow sense: the "secular religions" that fought each other so bloodily in the last century are now dead. But Fukuyama failed to note that nationalism remains very much alive. Moreover, he ignored the explosive potential of religious wars that has extended to a large part of the Islamic world.

Fukuyama's academic mentor, the political scientist Samuel Huntington, provided a few years later a gloomier account that saw a very different world. Huntington predicted that violence resulting from international anarchy and the absence of common values and institutions would erupt among civilizations rather than among states or ideologies. But Huntington's conception of what constitutes a civilization was hazy. He failed to take into account sufficiently conflicts within each so-called civilization, and he overestimated the importance of religion in the behavior of non-Western elites, who are often secularized and Westernized. Hence he could not clearly define the link between a civilization and the foreign policies of its member states.

Other, less sensational models still have adherents. The "realist" orthodoxy insists that nothing has changed in international relations since Thucydides and Machiavelli: a state's military and economic power determines its fate; interdependence and international institutions are secondary and fragile phenomena; and states' objectives are imposed by the threats to their survival or security. Such is the world described by Henry Kissinger. Unfortunately, this venerable model has trouble integrating change, especially globalization and the rise of nonstate actors. Moreover, it overlooks the need for international cooperation that results from such new threats as the proliferation of weapons of mass destruction (WMD). And it ignores what the scholar Raymond Aron called the "germ of a universal consciousness": the liberal, promarket norms that developed states have come to hold in common.

Taking Aron's point, many scholars today interpret the world in terms of a triumphant globalization that submerges borders through new means of information and communication. In this universe, a state choosing to stay closed invariably faces decline and growing discontent among its subjects, who are eager for material progress. But if it opens up, it must accept a reduced role that is mainly limited to social protection, physical protection against aggression or civil war, and maintaining national identity. The champion of this epic without heroes is *The New York Times* columnist Thomas Friedman. He contrasts barriers with open vistas, obsolescence with modernity, state control with free markets. He sees in globalization the light of dawn, the "golden straitjacket" that will force contentious publics to understand that the logic of globalization is that of peace (since war would interrupt globalization and therefore progress) and democracy (because new technologies increase individual autonomy and encourage initiative).

BACK TO REALITY

These MODELS come up hard against three realities. First, rivalries among great powers (and the capacity of smaller states to exploit such tensions) have most certainly not disappeared. For a while now, however, the existence of nuclear weapons has produced a certain degree of prudence among the powers that have them. The risk of destruction that these weapons hold has moderated the game and turned nuclear arms into instruments of last resort. But the game could heat up as more states seek other WMD as a way of narrowing the gap between the nuclear club and the other powers. The sale of such weapons thus becomes a hugely contentious issue, and efforts to slow down the spread of all WMD, especially to dangerous "rogue" states, can paradoxically become new causes of violence.

Second, if wars between states are becoming less common, wars within them are on the rise—as seen in the former Yugoslavia, Iraq, much of Africa, and Sri Lanka. Uninvolved states first tend to hesitate to get engaged in these complex conflicts, but they then (sometimes) intervene to prevent these conflicts from turning into regional catastrophes. The interveners, in turn, seek the help of the United Nations or regional organizations to rebuild these states, promote stability, and prevent future fragmentation and misery.

Third, states' foreign policies are shaped not only by realist geopolitical factors such as economics and military power but by domestic politics. Even in undemocratic regimes, forces such as xenophobic passions, economic grievances, and transnational ethnic solidarity can make policymaking far more complex and less predictable. Many states—especially the United States—have to grapple with the frequent interplay of competing government branches. And the importance of individual leaders and their personalities is often underestimated in the study of international affairs.

For realists, then, transnational terrorism creates a formidable dilemma. If a state is the victim of private actors such as terrorists, it will try to eliminate these groups by depriving them of sanctuaries and punishing the states that harbor them. The national interest of the attacked state will therefore require either armed interventions against governments supporting terrorists or a course of prudence and discreet pressure on other governments to bring these terrorists to justice. Either option requires a questioning of sovereignty—the holy concept of realist theo-

ries. The classical realist universe of Hans Morgenthau and Aron may therefore still be very much alive in a world of states, but it has increasingly hazy contours and offers only difficult choices when it faces the threat of terrorism.

At the same time, the real universe of globalization does not resemble the one that Friedman celebrates. In fact, globalization has three forms, each with its own problems. First is economic globalization, which results from recent revolutions in technology, information, trade, foreign investment, and international business. The main actors are companies, investors, banks, and private services industries, as well as states and international organizations. This present form of capitalism, ironically foreseen by Karl Marx and Friedrich Engels, poses a central dilemma between efficiency and fairness. The specialization and integration of firms make it possible to increase aggregate wealth, but the logic of pure capitalism does not favor social justice. Economic globalization has thus become a formidable cause of inequality among and within states, and the concern for global competitiveness limits the aptitude of states and other actors to address this problem.

Optimism regarding globalization rests on very fragile foundations.

Next comes cultural globalization. It stems from the technological revolution and economic globalization, which together foster the flow of cultural goods. Here the key choice is between uniformization (often termed "Americanization") and diversity. The result is both a "disenchantment of the world" (in Max Weber's words) and a reaction against uniformity. The latter takes form in a renaissance of local cultures and languages as well as assaults against Western culture, which is denounced as an arrogant bearer of a secular, revolutionary ideology and a mask for U.S. hegemony.

Finally there is political globalization, a product of the other two. It is characterized by the preponderance of the United States and its political institutions and by a vast array of international and regional organizations and transgovernmental networks (specializing in areas such as policing or migration or justice). It is also marked by private institutions that are neither governmental nor purely national—say, Doctors Without Borders or Amnesty International. But many of these agencies lack democratic accountability and are weak in scope, power, and authority. Furthermore, much uncertainty hangs over the fate of American hegemony, which faces significant resistance abroad and is affected by America's own oscillation between the temptations of domination and isolation.

The benefits of globalization are undeniable. But Friedman-like optimism rests on very fragile foundations. For one thing, globalization is neither inevitable nor irresistible. Rather, it is largely an American creation, rooted in the period after World War II and based on U.S. economic might. By extension, then, a deep and protracted economic crisis in the United States could have as devastating an effect on globalization as did the Great Depression.

Second, globalization's reach remains limited because it excludes many poor countries, and the states that it does transform react in different ways. This fact stems from the diversity of economic and social conditions at home as well as from partisan politics. The world is far away from a perfect integration of markets, services, and factors of production. Sometimes the simple existence of borders slows down and can even paralyze this integration; at other times it gives integration the flavors and colors of the dominant state (as in the case of the Internet).

Third, international civil society remains embryonic. Many nongovernmental organizations reflect only a tiny segment of the populations of their members' states. They largely represent only modernized countries, or those in which the weight of the state is not too heavy. Often, NGOs have little independence from governments.

Fourth, the individual emancipation so dear to Friedman does not quickly succeed in democratizing regimes, as one can see today in China. Nor does emancipation prevent public institutions such as the International Monetary Fund, the World Bank, or the World Trade Organization from remaining opaque in their activities and often arbitrary and unfair in their rulings.

Fifth, the attractive idea of improving the human condition through the abolition of barriers is dubious. Globalization is in fact only a sum of techniques (audio and videocassettes, the Internet, instantaneous communications) that are at the disposal of states or private actors. Self-interest and ideology, not humanitarian reasons, are what drive these actors. Their behavior is quite different from the vision of globalization as an Enlightenment-based utopia that is simultaneously scientific, rational, and universal. For many reasons—misery, injustice, humiliation, attachment to traditions, aspiration to more than just a better standard of living—this "Enlightenment" stereotype of globalization thus provokes revolt and dissatisfaction.

Another contradiction is also at work. On the one hand, international and transnational cooperation is necessary to ensure that globalization will not be undermined by the inequalities resulting from market fluctuations, weak state-sponsored protections, and the incapacity of many states to improve their fates by themselves. On the other hand, cooperation presupposes that many states and rich private players operate altruistically—which is certainly not the essence of international relations—or practice a remarkably generous conception of their long-term interests. But the fact remains that most rich states still refuse to provide sufficient development aid or to intervene in crisis situations such as the genocide in Rwanda. That reluctance compares poorly with the American enthusiasm to pursue the fight against al Qaeda and the Taliban. What is wrong here is not patriotic enthusiasm as such, but the weakness of the humanitarian impulse when the national interest in saving non-American victims is not self-evident.

IMAGINED COMMUNITIES

AMONG the many effects of globalization on international politics, three hold particular importance. The first concerns in-

stitutions. Contrary to realist predictions, most states are not perpetually at war with each other. Many regions and countries live in peace; in other cases, violence is internal rather than state-to-state. And since no government can do everything by itself, interstate organisms have emerged. The result, which can be termed "global society," seeks to reduce the potentially destructive effects of national regulations on the forces of integration. But it also seeks to ensure fairness in the world market and create international regulatory regimes in such areas as trade, communications, human rights, migration, and refugees. The main obstacle to this effort is the reluctance of states to accept global directives that might constrain the market or further reduce their sovereignty. Thus the UN's powers remain limited and sometimes only purely theoretical. International criminal justice is still only a spotty and contested last resort. In the world economy—where the market, not global governance, has been the main beneficiary of the state's retreat—the network of global institutions is fragmented and incomplete. Foreign investment remains ruled by bilateral agreements. Environmental protection is badly ensured, and issues such as migration and population growth are largely ignored. Institutional networks are not powerful enough to address unfettered short-term capital movements, the lack of international regulation on bankruptcy and competition, and primitive coordination among rich countries. In turn, the global "governance" that does exist is partial and weak at a time when economic globalization deprives many states of independent monetary and fiscal policies, or it obliges them to make cruel choices between economic competitiveness and the preservation of social safety nets. All the while, the United States displays an increasing impatience toward institutions that weigh on American freedom of action. Movement toward a world state looks increasingly unlikely. The more state sovereignty crumbles under the blows of globalization or such recent developments as humanitarian intervention and the fight against terrorism, the more states cling to what is left to them.

Second, globalization has not profoundly challenged the enduring national nature of citizenship. Economic life takes place on a global scale, but human identity remains national—hence the strong resistance to cultural homogenization. Over the centuries, increasingly centralized states have expanded their functions and tried to forge a sense of common identity for their subjects. But no central power in the world can do the same thing today, even in the European Union. There, a single currency and advanced economic coordination have not yet produced a unified economy or strong central institutions endowed with legal autonomy, nor have they resulted in a sense of postnational citizenship. The march from national identity to one that would be both national and European has only just begun. A world very partially unified by technology still has no collective consciousness or collective solidarity. What states are unwilling to do the world market cannot do all by itself, especially in engendering a sense of world citizenship.

Third, there is the relationship between globalization and violence. The traditional state of war, even if it is limited in scope, still persists. There are high risks of regional explosions in the Middle East and in East Asia, and these could seriously affect relations between the major powers. Because of this threat, and because modern arms are increasingly costly, the "anarchical society" of states lacks the resources to correct some of globalization's most flagrant flaws. These very costs, combined with the classic distrust among international actors who prefer to try to preserve their security alone or through traditional alliances, prevent a more satisfactory institutionalization of world politics—for example, an increase of the UN's powers. This step could happen if global society were provided with sufficient forces to prevent a conflict or restore peace—but it is not.

Globalization, far from spreading peace, thus seems to foster conflicts and resentments. The lowering of various barriers celebrated by Friedman, especially the spread of global media, makes it possible for the most deprived or oppressed to compare their fate with that of the free and well-off. These dispossessed then ask for help from others with common resentments, ethnic origin, or religious faith. Insofar as globalization enriches some and uproots many, those who are both poor and uprooted may seek revenge and self-esteem in terrorism.

GLOBALIZATION AND TERROR

TERRORISM is the poisoned fruit of several forces. It can be the weapon of the weak in a classic conflict among states or within a state, as in Kashmir or the Palestinian territories. But it can also be seen as a product of globalization. Transnational terrorism is made possible by the vast array of communication tools. Islamic terrorism, for example, is not only based on support for the Palestinian struggle and opposition to an invasive American presence. It is also fueled by a resistance to "unjust" economic globalization and to a Western culture deemed threatening to local religions and cultures.

If globalization often facilitates terrorist violence, the fight against this war without borders is potentially disastrous for both economic development and globalization. Antiterrorist measures restrict mobility and financial flows, while new terrorist attacks could lead the way for an antiglobalist reaction comparable to the chauvinistic paroxysms of the 1930s. Global terrorism is not the simple extension of war among states to nonstates. It is the subversion of traditional ways of war because it does not care about the sovereignty of either its enemies or the allies who shelter them. It provokes its victims to take measures that, in the name of legitimate defense, violate knowingly the sovereignty of those states accused of encouraging terror. (After all, it was not the Taliban's infamous domestic violations of human rights that led the United States into Afghanistan; it was the Taliban's support of Osama bin Laden.)

But all those trespasses against the sacred principles of sovereignty do not constitute progress toward global society, which has yet to agree on a common definition of terrorism or on a common policy against it. Indeed, the beneficiaries of the antiterrorist "war" have been the illiberal, poorer states that have lost so much of their sovereignty of late. Now the crackdown on terror allows them to tighten their controls on their own people, products, and money. They can give themselves new reasons to violate individual rights in the name of common defense against insecurity—and thus stop the slow, hesitant march toward international criminal justice.

Another main beneficiary will be the United States, the only actor capable of carrying the war against terrorism into all corners of the world. Despite its power, however, America cannot fully protect itself against future terrorist acts, nor can it fully overcome its ambivalence toward forms of interstate cooperation that might restrict U.S. freedom of action. Thus terrorism is a global phenomenon that ultimately reinforces the enemy—the state—at the same time as it tries to destroy it. The states that are its targets have no interest in applying the laws of war to their fight against terrorists; they have every interest in treating terrorists as outlaws and pariahs. The champions of globalization have sometimes glimpsed the "jungle" aspects of economic globalization, but few observers foresaw similar aspects in global terrorist and antiterrorist violence.

Finally, the unique position of the United States raises a serious question over the future of world affairs. In the realm of interstate problems, American behavior will determine whether the nonsuperpowers and weak states will continue to look at the United States as a friendly power (or at least a tolerable hegemon), or whether they are provoked by Washington's hubris into coalescing against American preponderance. America may be a hegemon, but combining rhetorical overkill and ill-defined designs is full of risks. Washington has yet to understand that nothing is more dangerous for a "hyperpower" than the temptation of unilateralism. It may well believe that the constraints of international agreements and organizations are not necessary, since U.S. values and power are all that is needed for world order. But in reality, those same international constraints provide far better opportunities for leadership than arrogant demonstrations of contempt for others' views, and they offer useful ways of restraining unilateralist behavior in other states. A hegemon concerned with prolonging its rule should be especially interested in using internationalist methods and institutions, for the gain in influence far exceeds the loss in freedom of action.

In the realm of global society, much will depend on whether the United States will overcome its frequent indifference to the costs that globalization imposes on poorer countries. For now, Washington is too reluctant to make resources available for economic development, and it remains hostile to agencies that monitor and regulate the global market. All too often, the right-leaning tendencies of the American political system push U.S. diplomacy toward an excessive reliance on America's greatest asset—military strength—as well as an excessive reliance on market capitalism and a "sovereigntism" that offends and alienates. That the mighty United States is so afraid of the world's imposing its "inferior" values on Americans is often a source of ridicule and indignation abroad.

ODD MAN OUT

FOR ALL THESE TENSIONS, it is still possible that the American war on terrorism will be contained by prudence, and that other governments will give priority to the many internal problems created by interstate rivalries and the flaws of globalization. But the world risks being squeezed between a new Scylla and Charybdis. The Charybdis is universal intervention, unilaterally decided by American leaders who are convinced that they have found a global mission provided by a colossal threat. Presentable as an epic contest between good and evil, this struggle offers the best way of rallying the population and overcoming domestic divisions. The Scylla is resignation to universal chaos in the form of new attacks by future bin Ladens, fresh humanitarian disasters, or regional wars that risk escalation. Only through wise judgment can the path between them be charted.

We can analyze the present, but we cannot predict the future. We live in a world where a society of uneven and often virtual states overlaps with a global society burdened by weak public institutions and underdeveloped civil society. A single power dominates, but its economy could become unmanageable or disrupted by future terrorist attacks. Thus to predict the future confidently would be highly incautious or naive. To be sure, the world has survived many crises, but it has done so at a very high price, even in times when WMD were not available.

Precisely because the future is neither decipherable nor determined, students of international relations face two missions. They must try to understand what goes on by taking an inventory of current goods and disentangling the threads of present networks. But the fear of confusing the empirical with the normative should not prevent them from writing as political philosophers at a time when many philosophers are extending their conceptions of just society to international relations. How can one make the global house more livable? The answer presupposes a political philosophy that would be both just and acceptable even to those whose values have other foundations. As the late philosopher Judith Shklar did, we can take as a point of departure and as a guiding thread the fate of the victims of violence, oppression, and misery; as a goal, we should seek material and moral emancipation. While taking into account the formidable constraints of the world as it is, it is possible to loosen them.

STANLEY HOFFMANN is Buttenwieser University Professor at Harvard University and a regular book reviewer for *Foreign Affairs*.

[The Hard Questions]

A NEW GRAND STRATEGY

For more than fifty years American foreign policy has sought to prevent the emergence of other great powers—a strategy that has proved burdensome, futile, and increasingly risky. The United States will be more secure, and the world more stable, if America now chooses to pass the buck and allow other countries to take care of themselves

BY BENJAMIN SCHWARZ
AND CHRISTOPHER LAYNE

Since the end of the Cold War, U.S. grand strategy has revolved around maintaining this country's overwhelming military, economic, and political preponderance. Until now most Americans have acquiesced in that strategy, because the costs seemed to be tolerably low. But the September 11 attacks have proved otherwise. Those assaults were neither random nor irrational. Those who undertook them acted with cool calculation to force the United States to alter specific policies—policies that largely flow from the global role America has chosen. The attacks were also a violent reaction to the very fact of America's pre-eminence.

Several tasks confront us. The most immediate is the one that rightly preoccupies the nation now: tracking down the al Qaeda terrorists and destroying their networks and their infrastructure, and waging war on the Taliban movement that harbors them. The larger task will take time, because it amounts to inventing a new American stance toward the world for the century ahead. We need to come to grips with an ironic possibility: that the very preponderance of American power may now make us not more secure but less secure. By the same token, it may actually be possible to achieve more of our ultimate foreign-policy goals by means of a diminished global presence.

Great powers have two basic strategic options: they can pursue geopolitical dominance (a "unipolar" strategy), or they can seek to maintain a rough balance of power among the strongest states in a region or around the world (a "multipolar" strategy). Since the late 1940s the United States has chosen the former course. True, even during the Cold War, when the world was essentially divided between the United States and the Soviet Union, a number of astute foreign-policy thinkers—including Walter Lippmann, George Kennan, and J. William Fulbright—argued that it was in America's interest to encourage Western Europe's and Japan's revival as independent great powers to relieve the United States of what Kennan called the "burdens of 'bi-polarity.'" But almost all American policymakers held that the United States had to contain its allies as much as it had to contain Moscow. By providing for the security of Britain, France, and (especially) Germany and Japan—by defending their access to far-flung economic and natural resources, and by enmeshing their foreign and military policies in alliances that America dominated—Washington prevented these former and potential great powers from embarking on independent, and (from the U.S. perspective) possibly destabilizing, foreign policies. This "reassurance strategy" (to use a term currently favored by policymakers) allowed for an unprecedented level of political and economic cooperation among the states of Western Europe and East Asia.

As noted, American policy since the end of the Cold War has aimed to ensure that the United States maintains its lofty perch. Every post-Cold War Pentagon assessment of national-security needs has insisted that America maintain its globe-girdling Cold War alliances, along with its Cold War defense-spending levels, even though the threat against which those alliances and budgets were ostensibly erected has disappeared. Some critics argue that this apparent stasis is born of bureaucratic inertia, or of a defense establishment's jealous guarding of its turf (and its trough). But in fact, given the logic behind American grand strategy, this continuity is entirely justifiable. The collapse of the Soviet Union hasn't altered the conviction among many American policymakers that a stable global economic and political order depends on Washington's maintaining preponderance (or, according to the official rhetoric, "leadership") over potential great powers. This means ameliorating their security problems.

The now infamous draft of the Pentagon's Defense Planning Guidance (prepared under the direction of the current undersecretary of defense for policy, Paul Wolfowitz), which was leaked to *The New York Times* in 1992, merely stated in undiplomatic language the logic that has long informed Washington's strategy. The United States, it argued, must continue to dominate the international system and thus to "discourage" the "advanced industrial nations from challenging our leadership or... even aspiring to a larger regional or global role." To accomplish this Washington must do nothing less than "retain the preeminent responsibility for addressing... those wrongs which threaten not only our interests, but those of our allies or friends, or which could seriously unsettle international relations." In other words, America must provide its allies with what one of the document's authors (now a special assistant to the President on the National Security Council) termed "adult supervision": the United States must not only impose a military protectorate over Europe and East Asia—regions composed of wealthy and technologically sophisticated states—but also safeguard Europe's and East Asia's worldwide interests, so that they need not develop military forces capable of "global power projection." (As Gabriel Robin, a former French representative to NATO, acknowledged, the U.S.-led alliance's "real function... is to serve as the chaperon of Europe." It is, Robin said, "the means to prevent [Europe] from establishing itself as an independent fortress and perhaps one day, a rival.") Those who argue that America's national-security spending is too high, given the end of the Cold War, often fail to appreciate the task Washington has assigned itself. The adult supervision of the world is an enormously expensive and complex undertaking, which perforce means that the United States must spend more on its military than do the next nine countries together—including Russia, Japan, China, France, Britain, and Germany.

An adult-supervision strategy entails a peculiar and recondite calculation of the world situation. For instance, although most Americans might believe that a reunified, democratic Korea would indisputably be in America's interests, the former National Security Advisor Zbigniew Brzezinski, in his 1997 book *The Grand Chessboard* (probably the fullest and frankest public exposition of America's post-Cold War global strategy), repeatedly explains how this development would in fact jeopardize America's unipolar strategy: it would, he argues, reflecting views long and widely held in policymaking circles, obviate the ostensible need for U.S. troops on the peninsula, which could lead to a U.S. pullback from East Asia, which could, in turn, lead to Japan's becoming "military, more self-sufficient," which would lead to political, military, and economic rivalry among the region's states. Thus the best situation is the status quo in Korea, which allows for U.S. forces to be stationed there indefinitely.

Similar—and more urgent, given the current war on terrorism—is the thinking underlying Washington's policy in the Persian Gulf. Why is the United States so deeply embroiled in this turbulent region? Many people, echoing a comment by Secretary of State James A. Baker during the Persian Gulf War, would probably answer with one word: oil. This answer is—and was—both true and misleading. America derives most of its oil from Alaska, Canada, the continental United States, Mexico, and Venezuela. About 25 percent of U.S. petroleum imports come from the Persian Gulf. If the United States adopted a national energy strategy, it could free itself from dependence on Persian Gulf oil. Nevertheless, Washington assumes responsibility for stabilizing the region because Western Europe and Japan are heavily dependent on its oil, and because soon China, owing to rapid economic growth, will be as well—and America wants to discourage those powers from developing the means to protect that resource for themselves. In an interview on National Public Radio early in October, Walter Russell Mead, a senior foreign-policy analyst at the Council on Foreign Relations, explained the basis of U.S. policy in the terms that NSC staffers, think-tank analysts, and State and Defense Department policy planners have used for years: "We do not get that large a percentage of our oil from the Middle East. Japan gets a lot more... And one of the reasons that we are sort of assuming this role of policeman of the Middle East, more or less, has more to do with making Japan and some other countries feel that their oil flow is assured... so that they don't then feel more need to create a great power, armed forces, and security doctrine, and you don't start getting a lot of great powers with conflicting interests sending their militaries all over the world."

Despite its sometimes esoteric logic, America's strategy of preponderance is seductive. In the abstract it makes sense that the United States should seek to amass as much power as possible. In this way the rationale behind U.S. strategy is analogous to that of a firm in an oligopolistic market, which drives its rivals out of business rather than risk its profits in a competitive environment. Theoretically, if a state can establish—and maintain—itself as the only great power in the international system, it will enjoy something very close to absolute security. But as history amply shows, when one state acquires too much power, others invariably fear that it will aggrandize itself at their expense. "Hegemonic empires," Henry Kissinger recently noted, "almost automatically elicit universal resistance, which is why all such claimants have sooner or later exhausted themselves."

More than two hundred years ago Edmund Burke warned his countrymen,

> Among precautions against ambition it may not be amiss to take one precaution against our *own*... I dread our being too much dreaded... We may say that we shall not abuse this astonishing and hitherto unheard-of power. But every nation will think we shall abuse it... Sooner or later, this state of things must produce a combination against us which may end in our ruin.

Like some optimistic Britons in the late eighteenth century, many American strategists today assert that the United States, the only superpower, is a "benevolent" hegemon, immunized from a backlash against its preponderance by what they call its "soft power"—that is, by the attractiveness of its liberal-democratic ideology and its open, syncretic culture. Washington also believes that others don't fear U.S. geopolitical pre-eminence because they know the United States will use its unprecedented power to promote the good of the international system rather than to advance its own selfish aims.

But states must always be more concerned with a predominant power's capabilities than with its intentions, and in fact well before September 11—indeed, throughout most of the past decade—other states have been profoundly anxious about the *im*balance of power in America's favor. This simmering mistrust of U.S. predominance intensified during the Clinton Administration, as other states responded to American hegemony by concerting their efforts against it. Russia and China, although long estranged, found common ground in a nascent alliance that opposed U.S. "hegemonism" and expressly aimed at re-establishing "a multipolar world." Arguing that the term "superpower" is inadequate to convey the true extent of America's economic and military pre-eminence, the French Foreign Minister Hubert Vedrine called the United States a "hyperpower." Even the Dutch Prime Minister declared that the European Union should make itself "a counterweight to the United States."

American intervention in Kosovo crystallized fears of U.S. hegemony, prompting the emergence of an anti-U.S. constellation of China, Russia, and India. Viewing the Kosovo war as a dangerous precedent establishing Washington's self-declared right to interfere in other countries' internal affairs, and asserting their support for a multipolar world, these three states increased their arms transfers and their sharing of military technology, specifically to counter American power. Also, the Kosovo conflict made apparent the disparity between America's geopolitical power and Europe's, inciting Europe to take its first serious steps toward redressing that disparity by acquiring—through the European Defense and Security Identity—the kinds of military capabilities it would need to act independent of the United States. If the European Union fulfills EDSI's longer-term goals, it will emerge as an unfettered strategic player in world politics. And that emergence will have been driven by the clear objective of investing Europe with the capability to act as a brake on America's aspirations.

Any remaining doubt that American hegemony could trigger a hostile reaction, whether reasonable or not, surely dissipated on September 11. The role the United States has assigned itself in the Persian Gulf has made it—not Japan, not the states of Western Europe, not China—vulnerable to a backlash. Iran, Iraq, and Afghanistan resent America's intrusion into regional affairs. The widespread perception within the region that the Middle East has long been a victim of "Western imperialism" of course exacerbates this animosity. Moreover, aggrieved groups throughout the Middle East contest the legitimacy of the regimes in Saudi Arabia, Kuwait, and the Gulf emirates which the United States is compelled to support, making America even more of a lightning rod for the politically disaffected. In this sense Osama bin Laden's brand of terrorism (which aims to compel the United States to remove its military forces from the Persian Gulf, and to replace America's client, the Saudi monarchy, with a fundamentalist Islamic government) dramatically illustrates U.S. vulnerability to the kind of "asymmetric warfare" of which some defense experts have warned.

The rise of new great powers is inevitable, and America's very primacy accelerates this process. If Washington continues to follow an adult-supervision strategy, which treats its "allies" as irresponsible adolescents and China and Russia as future enemies to be suppressed, its relations with these emerging great powers will be increasingly dangerous, as they coalesce against what they perceive as an American threat. But that is not even the worst conceivable outcome. What if a sullen and resentful China were to align itself with Islamic fundamentalist groups? Such a situation is hardly beyond the realm of possibility; partners form alliances not because they are friends, or because they have common values, but because they fear someone else more than they fear each other.

A strategy of preponderance is burdensome, Sisyphean, and profoundly risky. It is therefore time for U.S. policymakers to adopt a very different grand strategy: one that might be called offshore balancing. Rather than fear multipolarity, this strategy embraces it. It recognizes that instability—caused by the rise and fall of great powers, great-power rivalries, and messy regional conflicts—is a geopolitical fact of life. Offshore balancing accepts that the United States cannot prevent the rise of new great powers, either within the present American sphere (the European Union, Germany, Japan) or outside it (China, a resurgent Russia). Instead of exhausting its resources and drawing criticism or worse by keeping these entities weak, the United States would allow them to develop their militaries to provide for their own national and regional security. Among themselves, then, these states would maintain power balances, check the rise of overly ambitious global and regional powers, and stabilize Europe, East Asia, and the Persian Gulf. It would naturally be in their interests to do so.

It's always safest and cheapest to get others to stabilize the turbulent regions of the globe. Historically, however, this has seldom been an option, because if one lives in a dangerous neighborhood, one must be prepared to protect oneself from troublemakers rather than relying on someone else to do so. In fact, the only two great powers in modern history that successfully devolved onto others the responsibility for maintaining regional stability are Britain during its great-power heyday (1700–1914) and the United States (until 1945). They were able to do so because they had moats—a narrow one for England, and two very big ones for the United States—that kept predatory Eurasian great powers at bay.

As offshore balancers, Britain and the United States reaped enormous strategic dividends. While they were shielded from threatening states by geography, London and Washington could afford to maintain militaries smaller than those of Continental powers, and concentrate instead on getting rich. Often they could stay out of Europe's turmoil entirely, gaining in strength as other great powers fought debilitating wars. And even in wartime offshore balancers have enjoyed advantages that Continental powers have not. Instead of sending big armies to fight costly Continental wars, Britain, for instance, relied on its navy to blockade those states bidding for mastery of Europe and on its financial power to underwrite coalitions against them, and stuck its allies with the greater part of the blood price of defeating those powers that aspired to dominate the Continent.

The United States, of course, followed a similar strategy during World War II. From 1940 to 1944 it confined its role in the European war to providing economic assistance and munitions to the Soviet Union and Britain and—after entering the war, in December of 1941—to relatively low-cost strategic air bombardment of Germany, and peripheral land campaigns in North Africa and Italy. The United States was more than happy to delay the invasion of Europe until June of 1944. By then the Red Army—which inflicted about 88 percent of the Wehrmacht's casualties throughout the war—had mortally weakened Germany, but at a staggering cost.

Taken together, the experiences of Britain and America highlight the central feature of the offshore balancing strategy: it allows for burden *shifting*, rather than burden sharing. Offshore balancers can afford to be bystanders in the opening stages of conflict. Because the security of others is most immediately at risk, an offshore balancer can be confident that those others will attempt to defend themselves. Often they will do so expeditiously, obviating the offshore balancer's intervention. If, on the other hand, a predominant power seems to be winning, an offshore balancer can intervene decisively to forestall its victory (as Britain did against Philip II, Louis XIV, and Napoleon). And if the offshore balancer must intervene, the state aspiring to dominance will already have been at least somewhat bloodied, and thus not as formidable as it was for those who had the geopolitical misfortune to constitute the first line of defense.

The same dynamics apply—or would, if the United States gave them a chance—in regional conflicts, although not quite as dramatically. Great powers that border restive neighbors, or that are economically dependent on unstable regions, have a much larger interest than does the United States in policing those areas. Most regional power balances (the relative positions of, say, Hungary and Romania, or of one sub-Saharan state and another) need not concern the United States. America must intervene only to prevent a single power from dominating a strategically crucial area—and then only if the efforts of great powers with a larger stake in that region have failed to redress the imbalance. So for an offshore balancing strategy to work, the world must be multipolar—that is, there must be several other great powers, and major regional powers as well, onto which the United States can shift the burden of maintaining stability in various parts of the world.

For America the most important grand-strategic issue is what relations it will have with these new great powers. In fostering a multipolar world—in which the foreign and national-security policies of the emerging great powers will be largely devoted to their rivalries with one another and to quelling and containing regional instability—an offshore balancing strategy is, of course, opportunistic and self-serving. But it also exercises restraint and shows geopolitical respect. By abandoning the "preponderance" strategy's extravagant objectives, the United States can minimize the risks of open confrontation with the new great powers.

Although jockeying for advantage is a fact of life for great powers, coexistence, and even cooperation between and among them, is not unusual. Offshore balancing seeks to promote America's relative power and security, but it also aims to maximize the opportunity for the United States to be on decent terms with the other great powers. In this sense the strategy has

much in common with Richard Nixon and Henry Kissinger's vision of détente. That policy was a significant departure from previous Cold War approaches, in that the United States explicitly recognized the Soviet Union as a collaborator in, rather than a challenger to, the effort to maintain the stability of the international system. To understand this dramatic shift, contrast the inaugural address of John F. Kennedy, in which that paragon of Cold War liberalism advanced the stirring but rather dangerous notion that in the struggle with communism the United States would "pay any price, bear any burden," with Nixon's first inaugural address, which promulgated the realistic but conciliatory message that "we cannot expect to make everyone our friend, but we can try to make no one our enemy." This was détente's animating sentiment.

Détente was based on the assumption (hardly contested at the time) that the USSR wouldn't go away. Because the superpower rivalry could not be resolved without destroying humanity, there was, as Kissinger declared, "no alternative to coexistence." Détente, then, was a strategy for managing a permanent relationship. In what Nixon and Kissinger hoped would evolve into a mature relationship, Moscow and Washington would acknowledge each other's legitimate interests and try not to allow disagreements to poison accommodations. Détente was a shift in style with substance—or, rather, a shift in style with substantive consequences. The Soviets and the Chinese were to be approached not as alien ideologues but as intelligent adults with whom the United States should find a substantial area of common interest.

Similarly, an offshore balancing strategy would dictate that in order to coexist with the emerging great powers, or even to enjoy cooperative ties with them (in efforts to combat Islamic terrorism, for instance), the United States must start treating such powers like fellow adults. This would mean both accepting them as peers and acknowledging the legitimacy of their national interests. In concrete terms, here is how an offshore balancing strategy would apply to particular cases.

- Today, not for the first time in its history, Russia is down and out as a great power. But it has come back before and probably will do so again. Moreover, for the United States, Russia is crucial as a potential ally in three regions: Europe (vis-à-vis a European "superstate," or Germany if the EU project fails), East Asia (vis-à-vis China), and the Persian Gulf and Central Asia. As an offshore balancer, the United States would abandon plans for NATO expansion (which Russia regards as a strategic threat), and if Washington decides to undertake a national missile-defense program, it would follow through on the recent agreement to make deep cuts in its strategic nuclear arsenal to reassure Russia and China that it doesn't seek to gain a first-strike advantage. It would allow Russia to supervise its legitimate sphere of influence—in Chechnya and in Central Asia, where it is combating Islamic fundamentalists, as well as in parts of Eastern Europe and in the states that formerly composed the Soviet Union. America's direct sphere of influence embraces the area from the Canadian Arctic to Tierra del Fuego and from Greenland to

Guam. Surely we can tolerate other great powers' enjoying spheres of influence in their own parts of the world.

- With respect to China, the United States would recognize that the Taiwan issue is an internal Chinese matter. Taiwan's unresolved status is a legacy of the civil war that ended on the mainland in 1949. It is worth recalling that before the outbreak of the Korean War, Secretary of State Dean Acheson advised that the United States should extricate itself from the unfinished business of the Chinese civil war and leave Taiwan to its fate. A half century later it is time for the United States finally to do so. Washington would also fundamentally re-examine its notion of what constitutes a "China threat." That China, the largest and potentially most powerful state in East Asia, would seek a more assertive political, economic, and military role in the region—and would even want to end America's current strategic superiority there—hardly meets that threshold (although it is no doubt alarming to China's immediate neighbors). The United States should also mute its criticism of China's human-rights policy. Washington simply can't transform China into a liberal, free-market democracy, and U.S. pressure only exacerbates the friction in Sino-American relations. Generally, an offshore balancing strategy would hold that fatalism should replace idealism in America's attitude toward what used to be called the internal arrangements of other countries. It would also hold that our pragmatic policy choices, born of self-interest, should be embraced as such, and not clothed in altruism or idealism. Seeking to engender changes in other nations' fundamental values so that they resemble America's is an unreasonable goal of foreign policy. An offshore balancing strategy would accept other nations for what they are, or what their history has made them.

- With respect to Europe, the United States would endorse the EU's efforts—which Washington now opposes—to acquire the military capabilities it needs to defend its interests independent of the United States. At the same time, the United States would begin a phased withdrawal from its European security commitments. To be sure, many U.S. policymakers have argued that the Europeans have demonstrated their incapacity (during the Balkan crises, for instance) to act effectively without U.S. "leadership." But these protests are hypocritical—who can blame the Europeans for their inability to assert themselves in security affairs when Washington has for decades repeatedly squelched European initiatives that would have made that assertion possible? An offshore balancing strategy would hold that America's strategic interest in Europe does not demand that Washington insure against every untoward event there. Disorder in the Balkans and other places on Europe's fringes should be a matter for Europeans, who have the wherewithal to combat it, quarantine it, or, if they choose, ignore it. The United States would follow a similar policy with respect to Japan. Washington would announce to Tokyo its intention to terminate the Mutual Cooperation and Security Treaty within a specified time period (say, five years), at the end of which Japan, for more than fifty years a politically stable state, would have developed whatever military means it believes necessary to func-

tion as an independent great power. An offshore balancing strategy would turn on a simple truth: other states have at least as much interest as the United States does in secure sea-lanes, access to resources, and regional stability. The less America does, and the less others expect it to do, the more other states will do to help themselves.

Recognizing the legitimacy of other great powers' spheres of influence offers the United States a further strategic advantage. The Persian Gulf and Central Asia show why. Russia and China both are profoundly concerned about the spread of Islamic fundamentalism on their peripheries. In Chechnya, in Central Asia (where Russian troops help to defend the former Soviet republics of Uzbekistan and Tajikistan), and in the Caucasus, Moscow has fought major military campaigns to protect its southern flank against militant Islam. China, too, is combating terrorism fomented by Islamic separatists, in the Xinjiang province. Last June, Beijing and Moscow entered into a security relationship, the Shanghai Cooperation Organization (which also embraces three of the former Soviet republics in Central Asia), to coordinate efforts to combat the common threat to their security posed by these Islamic fundamentalist terrorist groups—groups linked to the Taliban and Osama bin Laden. Similarly, India, a possible future great power, has been battling Islamic terrorists who are waging a proxy war on Pakistan's behalf to wrest the disputed province of Kashmir away from New Delhi. Simply put, for reasons of security and access to oil, Russia, China, India, Western Europe, and Japan have strong reasons—stronger than America's—to pacify Central Asia and the Persian Gulf. By adopting an offshore balancing strategy, the United States will compel them to do so.

Passing the buck would help the United States out of the impasse that securing Afghanistan promises to be. The political and military challenges the war poses underscore how difficult and costly will be the effort to restore order in the country and the region when the fighting stops. When the United States has achieved its military goals in Afghanistan, it should announce a phased withdrawal from its security commitments in the region, shifting to others the hard job of stabilizing it.

The complexities involved in that job are numerous. Washington's very strategy of primacy, and America's concomitant military presence in the region, are in themselves a source of instability, especially for the regimes on which the United States relies. The regimes in Saudi Arabia and Pakistan, for instance, face doubtful prospects precisely because their close connection to Washington intensifies radical nationalist and Islamic fundamentalist opposition within those countries. For this reason none of the regional regimes in the current coalition can be especially dependable allies. Only with enormous pressure did a few of them even allow American forces to conduct offensive strikes on Afghanistan from bases on their territory. And fearing that popular anger at the U.S. military campaign will trigger domestic political explosions, many of these states pressed Washington to bring an early end to the war.

If America remains in the region indefinitely, it will have to prop up these unpopular or failing regimes. In Saudi Arabia the

United States could easily find itself militarily involved if internal upheaval threatens the monarchy's hold on power. To forestall economic collapse in Pakistan, Washington will have to donate billions of dollars in direct and indirect assistance. Finally, if the United States continues to play the role of regional gendarme, it will assume the thankless—and probably hopeless—burden of trying to put Afghanistan together again. Divided along ethnic, linguistic, and clan fault lines, the various factions inside Afghanistan cannot agree on that country's future political organization. (The forces making up the anti-Taliban contingent seem only to agree that they resent U.S. bombing of their country.) That the outside powers have conflicting goals for Afghanistan's future further complicates any sorting out of Afghanistan's political structure. If ever there was a place where America should devolve security responsibilities to others, it is the Persian Gulf and Southwest Asia region. Again, Western Europe, Japan, Russia, China, and India all have greater security and economic interests in the region than does the United States, and if America pulls out, they will police it because they must.

Rather than attempt to impose a Pax Americana on this endemically turbulent area, the United States should devote the resources it currently spends on this costly and dangerous job to rendering the region economically and strategically irrelevant. That is, America should pursue a national energy policy that would develop alternative sources of energy for the United States and, more important, the rest of the industrialized world. This colossal scientific and industrial effort should be our highest national-security priority (see "Mideast Oil Forever?," by Joseph J. Romm and Charles B. Curtis, April, 1996, *Atlantic*). If the United States shifts responsibility for stabilizing the region to the other great powers, the real price of Persian Gulf oil will become extremely high for them. It would then be in their interests to pool resources and expertise with America in what would amount to an international Manhattan Project to obviate the need for that oil—thus dramatically reducing the revenue streams to the regimes in Iran, Iraq, and Saudi Arabia. Doing so is surely a common international interest. If Washington were to spend the approximately $106 billion that—according to Earl Ravenal, a former Pentagon analyst—it is devoting this year to defending the Persian Gulf region, and if Western Europe, Japan, China, and Russia were to kick in what they would otherwise spend on policing the region, it's hard to imagine that this goal couldn't be achieved.

Some will assert, correctly, that if it abjures a strategy of preponderance, America will sacrifice some of the awe with which it is viewed by the world. But less awe and less influence will bring the United States more security. Some will object that the policy we advocate shuns the inspiring role of America as "the indispensable nation." But such a grandiose vision, while pleasing to our image of ourselves, is the antithesis of statecraft, which must be guided by discrimination on the basis of power, interest, and circumstance. Historically, the most imaginative statesmen and policies have hardly been visionary. For centuries, with flexibility and subtlety, British diplomats pursued a grand strategy that aimed at nothing more inspiring than ensuring a balance of power among the states of Western Europe. This was really just tactical fine-tuning on a grand scale, and so aroused the consternation of idealists of every stripe. For their part, America's nineteenth-century statesmen could not have been less idealistic or more pragmatic as they, by adroitly exploiting European great-power rivalries, maneuvered the British, the French, and the Spanish out of North America and established American predominance in the Western Hemisphere—probably the most stunning diplomatic achievement of modern history, and the very model of a successful multipolar strategy.

The policy we advocate is informed by the conviction that history is "just one damned thing after another"; we see no end to power politics. And we hold that the purpose of grand strategy isn't the pursuit of new world orders but simply making the best of bad choices—to use the political philosopher Michael Oakeshott's metaphor, keeping afloat in "a boundless and bottomless sea; there is neither harbor for shelter nor floor for anchorage, neither starting point nor appointed destination." Ours is a grand strategy for the long haul—and so, by the lights of visionaries who see foreign policy as a means of pursuing millennialist goals, not a very grand one. But the grander its foreign-policy vision, the more a state is trapped in the tyranny of its own construct: although recent administrations display an odd compulsion to devise and promulgate such visions of America's role in the world, those visions are in fact incompatible with the push and pull of strategy.

Finally, although some might characterize an offshore balancing strategy as isolationist, it emphatically is not. Rather, its guiding principle is a clear-eyed realism. It is a workaday policy—pragmatic, flexible, and opportunistic. But it will also bring America into a more respectful and natural relationship with the other great powers, as the United States forsakes the temptations of hegemony. "A mature great power will make measured and limited use of its power," Walter Lippmann wrote in 1965.

> It will eschew the theory of a global and universal duty, which not only commits it to unending wars of intervention, but intoxicates its thinking with the illusion that it is a crusader for righteousness… I am in favor of learning to behave like a great power, of getting rid of the globalism, which would not only entangle us everywhere, but is based on the totally vain notion that if we do not set the world in order, no matter what the price, we cannot live in the world safely… In the real world, we shall have to learn to live as a great power which defends itself and makes its way among other great powers.

MR. ORDER
Meets
MR. CHAOS

We live in an era of unprecedented prosperity, but when the financial bubble bursts we'll plunge into a world depression. Nations no longer go to war, but civil wars are booming. Humanity has embraced the idea of environmental interdependence, but the global ecosystem is in terminal crisis. Depending on your perspective, we stand either on the verge of a golden age or at the brink of disaster. Robert Wright and Robert Kaplan, two of the United States' most perceptive observers of world affairs and the human condition, met recently in Washington, D.C., to offer conflicting views of the path of history.

My Minivan and World Peace

By Robert Wright

Anyone who knows me would be surprised to find me cast as an optimist, but when you're juxtaposed with Robert Kaplan, it's not hard to come off looking pretty chipper and upbeat about the world.

What is the basis for my relative optimism? My prescription and diagnosis are built upon the notion of the non-zero-sum game, which is a reference to game theory. A zero-sum game is what you see in an athletic event like tennis: Every point in the match is good for one player and bad for the other. So the fates of the players are inversely correlated. In a non-zero-sum game, the fortunes can be positively correlated; the outcome can be win-win or lose-lose, depending on how competitors play the game. And, in fact, in a tennis doubles match the players on the same team have a highly non-zero-sum relationship because they'll both win or they'll both lose.

Nowadays we're all embedded in lots of non-zero-sum relationships that we really don't even think about. For instance, when I bought my Honda minivan I was in a non-zero-sum relationship with workers in various countries. The deal was I paid a tiny hit of their wages and they

built me a car. It is characteristic of globalization that it embeds us in these non-zero-sum relationships. It makes our fates more correlated with the fates of people at great distances. It's a subtle process that we usually don't think about, but every once in a while this correlation of fortunes becomes glaringly evident, as was the case with the Asian crisis when we realized that a financial downturn can instantly spread around the world; or when a virus spreads across the Internet and you realize that computer users on different continents are all vulnerable, their fates are correlated.

In theory, as globalization makes relations among nations more and more non-zero-sum, you would expect to see more in the way of institutionalized cooperation to address these problems. That is not a pathbreaking insight. For some time now, political scientists have been talking about the growing interdependence of nations and the growing logic behind cooperation. But I believe that this process is now moving so fast that, much sooner than most people expect, we're going to reach a system of institutionalized cooperation among nations that is so thorough it qualifies as world governance. I don't mean world government, a single centralized authority. I imagine a looser mix of global and regional organizations. But still I'm imagining some very significant sacrifices of national sovereignty to supranational bodies. We've already seen a little of this surrender of national sovereignty with the World Trade Organization, and I would argue there was a little bit of surrender (a well-advised surrender) when 174 nations signed the Chemical Weapons Convention.

Wright: "History is not just one damn thing after another, it's a process with a direction; it has an arrow."

I fully expect this trend toward global governance to continue, although I'm much more confident about it happening in the long run than in the short run. The zone of non-zero-sumness has been expanding for a very long time: You can go back to the Stone Age when the most complex polity on earth was a hunter-gatherer village and chart the evolution to the level of the chiefdom—a multivillage polity—and then to the level of the ancient state, and then to the system of modern nation-states, and so on. The key element that has driven the evolution of social complexity and of governance to higher levels is technology. Sometimes it is information technology, as when the invention of writing often accompanied the evolution of the first ancient states. Sometimes it is transportation technology and sometimes, ironically enough, it is weapons technology. Weapons technologies can make relations much more non-zero-sum—certainly nuclear weapons make war a very non-zero-sum endeavor in the sense of making it a lose-lose game, wherein the object of the game is never to play. Nuclear weapons thus strengthen the argument for a system of collective security pursued through some supranational institution such as the United Nations.

We don't know in detail what the future of technological evolution will be, but we have a pretty good idea. Information technologies will continue to evolve and enmesh people in webs of transactions, interactions, and interdependence. Weapons technologies will evolve, but perhaps more important, the information about how to build very lethal weapons of mass destruction will likely be accessible to more and more people. Thus, almost all nations share a common interest in controlling the development and use of these weapons. Technological evolution will continue doing what it has done for the broad sweep of history, which is expanding the realm of non-zero-sumness, making the fates of peoples and nations more correlated, and in the process driving governance to a higher level, to the global level.

Kaplan: "You're right... but neither is it on a direct, predetermined course... The course of history [is] just a gradual improvement, punctuated with a lot of ups and downs."

That the fates of the world's people have grown more and more correlated over time is not by itself especially good news. As you may have noticed, many examples of non-zero-sum dynamics are actually negative-sum games, lose-lose games, where the object of the game is to break even. Global warming is an example of such a negative-sum game—where we just want to fend off the bad outcome—that I think calls for institutionalized cooperation and some real, if small, sacrifice of national sovereignty.

So when I argue that history features more and more of this non-zero-sumness, that statement isn't by itself good or bad, it just is. It's just something we have to reckon with. But there is one feature of the direction of human history that is at least mildly upbeat, in some ways redeeming. It's what I call the expanding moral compass. Philosopher Peter Singer has written about this. If you go back to ancient Greece, there was a time when members of one Greek city-state considered members of another Greek city-state literally subhuman. They would slaughter and pillage without any compunction whatsoever. Then the Greeks underwent a process of enlightenment and they decided that actually other Greeks are humans, too. It's just the Persians who aren't humans. (Okay, it was limited progress, but it was progress.) And today I think we've made more progress, especially in economically developed nations. I think almost everyone in such countries would say that people everywhere, regardless of race, creed, or color, deserve at least minimal respect.

If you ask why that has happened, I argue that it gets back to this basic dynamic of history, this growth of non-zero-sumness. If you look at Greece at the time of their limited enlightenment, relations were growing more non-zero-sum among Greek city-states because they were fighting a war together against the Persians. They needed each other more, they were in the same boat, and to cooperate they had to accord each other at least minimal respect. And if you ask why an ethos of moral universalism now prevails in economically advanced, globally integrated nations, I would say it's the same answer. If you ask me why don't I think it's a good idea to bomb the Japanese, I'd say, "For one thing, because they built my minivan." I'm proud to say I have some more high-minded reasons as well, but I do think this basic, concrete interdependence forces people to accord one another at least minimal respect, to think a little about the welfare of people halfway around the world. I expect this dynamic to grow and persist in the future because in a world where disease can spread across borders in no time at all, it's in the interest of Americans to worry about the health of people in Africa or Asia. In a world where terrorists can wield unprecedentedly lethal technologies, it's in the interests of Americans to worry about political grievances before they fester to the point of terrorism. One feature of a globalized society is that disaster can happen at the global level, so we're now in this process where either we grasp the moral and political implications of this increasingly shared fate we have with other people or very bad things will happen.

The modern world is in many ways a disoriented and disturbing place. Things are changing very fast, but I think if you look at the broad sweep of the past it offers a way to orient ourselves. History is not just one damn thing after another, it's a process with a direction; it has an arrow. And I think if we use that arrow to orient ourselves then I would predict that the coming decades will not be characterized by chaos.

Robert Wright is author of Nonzero: The Logic of Human Destiny *(New York: Pantheon Books, 2000) and a visiting scholar at the University of Pennsylvania.*

Hope for the Best, Expect the Worst

Robert Kaplan

Well, Bob, while you've been looking ahead to discern the broad, cosmic sweep of history, I've been looking ahead just 10 or 15 years in terms of foreign policy—which is often most effective when it's conceived of in light of worst-case scenarios, in the hope that those scenarios don't occur. I should remind you that constructive pessimism is profoundly in the American tradition. It's the basis for the U.S. Constitution. If you read *The Federalist Papers*, you can see that Americans have become a country of optimists over 225 years precisely because we've had the good fortune of having our systems of government founded by pessimists. The French Revolution conversely was founded on optimism, on the belief that elites could engineer positive results from above, and it devolved into the guillotine and Napoleon's dictatorship. Alexander Hamilton, whom I consider the greatest of the Founding Fathers, said don't think there will be fewer wars in the world simply because there will be more democracies. In *Federalist Number Six* he said there are as many wars from commercial motives as from territorial aggrandizement. So it is in that spirit of *The Federalist Papers* that I'm going to present a scenario about what worries me over the next 10 or 15 years.

I wrote in 1994 that even as part of the globe was moving toward economic prosperity, another part—containing much of the population—was marching in another direction due to issues such as demography, resource scarcity and disease. So let me tell you how I see things now, seven years later. The European colonialists did a lot of terrible things, but they did bring a certain degree of order to much of sub-Saharan Africa, South Asia, and Central Asia. That colonial grid work of states started dissolving in the 1990s when we saw the weakening or outright collapse of several marginal places. I use the term "marginal" not because their well-being wasn't important, but because they had low populations, their

economies were small, and they didn't really affect the region around them all that much. Somalia, Sierra Leone, Tajikistan, Haiti, and Rwanda were not core regional states in any sense, but look at how they disrupted the international community.

Wright: "We're going to reach a system of institutionalized cooperation among nations that is so thorough it qualifies as world governance."

I believe that, for a number of reasons, we're going to see the weakening, dilution, and perhaps even crackup of larger, more complex, modern societies in the next 10 or 15 years in places such as Nigeria, Ivory Coast, and Pakistan. And we're going to see severe crises in countries like Brazil and India. This dissolution of the colonial grid work is going to create the kind of crises where there will be no intervention scenarios, or the intervention scenarios will be far worse than they were in Bosnia or Sierra Leone. The problem is not that these places have particularly bad governments. They're coping as best as any could. The reasons are far more complex and intractable.

First of all, these societies are modernizing. Although history teaches us that modern democratic institutions provide stability, history also reveals that the process of creating and developing modern democratic institutions is very destabilizing. As free-market democracies develop, more and more people are brought into the political process. And all of these people are full of yearning, ambitions, and demands that governing institutions very often cannot keep pace with. So things start to break down here and there. It is economic growth that typically fuels political upheavals, not poverty.

The other challenge to the stability of the nation-state is demography. You hear a lot about how the world population is aging, but that's over the long term and throughout the world as a whole [see "The Population Implosion," FOREIGN POLICY, March/April 2001]. But when you look ahead at just 20 or 30 countries over the next 10 or 20 years, you see dramatic rises in the youth population (what demographers call "youth bulges"). When you watch your television and you see unrest or rioting in Indonesia, Ivory Coast, Gaza, and the West Bank, what's similar about all of them? All of the violence is typically conducted by young men, ages 15 to 29, who are unemployed and frustrated. The sector of the young male population within this age group is going to grow dramatically in the countries that already have tremendous unrest and are already on the edge. In other words, the places that will have a population pyramid that is bottom-heavy with the youngest members of society are the ones that can least afford it.

And if that isn't enough, you've got urbanization. The 21st century is going to be the first century in world history when more than half of humanity will live in cities. Even sub-Saharan Africa is almost 50 percent urban. Urban societies are much more challenging to govern than rural societies. In rural societies people can grow their own food, so they are less susceptible to price increases for basic commodities. Rural societies don't require the complex infrastructure of sewage, potable water, electricity, and other things that urban societies have. Urbanization widens the scope of error for leaders in the developing world while simultaneously narrowing the scope for success. It is harder to satisfy an urban population than a rural population, especially when that population is growing in such leaps and bounds that governing institutions simply cannot keep pace.

Then you have resource scarcity, particularly water. I spent the summer in a small village in Portugal where we only had running water about eight hours a week. We had to drive about half a mile to a local fountain to fill pitchers of water. Anyone who has not gone without water has no idea what it's like not to be able to flush your toilet or take a bath. There's been a drought for the last four years across a swath of South Asia from Afghanistan, Pakistan, and into India. Dams are low, so there is not enough water for drinking or generating electricity. So in these hot cities of the subcontinent you have less and less air conditioning in the summer. This kind of stuff doesn't necessarily cause political crises, but it's all part of the background noise that aggravates existing crises. This frustration worsens ethnic tension and makes social divides harder to resolve. In short, people get angry. There was a spate of riots in Karachi, Pakistan, not long ago that was preceded by an extended period when there was very little electricity due to water shortages.

Kaplan: "We're going to see the weakening, dilution, and perhaps even crackup of larger, more complex, modern societies in the next 10 or 15 years..."

Then there's the issue of climate change. Let's just say for the sake of argument that this whole global warming issue has been exaggerated, that it really doesn't exist, that it's not going to be a problem. Well, even if you factor out global warming, the normal climatic variations of the earth during the next few decades will still ensure devastating floods and other upheavals because, for the first time in world history, you have hundreds of millions of human beings living in environmentally fragile terrain— where perhaps human beings were never meant to live at all. So even without global warming you're going to have natural events that can spark political upheaval.

And finally, the other factor that's going to spark serious institutional crises in a lot of states is democracy. Everyone wants to be democratic, no use denying it. But democracy tends to emerge best when it emerges last. It should be the capstone to all other types of development, when you already have middle classes that pay income taxes, when you already have institutions run by literate bureaucrats, when the major issues of a society (such as territorial borders) are all resolved and you already have a functioning polity. Then, and only then, can a society cope with weak minority governments. Then, and only then, can democracy unleash a nation's full potential. Right now, we're seeing democracy evolve in many places around the earth accompanied by unemployment and inflation rates every bit as dire as Germany in the 1930s, when Hitler emerged under democratic conditions, and in Italy, when Mussolini came to power in the early 1920s. I'm not arguing against democracy, but I believe democracy will be another destabilizing factor.

If it seems like I'm deliberately cultivating a sense of the tragic it's because that's how you avoid tragedy in the first place. Remember that Klemens von Metternich was so brilliant in creating a post-Napoleonic order that Europe saw decades of peace and prosperity—so much so that politicians in France and England lost their sense of the tragic. All they saw ahead were optimistic scenarios and, as such, they stumbled and miscalculated their way into World War I. Take my concern in that spirit.

Robert Kaplan is author of The Coming Anarchy: Shattering the Dreams of the Post Cold War *(New York: Vintage Books, 2000) and a senior fellow at the New America Foundation.*

In the Long Run, We're All Interdependent

Robert Wright responds.

Well Bob, I'm actually something of a fan of pessimism myself. I think it focuses us on the problems that need our attention. I find it particularly heartening that your books have a sizable American readership, since that suggests that Americans increasingly realize their fates are intertwined with the fates of people around the world. But I don't want to overdo the pessimism. And in particular I don't want to make it sound like globalization and its attendant technological fluctuations are part of some kind of uniformly bad force. I'm actually something of a cheerleader for globalization. It has problems, but I think on balance it's a good thing.

You said that the world was increasingly dividing into two parts, echoing the common refrain that globalization exacerbates income inequality worldwide. But that conclusion actually depends on how you examine the data. If you look at the number of rich versus poor nations, then you can certainly make that argument. But if you look at the total number of people in the world, ignoring where the borders fall, then what's happening in absolute terms is that there are fewer poor people than there used to be. And even in relative terms, it's far from clear that income inequality is growing, and a number of people have argued that the income gap is actually shrinking worldwide. It turns out that many of the world's poor people are concentrated in a few very large countries (like China and India) that have seen more progress than some of the smaller countries (notably those in Africa). But even in Africa, globalization has seen a kind of vindication: The countries that have seen the most economic advancement are the ones that are most open to trade and investment.

Another virtue of globalization is that it is basically an antiwar activity. I think as peoples and nations become more economically intertwined, war becomes more of a lose-lose kind of non-zero-sum game that it doesn't make sense to play. There still are wars in the world, but there is a very interesting feature of the modern world that is insufficiently noted: We increasingly think of wars between nations as something that poor countries do. Nobody expects any of the most economically advanced nations to go to war with one another, which represents a real shift of mind-set. If you look back at most of history it was really standard procedure for the most powerful polities to go to war with one another. Nowadays, most interstate fighting breaks out in parts of the world that could be termed "underglobalized" areas. I don't mean that pejoratively. It's not their fault that they're underglobalized. There are various quirks of history or geographical circumstance that explain why some parts of the world have advanced faster economically than others. But the fact is that wars are mostly a threat in the poorest parts of the world.

Now, when you get to subnational conflict, war within nations, I agree, Bob, that's a problem that may grow more serious. You argue that conflict is often exacerbated by economic development. I'd add another way in which modernization has given rise to intranational conflict, and that is through the propagation of information technology. As I suggested earlier, information technology has certain globalizing effects, but it also has fragmenting effects because whenever you lower the cost of communication you make it easier for small groups with meager resources to organize. It's no coincidence that the Protestant Reformation roughly coincided with the invention of the printing press. After Martin Luther had tacked up his 95 Theses, printers took it upon themselves to start printing them in various cities. That is how Luther first organized the masses, because printing was suddenly so cheap.

You're seeing the same thing in the modern world thanks to the Internet. Inevitably, information technology is going to empower separatist groups such as Muslims in the west of China and Basques in Spain. But, in the long run, you can imagine this secessionist frenzy working itself out, because as some of these subnational groups choose to drop out of nations they can at the same time cement themselves into supranational bodies. In fact, the Quebec separatists have said they plan to join the North American Free Trade Agreement (NAFTA) as soon as they get out of Canada, and I would expect that European separatist groups would be strongly tempted to join the European Union. So, I certainly agree that globalization presents us with all kinds of short-term difficulties, but I do still think it's a process that is fundamentally beneficial and will lead to a new equilibrium in the long run.

Passion Play

Robert Kaplan responds.

Bob, let me draw some distinctions here, just in the spirit of argument. You tend to put a lot of emphasis on the ability of people to make good, rational choices. But if you think that people are always going to behave accord-

The Clash of Interpretations

Samuel Huntington, cofounder of FOREIGN POLICY *and one of the world's most influential political scientists, incited widespread debate during the 1990S with his "Clash of Civilizations" thesis, which maintained that cultural fault lines would dominate the post-Cold War world. How do Huntington's views fare when examined through the prisms of order and chaos?*

Robert Wright:

I thought Dr. Huntington somewhat overdid the fissures between civilizations. The world is full of examples where nations with very different cultural heritages are on very good and stable terms, the United States and Japan to name just one example. At the same time, I realize that cultural history matters and it can be a source of tension. But to the extent that he was right about that, what bothered me was his prescription that we should be true to our cultural spirit. He said, for instance, that Australia is not really an Asian nation, so it should not be in a trade bloc with Asian nations, it should be part of an expanded NAFTA. It seems to me that the moral of the story is exactly the opposite: If indeed fissures among civilizations are deep and threatening (and sometimes they are), then the effort should be to bridge them with supranational organizations like trade blocs.

Robert Kaplan:

I thought Sam Huntington's thesis was brilliant, which is proven by the fact that it got everybody angry. Just look at the three countries that were brought into NATO: Hungary, Poland, and the Czech Republic were all part of Western Christendom, and countries that were left out (Bulgaria and Romania) were part of the Eastern Orthodox world. When you look at the current borders of NATO, it's basically a variation of the Holy Roman Empire in the 11th century. If you travel through the Middle East you see increasing tensions in villages between Christians and Muslims. All the corporate mergers are between American and European companies. It's true they have different corporate cultures, but they're similar compared to corporate cultures in other parts of the world. And our main foreign-policy challenge of the next 20 years will be managing a relationship with China. Probably never before have we had a major adversary where there is such a great chance for cultural misunderstanding.

But it's important to recall that Huntington's scenario was a paradigm. It was just a big abstract argument that you judge on the basis of whether it is better than any other generalizing abstract argument. I think on that basis he succeeded very well, but I would agree with Bob that the solution is maybe to bring Romania into NATO, to bridge some of these gaps so that NATO doesn't evolve into a bastion of Western Christendom.

ing to their best, rational self-interest, read *Mein Kampf*. As Hamilton said, "the passions of men will not conform to the dictates of reason and justice, without constraint." The U.S. Constitution was established to slyly organize and control our passions. I'm not convinced that we're going to act any more rationally than we have in the past. It is true that there is a movement toward world governance, but a single, unifying thread is not necessarily a good thing. For instance, the European Union could readily devolve into a benign bureaucratic despotism that will ignore the interests of the lower middle classes. I think the nationalist movements popping up throughout Europe are already a reaction to this benign bureaucratic despotism from Brussels. If there is to be world governance, it has to be a kind that doesn't only appeal to the elites.

And those who feel marginalized have resources at their disposal that go way beyond the Internet. The Industrial Revolution was about bigness—big aircraft carriers, tanks, and railway grids—so that only large states could take advantage of the power the Industrial Revolution had to offer. But when you're talking about cyberwarfare, biological weapons, and this whole new gamut of weaponry in the post-Industrial Revolution, when you live in a world where just a telephone jack and a petri dish give you power, then it's not just large nations that can benefit. Nonstate actors who feel shut out can also magnify their power through this new technology. Power relationships are going to be more complex than ever. You were right when you said that technology drives history, but it doesn't necessarily do so in an orderly manner.

As for the moral universalism that you mention, I think we have to be a bit careful because the West is now using the term "global community" in the way we used to use the term "free world." We're trying to define the whole world in terms of our own moral outlook and what we want. There may be other powers and other cultures that have different views of how the world should be organized, so we have to be careful not to sound triumphalist.

And although inequality might be decreasing, I believe the most significant form of inequality is not what we see between the United States and sub-Saharan Africa, but the income gap you see between the wealthy coastal community and poor interior of a place such as Ghana. The biggest divides are between these globalized communities within the poorest countries—with their own electricity generators, their own water wells, and their own private security guards—that are hooked up to the world economy and surrounded by people with whom they have less and less in common.

You're right Bob, history is not one damn thing after another, but neither is it on a direct, predetermined course à la Karl Marx. The philosopher who captures it all best is Charles de Montesquieu who, in *The Spirit of the Laws*, sees the course of history as just a gradual improvement, punctuated with a lot of ups and downs.

But, lest I sound too contrarian, allow me to point out that I've been concentrating on the zero-sum games that occur within your vast non-zero-sum game. So, in that sense, there is no contradiction between us.

Coffee, Tea, or Apocalypse

Robert Wright responds.

I would hate to let stand the accusation that I think people behave rationally. I've long argued that people are really quite spotty on this particular front, and the saving grace of history has been that whenever people screwed up in one part of the world, there were people in another part of the world who picked up the torch. So when the Roman emperors messed up and began exhibiting the sort of increasingly autocratic rule that people are prone to given the opportunity, and the barbarian hordes did us the service of dismantling the Roman Empire, there were other empires that could continue to grow and thrive.

But I agree that once we reach the global level of organization, we face exactly the threat you describe in that there is no longer a plurality of experiments going on around the world. Increasingly it's one big experiment, and if it collapses that is bad news on a very large scale. That is one reason why I always say that history almost reads better than any novel. The protagonist of this story, that is to say the human species, has been driven more or less inexorably to a moment of fundamental moral and political choice. Making the right decision really depends on our level of moral enlightenment, an understanding of the commonality of all human beings. And if we fail to make the right moral choice, then apocalypse could well ensue. We live at a time of great drama and I absolutely acknowledge that chaos is one of the prospects we face.

A Whiff of Medievalism

Robert Kaplan responds.

Bob, I think chaos is more than a prospect. As divisions within societies and nation-states become greater, chaos might very well be inevitable. A new global community is taking root, but it is doing so at the top. Right now that community is so small that it's still a Potemkin village, but it's nor going to be that way forever. The middle and upper middle classes, what I call the "nouvelle cuisine classes," are merging together at the pinnacle. If you've ever been to the annual Davos conference, you are struck by how medieval it has become: You see the world's elites gathered together, just like the aristocrats of Germany, France, and England 200 or 300 years ago, who had more in common with each other than with their own peasants at home. Our sense of identity is being driven more by what economic class we belong to than what country we live in. The world is moving very slowly and inexorably out of the nation-state phase. That may ultimately lead to something better, but the process of leading to something better is very chaotic.

Want to Know More?

This debate is based on a dialogue between Robert Wright and Robert Kaplan that took place at the **Meridian International Center** in Washington, D.C., on March 13, 2001. Ambassador Walter Cutler moderated the event.

Robert Wright Recommends:

The great historian William McNeill's books *The Rise of the West: A History of the Human Community* (Chicago: University of Chicago Press, 1999) and *A World History* (New York: Oxford University Press,, 1999) convey the directional nature of history without sacrificing nuance. But seeing direction in history is so unpopular among historians that the task is often left to nonhistorians, such as political scientist Francis Fukuyama. His justly famous *The End of History and the Last Man* (New York: Free Press, 1992) differs from my *Nonzero: The Logic of Human Destiny* (New York: Pantheon Books, 2000) in that mine is a more materialist (in a somewhat Marxist sense) account of history's driving force. Peter Singer's *The Expanding Circle: Ethics and Sociobiology* (Oxford: Oxford University Press, 1981) acutely noted history's moral direction—the erratic movement over the last few millenniums toward moral universalism (though, again, my explanation is more materialist than his). Thomas Friedman's bestseller *The Lexus and the Olive Tree: Understanding Globalization* (New York: Farrar, Straus & Giroux, 1999) gives due weight to globalization's positive side without being Panglossian. His term the "superempowered angry man" captures the growing ability of small groups to wreak massive damage. For an abstract but fascinating account of how non-zero-sumness can beget more non-zero-sumness as the growth of cooperation strengthens the logic of further cooperation, see Robert Axelrod's classic *The Evolution of Cooperation* (New York: Penguin Books, 1990).

Robert Kaplan Recommends:

Paul Kennedy's *Preparing for the 21st Century*. (New York: Random House, 1993) is the best primer available on how demography, the environment, technology, and other related factors will test humankind's political skills in the future. Martin van Creveld's *The Transformation of War* (New York: Free Press, 1991) is a brilliant work of sustained, abstract thinking (perhaps the greatest since Carl von Clausewitz's *On War*) that shows how technology will be used toward primitive ends in future wars. In *The Clash of Civilizations and the Remaking of World Order* (New York: Simon & Schuster, 1996) Harvard professor Samuel Huntington—who, since the 1950s, has been a controversial but uncannily accurate forecaster—offers a prognosis of a future where tangible cultural divides will replace artificial, ideological ones. My own *The Coming Anarchy: Shattering the Dreams of the Post Cold War* (New York: Vintage Books, 2000) identifies the demographic, environmental, and political problems ahead while finding hope in ancient and modern philosophy.

Among the greatest works of philosophy that demonstrate how and why order must precede democracy—and how the search for order stems from the human need to be protected from others—is Thomas Hobbes' *Leviathan* (New York: W. W. Norton & Company, 1997). Arnold J. Toynbee's *A Study of History* (Oxford: Oxford University Press, 1987) presents an epic pageant explaining how culturally based states have arisen out of the ashes of preceding ones.

• For links to relevant Web sites, as well as a comprehensive index of related FOREIGN POLICY articles, access **www.foreignpolicy.com**.

UNIT 2
Population and Food Production

Unit Selections

5. **The Big Crunch**, Jeffrey Kluger
6. **Breaking *Out* or Breaking *Down***, Lester R. Brown and Brian Halweil
7. **Bittersweet Harvest: The Debate Over Genetically Modified Crops**, Honor Hsin

Key Points to Consider

- What are the basic characteristics and trends of the world's population? How many people are there? How long do people typically live?

- How fast is the world's population growing? What are the reasons for this growth? How do population dynamics vary from one region to the next?

- How does rapid population growth affect the quality of the environment, social structures, and the ways in which humanity views itself?

- There is a growing debate about genetically modified food. What are the differing perspectives on this debate and what issues are likely to be contested in the near future?

- How can economic and social policies be changed in order to reduce the impact of population growth on environmental quality?

- In an era of global interdependence, how much impact can individual governments have on demographic changes?

 **Links: www.dushkin.com/online/**
These sites are annotated in the World Wide Web pages.

The Hunger Project
   http://www.thp.org
Penn Library: Resources by Subject
   http://www.library.upenn.edu/cgi-bin/res/sr.cgi
World Health Organization
   http://www.who.int
WWW Virtual Library: Demography & Population Studies
   http://demography.anu.edu.au/VirtualLibrary/

After World War II, the world's population reached an estimated 2 billion people. It had taken 250 years to triple to that level. In the 55 years since the end of World War II, the population tripled again to 6 billion. When the typical reader of this book reaches the age of 50, experts estimate that the global population will have reached 8½ billion! By 2050, or about 100 years after World War II, some experts forecast that 10 to 12 billion people may populate the world. A person born in 1946 (a so-called baby boomer) who lives to be 100 could see a six-fold increase in population.

Nothing like this has ever occurred before. To state this in a different way: In the next 50 years there will have to be twice as much food grown, twice as many schools and hospitals available, and twice as much of everything else just to maintain the current and rather uneven standard of living. We live in an unprecedented time in human history.

One of the most interesting aspects of this population growth is that there is little agreement about whether this situation is good or bad. The government of China, for example, has a policy that encourages couples to have only one child. In contrast, there are a few governments that use various financial incentives to promote large families.

Some experts view population growth as the major problem facing

the world, while others see it as secondary to social, economic, and political problems. The theme of conflicting views, in short, has been carried forward from the introductory unit of this book to the more specific discussion of population.

As the world celebrates the new millennium, there are many population issues that transcend numerical or economic considerations. The disappearance of indigenous cultures is a good example of the pressures of population growth on people who live on the margins of modern society. Finally, while demographers develop various scenarios forecasting population growth, it is important to remember that there are circumstances that could lead not to growth but to a significant decline in global population. The spread of AIDS and other infectious diseases reveals that confidence in modern medicine's ability to control these scourges may be premature. Nature has its own checks and balances to the population dynamic that are not policy instruments of some international organization. This factor is often overlooked in an age of technological optimism.

The lead article in this section provides an overview of the general demographic trends of the contemporary world. The unit continues with a more focused discussion of adverse situations that are creating a sudden reversal in the general trend to longer life.

There is no greater check on population growth than the ability to produce an adequate food supply. Some experts question whether current technologies are sustainable over the long run. How much food are we going to need in the decades to come, and how are farmers and fishers going to produce it?

Making predictions about the future of the world's population is a complicated task, for there are a variety of forces at work and considerable variation from region to region. The danger of oversimplification must be overcome if governments and international organizations are going to respond with meaningful policies. Perhaps one could say that there is not a global population problem but rather many population challenges that vary from country to country and region to region.

THE BIG CRUNCH

Birthrates are falling, but it may be a half-century before the number of people—and their impact—reaches a peak

By Jeffrey Kluger

ODDS ARE YOU'LL NEVER MEET ANY OF THE ESTIMATED 247 HUMAN BEINGS WHO WERE BORN IN THE PAST MINUTE. IN A POPULATION OF 6 BILLION, 247 IS A DEMOGRAPHIC HICCUP. IN THE MINUTE BEFORE LAST, HOWEVER, THERE WERE ANOTHER 247. IN THE MINUTES TO COME THERE WILL be another, then another, then another. By next year at this time, all those minutes will have produced nearly 130 million newcomers to the great human mosh pit. That kind of crowd is awfully hard to miss.

For folks inclined to fret that the earth is heading for the environmental abyss, the population problem has always been one of the biggest causes for worry—and with good reason. The last time humanity celebrated a new century there were 1.6 billion people here for the party—or a quarter as many as this time. In 1900 the average life expectancy was, in some places, as low as 23 years; now it's 65, meaning the extra billions are staying around longer and demanding more from the planet. The 130 million or so births registered annually—

even after subtracting the 52 million deaths—is still the equivalent of adding nearly one new Germany to the world's population each year.

But things may not be as bleak as they seem. Lately demographers have come to the conclusion that the population locomotive—while still cannonballing ahead—may be chugging toward a stop. In country after country, birthrates are easing, and the population growth rate is falling.

To be sure, this kind of success is uneven. For every region in the world that has brought its population under control, there's another where things are still exploding. For every country that has figured out the art of sustainable agriculture, there are others that have worked their land to exhaustion. The population bomb may yet go off before governments can snuff the fuse, but for now, the news is better than it's been in a long time. "We could have an end in sight to population growth in the next century," says Carl Haub, a demographer with the nonprofit Population Research Bureau. "That's a major change."

Cheering as the population reports are becoming today, for much of the past 50 years, demographers were bearers of mostly bad tidings. In census after census, they reported that humanity was not just settling the planet but smothering it. It was not until the century was nearly two-thirds over that scientists and governments finally bestirred themselves to do something about it. The first great brake on population growth came in the early 1960s, with the development of the birth-control pill, a magic pharmacological bullet that made contraception easier—not to mention tidier—than it had ever been before. In 1969 the United Nations got in on the population game, creating the U.N. Population Fund, a global organization dedicated to bringing family-planning techniques to women who would not otherwise have them. In the decades that followed, the U.N. increased its commitment, sponsoring numerous global symposiums to address the population problem further. The most significant was the 1994 Cairo conference, where attendees pledged $5.7 billion to reduce birthrates in the developing world and acknowledged that giving women more education and reproductive freedom was the key to accomplishing that goal. Even a global calamity like AIDS has yielded unexpected dividends, with international campaigns to promote condom use and abstinence helping to prevent not only disease transmission but also conception.

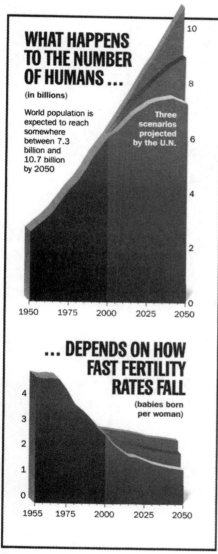

WHAT HAPPENS TO THE NUMBER OF HUMANS ...

(in billions)

World population is expected to reach somewhere between 7.3 billion and 10.7 billion by 2050

Three scenarios projected by the U.N.

10
8
6
4
2
0

1950 1975 2000 2025 2050

... DEPENDS ON HOW FAST FERTILITY RATES FALL

(babies born per woman)

4
3
2
1
0

1955 1975 2000 2025 2050

Source: United Nations

Such efforts have paid off in a big way. According to U.N. head counters, the average number of children produced per couple in the developing world—a figure that reached a whopping 4.9 earlier this century—has plunged to just 2.7. In many countries, including Spain, Slovenia, Greece and Germany, the fertility rate is well below 1.5, meaning parents are pro-

ducing 25% fewer offspring than would be needed to replace themselves—in effect, throwing the census into reverse. A little more than 30 years ago, global population growth was 2.04% a year, the highest in human history. Today it's just 1.3%. "It was a remarkable century," says Joseph Chamie of the U.N. Population Division. "We quadrupled the population in 100 years, but that's not going to happen again."

Sunny as the global averages look, however, things get a lot darker when you break them down by region. Even the best family-planning programs do no good if there is neither the money nor governmental expertise to carry them out, and in less-developed countries—which currently account for a staggering 96% of the annual population increase—both are sorely lacking. In parts of the Middle East and Africa, the fertility rate exceeds seven babies per woman. In India, nearly 16 million births are registered each year, for a growth rate of 1.8%. While Europe's population was three times that of Africa in 1950, today the two continents have about the same count. At the current rate, Africa will triple Europe in another 50 years.

Many of the countries in the deepest demographic trouble have imposed aggressive family-planning programs, only to see them go badly—even criminally—awry. In the 1970s, Indian Prime Minister Indira Gandhi tried to reduce the national birthrate by offering men cash and transistor radios if they would undergo vasectomies. In the communities in which those sweeteners failed, the government resorted to coercion, putting millions of males—from teenage boys to elderly men—on the operating table. Amid the popular backlash that followed, Gandhi's government was turned out of office, and the public rejected family planning.

China's similarly notorious one-child policy has done a better job of slowing population growth but not without problems. In a country that values boys over girls, one-child rules have led to abandonments, abortions and infanticides, as couples limited to a single offspring keep spinning the reproductive wheel until it comes up male. "We've learned that there is no such thing as 'population control,'" says Alex Marshall of the U.N. Population Fund. "You don't control it. You allow people to make up their own mind."

That strategy has worked in many countries that once had runaway population growth. Mexico, one of Latin America's population success stories, has

made government-subsidized contraception widely available and at the same time launched public-information campaigns to teach people the value of using it. A recent series of ads aimed at men makes the powerful point that there is more machismo in clothing and feeding offspring than in conceiving and leaving them. In the past 30 years, the average number of children born to a Mexican woman has plunged from seven to just 2.5. Many developing nations are starting to recognize the importance of educating women and letting them—not just their husbands—have a say in how many children they will have.

But bringing down birthrates loses some of its effectiveness as mortality rates also fall. At the same time Mexico reduced its children-per-mother figure, for example, it also boosted its average life expectancy from 50 years to 72—a wonderful accomplishment, but one that offsets part of the gain achieved by reducing the number of births.

When people live longer, populations grow not just bigger but also older and frailer. In the U.S. there has been no end of hand wringing over what will happen when baby boomers—who owe their very existence to the procreative free-for-all that followed World War II—retire, leaving themselves to be supported by the much smaller generation they produced. In Germany there are currently four workers for every retired person. Before long that ratio will be down to just 2 to 1.

STATE OF THE PLANET

Humans already use 54% of Earth's rainfall, says the U.N. report, and 70% of that goes to agriculture

For now, the only answer may be to tough things out for a while, waiting for the billions of people born during the great population booms to live out their long life, while at the same time continuing to reduce birthrates further so that things don't get thrown so far out of kilter again. But there's no telling if the earth—already worked to exhaustion feeding the 6 billion people currently here—can take much more. People in the richest countries consume a disproportionate share of the world's resources, and as poorer nations push to catch up, pressure on the planet will keep growing. "An ecologist looks at the population size relative to the carrying capacity of Earth," says Lester Brown, president of the Worldwatch Institute. "Looking at it that way, things are much worse than we expected them to be 20 years ago."

How much better they'll get will be decided in the next half-century (*see chart*). According to three scenarios published by the U.N., the global population in the year 2050 will be somewhere between 7.3 billion and 10.7 billion, depending on how fast the fertility rate falls. The difference between the high scenario and the low scenario? Just one child per couple. With the species poised on that kind of demographic knife edge, it pays for those couples to make their choices carefully.

—Reported by William Dowell/New York, Meenakshi Ganguly/New Delhi and Dick Thompson/Washington

Breaking Out or Breaking Down

In some parts of the world, the historic trend toward longer life has been abruptly reversed.

by Lester R. Brown and Brian Halweil

On October 12 of this year, the world's human population is projected to pass 6 billion. The day will be soberly observed by population and development experts, but media attention will do nothing to immediately slow the expansion. During that day, the global total will swell by another 214,000—enough people to fill two of the world's largest sports stadiums.

Even as world population continues to climb, it is becoming clear that the several billion additional people projected for the next half century are not likely to materialize. What is not clear is how the growth will be curtailed. Unfortunately, in some countries, a slowing of the growth is taking place only partly because of success in bringing birth rates down—and increasingly because of newly emergent conditions that are raising death rates.

Evidence of this shift became apparent in late October, 1998, when U.N. demographers released their biennial update of world population projections, revising the projected global population for 2050. Instead of rising in the next 50 years by more than half, to 9.4 billion (as computed in 1996), the 1998 projection rose only to 8.9 billion. The good news was that two-thirds of this anticipated slow-down was expected to be the result of falling fertility—of the decisions of more couples to have fewer children. But the other third was due to rising death rates, largely as a result of rising mortality from AIDS.

This rather sudden reversal in the human death rate trend marks a tragic new development in world demography, which is dividing the developing countries into two groups. When these countries embarked on the development journey a half century or so ago, they followed one of two paths. In the first, illustrated by the East Asian

nations of South Korea, Taiwan, and Thailand, early efforts to shift to smaller families set in motion a positive cycle of rising living standards and falling fertility. Those countries are now moving toward population stability.

In the second category, which prevails in sub-Saharan Africa (770 million people) and the Indian subcontinent (1.3 billion), fertility has remained high or fallen very little, setting the stage for a vicious downward spiral in which rapid population growth reinforces poverty, and in which some segments of society eventually are deprived of the resources needed even to survive. In Ethiopia, Nigeria, and Pakistan, for example, demographers estimate that the next half-century will bring a doubling or near-tripling of populations. Even now, people in these regions each day awaken to a range of daunting conditions that threatens to drop their living standards below the level at which humans can survive.

We now see three clearly identifiable trends that either are already raising death rates or are likely to do so in these regions: the spread of the HIV virus that causes AIDS, the depletion of aquifers, and the shrinking amount of cropland available to support each person. The HIV epidemic is spiraling out of control in sub-Saharan Africa. The depletion of aquifers has become a major threat to India, where water tables are falling almost everywhere. The shrinkage in cropland per person threatens to force reductions in food consumed per person, increasing malnutrition—and threatening lives—in many parts of these regions.

Containing one-third of the world's people, these two regions now face a potentially dramatic shortening of life expectancy. In sub-Saharan Africa, mortality rates are already rising, and in the Indian subcontinent they could begin rising soon. Without clearly defined national strategies for quickly lowering birth rates in these countries, and without a commitment by the international community to support them in their efforts, one-third of humanity could slide into a demographic black hole.

Birth and Death

Since 1950, we have witnessed more growth in world population than during the preceding 4 million years since our human ancestors first stood upright. This post-1950 explosion can be attributed, in part, to several developments that reduced death rates throughout the developing world. The wider availability of safe drinking water, childhood immunization programs, antibiotics, and expanding food production sharply reduced the number of people dying of hunger and from infectious diseases. Together these trends dramatically lowered mortality levels.

But while death rates fell, birth rates remained high. As a result, in many countries, population growth rose to 3 percent or more per year—rates for which there was no historical precedent. A 3 percent annual increase in population leads to a twenty-fold increase within a century. Ecologists have long known that such rates of population growth—which have now been sustained for close to half a century in many countries—could not be sustained indefinitely. At some point, if birth rates did not come down, disease, hunger, or conflict would force death rates up.

Projected Population Growth in Selected Developing Countries, 1999 to 2050

	1999 (millions)	2050 (millions)	Growth From 1999 to 2050 (millions)	(percent)
Developing Countries That Have Slowed Population Growth:				
South Korea	46	51	5	+11
Taiwan	22	25	3	+14
Thailand	61	74	13	+21
Developing Countries Where Rapid Population Growth Continues:				
Ethiopia	61	169	108	+177
Nigeria	109	244	135	+124
Pakistan	152	345	193	+127

Source: *United Nations, Global Population Projections, 1998.*

Although most of the world has succeeded in reducing birth rates to some degree, only some 32 countries—containing a mere 12 percent of the world's people—have achieved population stability. In these countries, growth rates range between 0.4 percent per year and minus 0.6 percent per year. With the exception of Japan, all of the 32 countries are in Europe, and all are industrial. Although other industrial countries, such as the United States, are still experiencing some population growth as a result of a persistent excess of births over deaths, the population of the industrial world as a whole is not projected to grow at all in the next century—unless, perhaps, through the arrival of migrants from more crowded regions.

Within the developing world, the most impressive progress in reducing fertility has come in East Asia. South Korea, Taiwan, and Thailand have all reduced their population growth rates to roughly one percent per year and are approaching stability. (See table, this page.) The biggest country in Latin America—Brazil—has reduced its population growth to 1.4 percent per year. Most other countries in Latin America are also making progress on this front. In contrast, the countries of sub-Saharan Africa and the Indian subcontinent have lagged in lowering growth rates, and populations are still rising ominously—at rates of 2 to 3 percent or more per year.

Graphically illustrating this contrast are Thailand and Ethiopia, each with 61 million people. Thailand is projected to add 13 million people over the next half century for a gain of 21 percent. Ethiopia, meanwhile, is projected to add 108 million for a gain of 177 percent. (The U.N.'s projections are based on such factors as the number of children per woman, infant mortality, and average life span in each country—factors that could change in time, but meanwhile differ sharply in the two countries.) The deep poverty among those living in sub-Saharan Africa and the Indian subcontinent has been a principal factor in their rapid population growth, as couples lack access to the kinds of basic social services and education that allow control over reproductive choices. Yet, the population growth, in turn, has only worsened their poverty—perpetuating a vicious cycle in which hopes of breaking out become dimmer with each passing year.

After several decades of rapid population growth, governments of many developing countries are simply being overwhelmed by their crowding—and are suffering from what we term "demographic fatigue." The simultaneous challenges of educating growing numbers of children, creating jobs for the swelling numbers of young people coming into the job market, and confronting such environmental consequences of rapid population growth as deforestation, soil erosion, and falling water tables, are undermining the capacity of governments to cope. When a major new threat arises, as has happened with the HIV virus, governments often cannot muster the leadership energy and fiscal resources to mobilize effectively. Social problems that are easily contained in industrial societies can become humanitarian disasters in many developing ones. As a result, some of the latter may soon see their population growth curves abruptly flattened, or even thrown into decline, not because of falling birth rates but because of fast-rising death rates. In some countries, that process has already begun.

Countries Where HIV Infection Rate Among Adults Is Greater Than Ten Percent

Country	Population	Share of Adult Population Infected
	(millions)	(percent)
Zimbabwe	11.7	26
Botswana	1.5	25
South Africa	43.3	22
Namibia	1.6	20
Zambia	8.5	19
Swaziland	0.9	18
Malawi	10.1	15
Mozambique	18.3	14
Rwanda	5.9	13
Kenya	28.4	12
Central African Republic	3.4	11
Cote d'Ivoire	14.3	10

Source: UNAIDS

Shades of the Black Death

Industrial countries have held HIV infection rates under 1 percent of the adult population, but in many sub-Saharan African countries, they are spiraling upward, out of control. In Zimbabwe, 26 percent of the adult population is infected; in Botswana, the rate is 25 percent. In South Africa, a country of 43 million people, 22 percent are infected. In Namibia, Swaziland, and Zambia, 18 to 20 percent are. (See table, this page.) In these countries, there is little to suggest that these rates will not continue to climb.

In other African nations, including some with large populations, the rates are lower but climbing fast. In both Tanzania, with 32 million people, and Ethiopia, with its 61 million, the race is now 9 percent. In Nigeria, the continent's largest country with 111 million people, the latest estimate now puts the infection rate also at 9 percent and rising.

What makes this picture even more disturbing is that most Africans carrying the virus do not yet know they are infected, which means the disease can gain enormous momentum in areas where it is still largely invisible. This, combined with the social taboo that surrounds HIV/AIDS in Africa, has made it extremely difficult to mount an effective control effort.

Barring a medical miracle, countries such as Zimbabwe, Botswana, and South Africa will lose at least 20 percent of their adult populations to AIDS within the next

decade, simply because few of those now infected with the virus can afford treatment with the costly antiviral drugs now used in industrial countries. To find a precedent for such a devastating region-wide loss of life from an infectious disease, we have to go back to the decimation of Native American communities by the introduction of small pox in the sixteenth century from Europe or to the bubonic plaque that claimed roughly a third of Europe's population in the fourteenth century (see table, next page).

Reversing Progress

The burden of HIV is not limited to those infected, or even to their generation. Like a powerful storm or war that lays waste to a nation's physical infrastructure, a growing HIV epidemic damages a nation's social infrastructure, with lingering demographic and economic effects. A viral epidemic that grows out of control is likely to reinforce many of the very conditions—poverty, illiteracy, malnutrition—that gave it an opening in the first place.

Using life expectancy—the sentinel indicator of development—as a measure, we can see that the HIV virus is reversing the gains of the last several decades. For example, in Botswana life expectancy has fallen from 61 years in 1990 to 44 years in 1999. By 2010, it is projected to drop to 39 years—a life expectancy more characteristic of medieval times than of what we had hoped for in the twenty-first century.

Beyond its impact on mortality, HIV also reduced fertility. For women, who live on average scarcely 10 years after becoming infected, many will die long before they have reached the end of their reproductive years. As the symptoms of AIDS begin to develop, women are less likely to conceive. For those who do conceive, the likelihood of spontaneous abortion rises. And among the reduced number who do give birth, an estimated 30 percent of the infants born are infected and an additional 20 percent are likely to be infected before they are weaned. For babies born with the virus, life expectancy is less than 2 years. The rate of population growth falls, but not in the way any family-planning group wants to see.

One of the most disturbing social consequences of the HIV epidemic is the number of orphans that it produces. Conjugal sex is one of the surest ways to spread AIDS, so if one parent dies, there is a good change the other will as well. By the end of 1997, there were already 7.8 million AIDS orphans in Africa—a new and rapidly growing social subset. The burden of raising these AIDS orphans falls first on the extended family, and then on society at large. Mortality rates for these orphans are likely to be much higher than the rates for children whose parents are still with them.

As the epidemic progresses and the symptoms become visible, health care systems in developing countries are being overwhelmed. The estimated cost of providing an-

Profiles of Major Epidemics Throughout Human History

Epidemic and Date	Mode of Introduction and Spread	Description of Plague and Its Effects on Population
Black Death in Europe, 14th century	Originating in Asia, the plague bacteria moved westward via trade routes, entering Europe in 1347; transmitted via rats as well as coughing and sneezing.	One fourth of the population of Europe was wiped out (an estimated 25 million deaths); old, young, and poor hit hardest.
Smallpox in the New World, 16th century	Spanish conquistadors and European colonists introduced virus into the Americas, where it spread through respiratory channels and physical contact.	Decimated Aztec, Incan, and native American civilizations, killing 10 to 20 million.
HIV/AIDS, worldwide, 1980 to present	Thought to have originated in Africa; a primate virus that mutated and spread to infect humans; transmitted by the exchange of bodily fluids, including blood, semen, and breast milk.	More than 14 million deaths worldwide thus far; an additional 33 million infected; one-fifth of adult population infected in several African nations; strikes economically active populations hardest.

Source: Jared Diamond, *Guns, Germs, and Steel: The Fates of Human Societies*, 1997; UNAIDS.

tiviral treatment (the standard regimen used to reduce symptoms, improve life quality, and postpone death) to all infected individuals in Malawi, Mozambique, Uganda, and Tanzania would be larger than the GNPs of those countries. In some hospitals in South Africa, 70 percent of the beds are occupied by AIDS patients. In Zimbabwe, half the health care budget now goes to deal with AIDS. As AIDS patients increasingly monopolize nurses' and doctors' schedules, and drain funds from health care budgets, the capacity to provide basic health care to the general population—including the immunizations and treatments for routine illnesses that have underpinned the decline in mortality and the rise in life expectancy in developing countries—begins to falter.

Worldwide, more than half of all new HIV infections occur in people between the ages of 15 and 24—an atypical pattern for an infectious disease. Human scourges have historically spread through respiratory exposure to coughing or sneezing, or through physical contact via shaking hands, food handling, and so on. Since nearly everyone is vulnerable to such exposure, the victims of most infectious diseases are simply those among society at large who have the weakest immune systems—generally the very young and the elderly. But with HIV, because the primary means of transmission is unprotected sexual activity, the ones who are most vulnerable to infection are those who are most sexually active—young, healthy adults in the prime of their lives. According to a UNAIDS report, "the bulk of the increase in adult death is in the younger adult ages—a pattern that is common in wartime and has become a signature of the AIDS epidemic, but that is otherwise rarely seen."

One consequence of this adult die-off is an increase in the number of children and elderly who are dependent on each economically productive adult. This makes it more difficult for societies to save and, therefore, to make the investments needed to improve living conditions. To make matters worse, in Africa it is often the better educated, more socially mobile populations who have the highest infection rate. Africa is losing the agronomists, the engineers, and the teachers it needs to sustain its economic development. In South Africa, for example, at the University of Durban-Westville, where many of the country's future leaders are trained, 25 percent of the students are HIV positive.

Countries where labor forces have such high infection levels will find it increasingly difficult to attract foreign investment. Companies operating in countries with high infection rates face a doubling, tripling, or even quadrupling of their health insurance costs. Firms once operating in the black suddenly find themselves in the red. What has begun as an unprecedented social tragedy is beginning to translate into an economic disaster. Municipalities throughout South Africa have been hesitant to publicize the extent of their local epidemics or scale up control efforts for fear of deterring outside investment and tourism.

The feedback loops launched by AIDS may be quite predictable in some cases, but could also destabilize societies in unanticipated ways. For example, where levels of unemployment are already high—the present situation in most African nations—a growing population of orphans and displaced youths could exacerbate crime. Moreover, a country in which a substantial share of the population suffers from impaired immune systems as a result of

AIDS is much more vulnerable to the spread of other infectious diseases, such as tuberculosis, and waterborne illness. In Zimbabwe, the last few years have brought a rapid rise in deaths due to tuberculosis, malaria, and even the bubonic plague—even among those who are not HIV positive. Even without such synergies, in the early years of the next century, the HIV epidemic is poised to claim more lives than did World War II.

Sinking Water Tables

While AIDS is already raising death rates in sub-Saharan Africa, the emergence of acute water shortages could have the same effect in India. As population grows, so does the need for water. Home to only 358 million people in 1950, India will pass the one-billion mark later this year. It is projected to overtake China as the most populous nation around the year 2037, and to reach 1.5 billion by 2050.

As India's population has soared, its demand for water for irrigation, industry, and domestic use has climbed far beyond the sustainable yield of the country's aquifers. According to the International Water Management Institute (IWMI), water is being pumped from India's aquifers at twice the rate the aquifers are recharged by rainfall. As a result, water tables are falling by one to three meters per year almost everywhere in the country. In thousands of villages, wells are running dry.

In some cases, wells are simply drilled deeper—if there is a deeper aquifer within reach. But many villages now depend on trucks to bring in water for household use. Other villages cannot afford such deliveries, and have entered a purgatory of declining options—lacking enough water even for basic hygiene. In India's western state of Gujarat, water tables are falling by as much as five meters per year, and farmers now have to drill their wells down to between 700 and 1200 feet to reach the receding supply. Only the more affluent can afford to drill to such depths.

Although irrigation goes back some 6,000 years, aquifer depletion is a rather recent phenomenon. It is only within the last half century or so that the availability of powerful diesel and electric pumps has made it possible to extract water at rates that exceed recharge rates. Little is known about the total capacity of India's underground supply, but the unsustainability of the current consumption is clear. If the country is currently pumping water at double the rate at which its aquifers recharge, for example, we know that when the aquifers are eventually depleted, the rate of pumping will necessarily have to be reduced to the recharge rate—which would mean that the amount of water pumped would be cut in half. With at least 55 percent of India's grain production now coming from irrigated lands, IWMI speculates that aquifer depletion could reduce India's harvest by one-fourth. Such a massive cutback could prove catastrophic for a nation where 53 percent of the children are already undernourished and underweight.

Impending aquifer depletion is not unique to India. It is also evident in China, North Africa and the Middle East, as well as in large tracts of the United States. However, in wealthy Kuwait or Saudi Arabia, precariously low water availability per person is not life-threatening because these countries can easily afford to import the food that they cannot produce domestically. Since it takes 1,000 tons of water to produce a ton of grain, the ability to import food is in effect an ability to import water. But in poor nations, like India, where people are immediately dependent on the natural-resource base for subsistence and often lack money to buy food, they are limited to the water they can obtain from their immediate surroundings—and are much more endangered if it disappears.

In India—as in other nations—poorer farmers are thus disproportionately affected by water scarcity, since they often cannot get the capital or credit to obtain bigger pumps necessary to extract water from ever-greater depths. Those farmers who can no longer deepen their wells often shift their cropping patterns to include more water-efficient—but lower-yielding—crops, such as mustard, sorghum, or millet. Some have abandoned irrigated farming altogether, resigning themselves to the diminished productivity that comes with depending only on rainfall.

When production drops, of course, poverty deepens. When that happens, experience shows that most people, before succumbing to hunger or starvation, will migrate. On Gujarat's western coast, for example, the overpumping of underground water has led to rapid salt-water intrusion as seawater seeps in to fill the vacuum left by the freshwater. The groundwater has become so saline that farming with it is impossible, and this has driven a massive migration of farmers inland in search of work.

Village communities in India tend to be rather insular, so that these migrants—uprooted from their homes—cannot take advantage of the social safety net that comes with community and family bonds. Local housing restrictions force them to camp in the fields, and their access to village clinics, schools, and other social services is restricted. But while attempting to flee, the migrants also bring some of their troubles along with them. Navroz Dubash, a researcher at the World Resources Institute who examined some of the effects of the water scarcity in Gujarat, notes that the flood of migrants depresses the local labor markets, driving down wages and diminishing the bargaining power of all landless laborers in the region.

In the web of feedback loops linking health and water supply, another entanglement is that when the *quantity* of available water declines, the *quality* of the water, too, may decline, because shrinking bodies of water lose their efficacy in diluting salts or pollutants. In Gujarat, water pumped from more than 700 feet down tends to have an unhealthy concentration of some inorganic elements, such as fluoride. As villagers drink and irrigate with this contaminated water, the degeneration of teeth and bones known as fluorosis has emerged as a major health threat. Similarly, in both West Bengal, India and Bangladesh, receding water tables have exposed arsenic-laden sediments to oxygen, converting them to a water-soluble form. According to UNDP estimates, at least 30 million people are exposed to health-impairing levels of arsenic in their drinking water.

AIDS attacks whole communities, but unlike other scourges it takes its heaviest toll on teenagers and young adults—the people most needed to care for children and keep the economy productive.

As poverty deepens in the rural regions of India—and is driven deeper by mutually exacerbating health threats and water scarcities—migration from rural to urban areas is likely to increase. But for those who leave the farms, conditions in the cities may be no better. If water is scarce in the countryside, it is also likely to be scarce in the squatter settlements or other urban areas accessible to the poor. And where water is scarce, access to adequate sanitation and health services is poor. In most developing nations, the incidence of infectious diseases, including waterborne microbes, tuberculosis, and HIV/AIDS, is considerably higher in urban slums—where poverty and compromised health define the way of life—than in the rest of the city.

In India, with so many of the children undernourished, even a modest decline in the country's ability to produce or purchase food is likely to increase child mortality. With India's population expected to increase by 100 million people per decade over the next half century, the potential losses of irrigation water pose an ominous specter not only to the Indian people now living but to the hundreds of millions more yet to come.

Shrinking Cropland Per Person

The third threat that hangs over the future of nearly all the countries where rapid population growth continues is the steady decline in the amount of cropland remaining per person—a threat both of rising population and of the conversion of cropland to other uses. In this analysis, we use grainland per person as a surrogate for cropland, because in most developing countries the bulk of land is used to produce grain, and the data are much more reliable. Among the more populous countries where this trend threatens future food security are Nigeria, Ethiopia, and Pakistan—all countries with weak family-planning programs.

As a limited amount of arable land continues to be divided among larger numbers of people, the average amount of cropland available for each person inexorably shrinks. Eventually, it drops below the point where people can feed themselves. Below 600 square meters of grainland per person (about the area of a basketball court), nations typically begin to depend heavily on imported grain. Cropland scarcity, like, water scarcity, can easily be translated into increased food imports in countries that can afford to import grain. But in the poorer nations of sub-Saharan Africa and the Indian subcontinent, subsistence farmers may not have access to imports. For them, land scarcity readily translates into malnutrition, hunger, rising mortality, and migration—and sometimes conflict. While most experts agree that resource scarcity alone is rarely the cause of violent conflict, resource scarcity has often compounded socioeconomic and political disruptions enough to drive unstable situations over the edge.

Thomas Homer-Dixon, director of the Project on Environment, Population, and Security at the University of Toronto, notes that "environmental scarcity is, without doubt, a significant cause of today's unprecedented levels of internal and international migration around the world." He has examined two cases in South Asia—a region plagued by land and water scarcity—in which resource constraints were underlying factors in mass migration and resulting conflict.

In the first case, Homer-Dixon finds that over the last few decades, land scarcity has caused millions of Bangladeshis to migrate to the Indian states of Assam, Tripura, and West Bengal. These movements expanded in the late 1970s after several years of flooding in Bangladesh, when population growth had reduced the grainland per person in Bangladesh to less than 0.08 hectares. As the average person's share of cropland began to shrink below the survival level, the lure of somewhat less densely populated land across the border in the Indian state of Assam became irresistible. By 1990, more than 7 million Bangladeshis had crossed the border, pushing Assam's population from 15 million to 22 million. The new immigrants in turn exacerbated land shortages in the Indian states, setting off a string of ethnic conflicts that have so far killed more than 5,000 people.

In the second case, Homer-Dixon and a colleague, Peter Gizewski, studied the massive rural-to-urban migration that has taken place in recent years in Pakistan. This migration, combined with population growth within the cities, has resulted in staggering urban growth rates of roughly 15 percent a year. Karachi, Pakistan's coastal capital, has seen its population balloon to 11 million. Urban services have been unable to keep pace with growth, especially for low-income dwellers. Shortages of water, sanitation, health services and jobs have become especially acute, leading to deteriorating public health and growing impoverishment.

"This migration… aggravates tensions and violence among diverse ethnic groups," according to Homer-

Dixon and Gizewski. "This violence, in turn, threatens the general stability of Pakistani society." The cities of Karachi, Hyderabad, Islamabad, and Rawalpindi, in particular, have become highly volatile, so that "an isolated, seemingly chance incident—such as a traffic accident or short-term breakdown in services—ignites explosive violence." In 1994, water shortages in Islamabad provoked widespread protest and violent confrontation with police in hard-hit poorer districts.

When people of parenting age die, the elderly are often left alone to care for the children. Meanwhile, poverty worsens with the loss of wage-earners. In other situations, poverty is worsened by declines in the amounts of productive land or fresh water available to each person and here, too, death may take an unnatural toll.

Without efforts to step up family planning in Pakistan, these patterns are likely to be magnified. Population is projected to grow from 146 million today to 345 million in 2050, shrinking the grainland area per person in Pakistan to a miniscule 0.036 hectares by 2050—less than half of what it is today. A family of six will then have to produce its food on roughly one-fifth of a hectare, or half an acre—the equivalent of a small suburban building lot in the United States.

Similar prospects are in the offing for Nigeria, where population is projected to double to 244 million over the next half century, and in Ethiopia, where population is projected to nearly triple. In both, of course, the area of grainland per person will shrink dramatically. In Ethiopia, if the projected population growth materializes, it will cut the amount of cropland per person to one-third of its current 0.12 hectares per person—a level at which already more than half of the country's children are undernourished. And even as its per capita land shrinks, its long-term water supply is jeopardized by the demands of nine other rapidly growing, water-scarce nations throughout the Nile River basin. But even these projections may underestimate the problem, because they assume an equitable distribution of land among all people. In reality, the inequalities in land distribution that exist in many African and South Asian nations mean that as the competition for declining resources becomes more intense, the poorer and more marginal groups face even harsher deprivations than the averages imply.

Moreover, in these projections we have assumed that the total grainland area over the next half-century will not change. In reality this may be overly optimistic simply because of the ongoing conversion of cropland to nonfarm uses and the loss of cropland from degradation. A steadily growing population generates a need for more homes, schools, and factories, many of which will be built on once-productive farmland. Degradation, which may take the form of soil erosion or of the waterlogging and salinization of irrigated land, is also claiming cropland.

Epidemics, resource scarcity, and other societal stresses thus do not operate in isolation. Several disruptive trends will often intersect synergistically, compounding their effects on public health, the environment, the economy, and the society. Such combinations can happen anywhere, but the effects are likely to be especially pernicious—and sometimes dangerously unpredictable—in such places as Bombay and Lagos, where HIV prevalence is on the rise, and where fresh water and good land are increasingly beyond the reach of the poor.

Regaining Control of Our Destiny

The threats from HIV, aquifer depletion, and shrinking cropland are not new or unexpected. We have known for at least 15 years that the HIV virus could decimate human populations if it is not controlled. In each of the last 18 years, the annual number of new HIV infections has risen, climbing from an estimated 200,000 new infections in 1981 to nearly 6 million in 1998. Of the 47 million people infected thus far, 14 million have died. In the absence of a low-cost cure, most of the remaining 33 million will be dead by 2005.

It may seem hard to believe, given the advanced medical knowledge of the late twentieth century, that a controllable disease is decimating human populations in so many countries. Similarly, it is hard to understand how falling water tables, which may prove an even greater threat to future economic progress, could be so widely ignored.

The arithmetic of emerging resource shortages is not difficult. The mystery is not in the numbers, but in our failure to do what is needed to prevent such threats from spiraling out of control.

Today's political leaders show few signs of comprehending the long-term consequences of persistent environmental and social trends, or of the interconnectedness of these trends. Despite advances in our understanding of the complex—often chaotic—nature of biological, ecological, and climatological systems, political thought continues to be dominated by reductionist thinking that fails to target the root causes of problems. As a result, political action focuses on responses to crises rather than prevention.

Leaders who are prepared to meet the challenges of the next century will need to understand that universal access to family planning not only is essential to coping with resource scarcity and the spread of HIV/AIDS, but is likely to improve the quality of life for the citizens they serve. Family planning comprises wide availability of contraception and reproductive healthcare, as well as improved access to educational opportunities for young women and men. Lower birth rates generally allow greater investment in each child, as has occurred in East Asia.

Overwhelmed by multiple attacks on its health, the society falls deeper into poverty and as the cycle continues, more of its people die prematurely.

Leaders all over the world—not just in Africa and Asia—now need to realize that the adverse effects of global population growth will affect those living in nations such as the United States or Germany, that seem at first glance to be relatively protected from the ravages now looming in Zimbabwe or Ethiopia. Economist Herman Daly observes that whereas in the past surplus labor in one nation had the effect of driving down wages only in that nation, "global economic integration will be the means by which the consequences of overpopulation in the Third World are generalized to the globe as a whole." Large infusions of job-seekers into Brazil's or India's work force that may lower wages there may now also mean large infusions into the global workforce, with potentially similar consequences.

As the recent Asian economic downturn further demonstrates, "localized instability" is becoming an anachronistic concept. The consequences of social unrest in one nation, whether resulting from a currency crisis or an environmental crisis, can quickly cross national boundaries. Several nations, including the United States, now recognize world population growth as a national security issue. As the U.S. Department of State Strategic Plan, issued in September 1997, explains, "Stabilizing population growth is vital to U.S. interests.... Not only will early stabilization of the world's population promote environmentally sustainable economic development in other countries, but it will benefit the United States by improving trade opportunities and mitigating future global crises."

One of the keys to helping countries quickly slow population growth, before it becomes unmanageable, is expanded international assistance for reproductive health and family planning. At the United Nations Conference on Population and Development held in Cairo in 1994, it was estimated that the annual cost of providing quality reproductive health services to all those in need in developing countries would amount to $17 billion in the year 2000. By 2015, the cost would climb to $22 billion.

Industrial countries agreed to provide one-third of the funds, with the developing countries providing the remaining two-thirds. While developing countries have largely honored their commitments, the industrial countries—and most conspicuously, the United States—have reneged on theirs. And in late 1998, the U.S. Congress—mired in the quicksand of anti-abortion politics—withdrew all funding for the U.N. Population Fund, the principal source of international family planning assistance. Thus was thrown aside the kind of assistance that helps both to slow population growth and to check the spread of the HIV virus.

In most nations, stabilizing population will require mobilization of domestic resources that may now be tied up in defense expenditures, crony capitalism or government corruption. But without outside assistance, many nations many still struggle to provide universal family planning. For this reason, delegates at Cairo agreed that the immense resources and power found in the First World are indispensable in this effort. And as wealth further consolidates in the North and the number living in absolute poverty increases in the South, the argument for assistance grows more and more compelling. Given the social consequences of one-third of the world heading into a demographic nightmare, failure to provide such assistance is unconscionable.

Lester Brown is president of the Worldwatch Institute and Brian Halweil is a staff researcher at the Institute.

From *World Watch*, September/October 1999, pp. 20-29. © 1999 by The Worldwatch Institute. Reprinted by permission.

Bittersweet Harvest

The Debate Over Genetically Modified Crops

Honor Hsin

In 1982 scientists on the 4th floor of the Monsanto Company U Building successfully introduced a foreign gene into a plant cell for the first time in history. These plants were genetically modified: they continued to express the new gene while exhibiting normal plant physiology and producing normal offspring. This breakthrough spawned the field of genetically modified (GM) crop production. Since the discovery, however, the international response to GM crops has been mixed. Along with the tremendous potential that lies vested in this technology, there are many risks and uncertainties involved as well. Arguments have centered on the health implications and environmental impact of cultivating GM crops and have raised disputes over national interests, global policy, and corporate agendas. Although there are many sides to this debate, discussions on GM crop regulation should be held within the context of scientific evidence, coupled with a careful weighing of present and future agricultural prospects.

Benefits and Costs

The possibility of environmental benefits first spurred the development of GM crops. The environmental issues at stake can be illustrated by one example of a potent genetic modification, the introduction of an endotoxin gene from *Bacillus thuringiensis* (Bt), a soil microorganism used for decades by organic growers as an insecticide, into soybeans, corn, and cotton. These GM crops promise to reduce the need to spray large amounts of chemicals into a field's ecosystem since the toxins are produced by

the plants themselves. The Bt crops pose environmental risks, however, and could possibly harm other organisms. Bt corn was shown to harm monarch butterfly caterpillars in the laboratory, although later studies performed with more realistic farming conditions found this result conclusively only with Syngenta Company's Bt maize, which expressed up to 40 percent more toxin than other brands. Another pertinent environmental issue is the possible evolution of Bt resistance in pests. Since the Bt toxin expressed by the crops is ubiquitous in the field, there is positive selection for resistance against it, which would quickly make Bt's effect obsolete. Experimentation has begun, however, that involves regulating the percentage of Bt crops in a field so that a balance can be achieved between high yields and survival of Bt-sensitive pests. Although there are still multiple layers of ecosystem complexity that need to be considered, careful scientific research can begin to address these questions.

Another potential area of risk that needs to be analyzed is the effect of GM crops on human health. A possible consequence of Bt expression in crops is the development of allergic reactions in farmers since the toxin is more highly concentrated in the crops than in the field. Furthermore, the method used to insert foreign genes into GM crops always risks manipulation of unknown genes in the plant, resulting in unforeseen consequences. The effects of GM crops on humans therefore must be tested rigorously. Fortunately, no solid evidence yet exists for adverse physiological reactions to GM crops in humans,

and some scientists argue that these same genetic-modification techniques are also currently being used in the development of pharmaceutical and industrial products.

A prevailing theme in the GM debate is that when discrepancies between scientific consensus and government policy result in unwanted consequences, the blame is often placed directly on GM crop technology itself. In 2000 about 300,000 acres of StarLink corn, a Bt crop produced by Aventis CropScience, were being cultivated in the United States. Since the US Environmental Protection Agency had declared its uncertainty over the allergenic potential of StarLink, the crops were grown with the understanding that they would be used solely as animal feed. Later that year news broke that StarLink corn had found its way into numerous taco food products around the world. This incident received wide press coverage and brought instant attention to the debate over GM crop safety. More at issue, though, were the United States' lax policies of GM crop approval and regulation. For nearly a decade, the US government made no distinction between GM crops and organically grown crops, and allergenicity safety tests were not mandatory. Only recently has the US Food and Drug Administration begun to reconsider its policies.

Canada is another leading producer of GM crops, with regulatory policies similar to those of the United States. Recent controversy surrounding Canada's cultivation of GM rapeseed, or canola, brought attention to another major environmental risk of GM crops. Unlike wheat and soybeans, which can self-pollinate to reproduce, the pollen of rapeseed plants spreads up to 800 meters beyond the field. There have been concerns in Ottawa over the government's refusal to reveal the location of ongoing GM wheat testing by Monsanto, resulting in fear of unwanted pollen spreading. This issue demonstrates one of the most potent risks of GM crops: uncontrolled breeding and the introduction of foreign genes into the natural ecosystem. An example of such an incident is Mexico's discovery of transgenic genes in non-GM strains of maize, although this result is still under scrutiny. More measures must be tested to restrain these possibilities. Current research on introducing the foreign genes into chloroplasts, which are only carried in the maternal line and not in pollen, offers a promising example.

Unfortunately, activist organizations rarely cite credible scientific evidence in their positions and have won much public sympathy by exploiting popular fears and misconceptions about genetic-engineering technology.

Europe's policy toward GM crops lies on the opposite end of the spectrum. In 1996 Europe approved the import of Monsanto's Roundup Ready soybeans and in 1997 authorized the cultivation of GM corn from Novartis. At around this time, however, there were rising concerns in Britain over BSE (bovine spongiform encephalopathy), or mad cow disease, which was thought to have killed more than two dozen people and cost the country the equivalent of billions of US dollars. The public was enraged over what it believed was a failure of government regulation, and in 1998 the European Commission voted to ban the import and cultivation of new GM crops. Besides the disappointment of private GM corporations like Monsanto, the United States claims to have lost US$600 million in corn exports to the

European Union. Recently, several European countries have considered lifting the ban contingent on the establishment of adequate labeling practices. The United States has complained to EU officials that labeling requirements discriminate against its agricultural exports, bringing the GM debate into the midst of a world trade dispute. In late January 2000, a tentative agreement was reached on the Montreal Biosafety Protocol in which the United States, Canada, Australia, Argentina, Uruguay, and Chile agreed to preliminary labeling of international exports and a precautionary principle allowing EU countries to reject imports if a scientific risk assessment of the imported crop is provided. This agreement, however, does not override decisions made by the World Trade Organization.

Corporate Control

The European public's anti-GM crop stance stems primarily from the success of environmental advocacy groups such as Greenpeace and Friends of the Earth. Numerous demonstrations have occurred throughout Britain, France, and other EU countries where GM crops have been uprooted and destroyed. Unfortunately, activist organizations rarely cite credible scientific evidence in their positions and have won much public sympathy by exploiting popular fears and misconceptions about genetic-engineering technology.

One issue they highlight that might prove significant, however, is the role of corporate interests in the GM-crop debate. A few years ago, Monsanto's attempt to acquire the "terminator" technology sparked tremendous controversy. This patent consisted of an elaborate genetically engineered control system designed to inhibit the generation of fertile seeds from crops. In essence, it was developed so that farmers would need to purchase new GM seeds each year, although arguments were raised that this technology could help prevent uncontrolled GM crop breeding. After much pressure from the nonprofit advocacy group Rural Advancement Foundation International, however, Monsanto announced in late 1999 that it would not market the "terminator" technology.

The "terminator" ordeal attracted so much attention because it placed Monsanto's corporate interest directly against the strongest argument in favor of genetic-engineering technology: potential cost savings and nutritional value of GM crops to developing countries. The UN Development Programme recently affirmed that GM crops could be the key to alleviating global hunger. Although the United Nations has expressed concern over precautionary testing of crops (through agencies like the World Health Organization), some contend that Western opposition to this technology ignores concerns of sub-Saharan and South Asian countries where malnutrition and poverty are widespread.

India is among those nations that could benefit from GM-crop technology. India's population has been growing by 1.8 percent annually; by 2025 India will need to produce 30 percent more grain per year to feed the twenty million new mouths added to its population. The need for higher food productivity is highlighted by incidents of poor farmers in Warangal and Punjab who have committed suicide when faced with devastated crops and huge debts on pesticides. The Indian government has approved several GM crops for commercial production, and testing has also commenced on transgenic cotton, rice, maize, tomato, and cauliflower, crops that would reduce the need for pesticides. A recent furor erupted over the discovery of around 11,000 hectares of illegal Bt cotton in Gujarat. The Gujarat administration responded immediately by ordering the fields stripped, the crops burned, and the seeds destroyed. There is still uncertainty over who will repay the farmers, who claim that Mahyco, a Monsanto subsidiary, is attempting to monopolize the distribution of Bt crops in India, and that the Indian government is also yielding to pressure from pesticide manufacturers. Corporate battles still abound in a nation where many farmers appear to be in need of agricultural change.

Feed the World

Many opponents of GM crops argue that the technology is not needed to help solve the problem of world hunger, with 800 million people who do not have enough to eat. They often argue that the world produces enough food to feed nine billion people while there are only six billion people today, implying that global hunger is simply a matter of distribution and not food productivity. Unfortunately, fixing the distribution problem is a complex issue. Purchasing power would need to increase in developing countries, coupled with increased food production in both developing and developed countries so that crops can be marketed at a price the underprivileged can afford. Since land for farming is limited, the remaining option for increasing crop productivity is to increase yield. While GM-crop technology is not the only method that can be used to achieve this end, it can contribute greatly toward it.

On Dr. Shiva's argument for supporting local knowledge in agricultural practices, Dr. Prakash argues that, from experience, "[local knowledge] is losing one third of your children before they hit the age of three. Is that the local knowledge that you want to keep reinforcing and keep perpetuating?"

Some consider GM crops part of a series of corporate attempts to control markets in developing countries and thus they brand GM technology another globalization "evil." Dr. Vandana Shiva of the Research Foundation for Science, Technology, and Ecology argues that globalization has pressured farmers in developing countries to grow monocultures—single-crop farming—instead of fostering sustainable agricultural diversity. Genetic engineering, in this view, is the next industrialization effort after chemical pesticides, and would also bear

no greater benefit than indigenous polycultural farming. The Food and Agricultural Organization of the United Nations also notes the leaning of research investment toward monocultures, spurred on by the profit potential of GM crops.

On the other hand, GM-crop technology serves to increase crop yield on land already in use for agricultural purposes, thereby preserving biodiversity in unused land. In the words of Dr. C. S. Prakash "using genetics helped [to] save so much valuable land from being under the plow." On Shiva's argument for supporting local knowledge in agricultural practices, Dr. Prakash argues that, from experience, "[local knowledge] is losing one third of your children before they hit the age of three. Is that the local knowledge that you want to keep reinforcing and keep perpetuating?"

Continuing along these lines and bringing GM technology in developing countries into the broader context of morality, leaders including Per Pinstrup-Andersen, director of the International Food Policy Research Institute, and Hassan Adamu, Nigeria's minister of agriculture, emphasize the importance of providing freedom of access, education, and choice in GM technology to the individual farmer himself. In Africa, for example, many local farmers have benefited from hybrid seeds obtained from multinational corporations. On a larger scale, however, Africa's agricultural production per unit area is among the lowest in the world, and great potential lies in utilizing GM crops to help combat pestilence and drought problems. On the issue of local knowledge, Dr. Florence Wambugu of the International Service for the Acquisition of Agribiotech Applications in Kenya (ISAAA) asserts that GM crops consist of "packaged technology in the seed" that can yield benefits without a change in local agricultural customs.

On another front of the world hunger debate, a promising benefit that GM-crop technology brings to developing countries is the introduction or enhancement of nutrients in crops. The first prod-

uct to address this was "golden rice," an engineered form of rice that expresses high levels of beta-carotene, a precursor of Vitamin A, which could be used to combat Vitamin A deficiency found in over 120 million children worldwide. Although many advocacy groups claim that the increased levels of Vitamin A from a golden rice diet are not high enough to fully meet recommended doses of Vitamin A, studies suggest that a less-than-full dose can still make a difference in an individual whose Vitamin A intake is already deficiently low. Currently the International Rice Research Institute is evaluating environmental and health concerns. After such tests are completed, however, there remains one final hurdle in the marketing process that advocates on both sides of the GM debate do agree on: multilateral access and sharing between public and private sectors. The International Undertaking on Plant Genetic Resources was established to foster such relationships for the world's key crops, but more discussions will have to take place on the intellectual-property rights of GM-crop patents.

Science First

Monsanto recently drafted a pledge of Five Commitments: Respect, Transparency, Dialogue, Sharing, and Benefits. These are qualities that all multinational organizations should bring to the debate over GM crops. In the meantime, the technology of genetic engineering has already emerged and bears promising potential. On the question of world hunger, GM crops are not the full solution, but they can play a part in one. There are possible risks which must be examined and compared to the risks associated with current agricultural conditions, and progress must not be sought too hastily. It is important to base considerations of the benefits and risks of GM crops on careful scientific research, rather than corporate interest or public fears.

HONOR HSIN, Staff Writer, *Harvard International Review*

UNIT 3

The Global Environment and Natural Resources Utilization

Unit Selections

8. **The Challenges We Face**, Jeffrey Kluger and Andrea Dorfman
9. **The Heat Is On**, Ralph Nader and Sallie Baliunas
10. **We *Can* Build a Sustainable Economy**, Lester R. Brown

Key Points to Consider

- What are the basic environmental challenges that confront both governments and individual consumers?

- What are the dimensions to the debate about global warming and the use of fossil fuel?

- What transformations will societies that are heavy users of fossil fuels undergo in order to meet future energy needs?

- Has the international community adequately responded to problems of pollution and threats to our common natural heritage? Why or why not?

- What is the natural resource picture going to look like 30 years from now?

- How is society, in general, likely to respond to the conflicts between lifestyle and resource conservation?

- If a sustainable economy were to be organized, what changes in behavior and values would be necessary?

 **Links: www.dushkin.com/online/**
These sites are annotated in the World Wide Web pages.

Friends of the Earth
http://www.foe.co.uk/index.html

National Geographic Society
http://www.nationalgeographic.com

National Oceanic and Atmospheric Administration (NOAA)
http://www.noaa.gov

Public Utilities Commission of Ohio (PUCO)
http://www.puc.state.oh.us/consumer/gcc/index.html

SocioSite: Sociological Subject Areas
http://www.pscw.uva.nl/sociosite/TOPICS/

United Nations Environment Programme (UNEP)
http://www.unep.ch

Beginning in the eighteenth century, the modern nation-state was conceived, and over many generations it has evolved to the point where it is now difficult to imagine a world without national governments. These legal entities have been viewed as separate, self-contained units that independently pursue their "national interests." Scholars often described the world as a political community of independent units that interact with each other (a concept that has been described as a billiard ball model).

This perspective of the international community as comprising self-contained and self-directed units has undergone major rethinking in the past 30 years. One of the reasons for this is the international dimensions of the demands being placed on natural resources. The Middle East, for example, contains a majority of the world's oil reserves. The United States, Western Europe, and Japan are very dependent on this vital source of energy. This unbalanced supply and demand equation has created an unprecedented lack of self-sufficiency for the world's major economic powers.

The increased interdependence of countries is further illustrated by the fact that air and water pollution often do not respect political boundaries. One country's smoke is often another country's acid rain. The concept that independent political units control their own destiny, in short, makes less sense than it may have 100 years ago. In order to more fully understand why this is so, one must first look at how Earth's natural resources are being utilized and how this may be affecting the global environment.

The initial articles in the unit examine the broad dimensions of the uses and abuses of natural resources. The central theme in these articles is whether or not human activity is in fact bringing about fundamental changes in the functioning of Earth's self-regulating ecological systems. In many cases an unsustainable rate of usage is under way, and, as a consequence, an alarming decline in the quality of the natural resource base is taking place.

An important conclusion resulting from this analysis is that contemporary methods of resource utilization often create problems that transcend national boundaries. Global climate changes, for example, will affect everyone, and if these changes are to be successfully addressed, international collaboration will be required. The consequences of basic human activities such as growing and cooking food are profound when multiplied billions of times every day. A single country or even a few countries working together cannot have a significant impact on redressing these problems. Solutions will have to be conceived that are truly global in scope. Just as there are shortages of natural resources, there are also shortages of new ideas for solving many of these problems.

The unit concludes with a discussion of the issues involved in moving from a perspective of the environment as simply an economic resource to be consumed to a perspective that has been defined as "sustainable development." This change is easily called for, but in fact it goes to the core of social values and basic economic activities. Developing sustainable practices, therefore, is a challenge of unprecedented magnitude.

Nature is not some object "out there" to be visited at a national park. It is the food we eat and the energy we consume. Human beings are joined in the most intimate of relationships with the natural world in order to survive from one day to the next. It is ironic how little time is spent thinking about this relationship. This lack of attention, however, is not likely to continue, for rapidly growing numbers of people and the increased use of energy-consuming technologies are placing unprecedented pressures on Earth's carrying capacity.

The Challenges We Face

By JEFFREY KLUGER and ANDREA DORFMAN

For starters, let's be clear about what we mean by "saving the earth." The globe doesn't need to be saved by us, and we couldn't kill it if we tried. What we do need to save—and what we have done a fair job of bollixing up so far—is the earth as we like it, with its climate, air, water and biomass all in that destructible balance that best supports life as we have come to know it. Muck that up, and the planet will simply shake us off, as it's shaken off countless species before us. In the end, then, it's us we're trying to save—and while the job is doable, it won't be easy.

The 1992 Earth Summit in Rio de Janeiro was the last time world leaders assembled to look at how to heal the ailing environment. Now, 10 years later, Presidents and Prime Ministers are convening at the World Summit on Sustainable Development in Johannesburg next week to reassess the planet's condition and talk about where to go from here. In many ways, things haven't changed: the air is just as grimy in many places, the oceans just as stressed, and most treaties designed to do something about it lie in incomplete states of ratification or implementation. Yet we're oddly smarter than we were in Rio. If years of environmental false starts have taught us anything, it's that it's time to quit seeing the job of cleaning up the world as a zero-sum game between industrial progress on the one hand and a healthy planet on the other. The fact is, it's development—well-planned, well-executed sustainable development—that may be what saves our bacon before it's too late.

Food
As we try to nourish 6 billion people, both bioengineering and organic farming will help

As the summiteers gather in Johannesburg, TIME is looking ahead to what the unfolding century—a green century—could be like. In this special report, we will examine several avenues to a healthier future, including green industry, green architecture, green energy, green transportation and even a greener approach to wilderness preservation. All of them have been explored before, but never so urgently as now. What gives such endeavors their new credibility is the hope and notion of sustainable development, a concept that can be hard to implement but wonderfully simple to understand.

Population
The tide of people may not ebb until the head count hits the 11 billion mark

With 6.1 billion people relying on the resources of the same small planet, we're coming to realize that we're drawing from a finite account. The amount of crops, animals and other biomatter we extract from the earth each year exceeds what the planet can replace by an estimated 20%, meaning it takes 14.4 months to replenish what we use in 12—deficit spending of the worst kind. Sustainable development works to reverse that, to expand the resource base and adjust how we use it so we're living off biological interest without ever touching principal. "The old environmental movement had a reputation of elitism," says Mark Malloch Brown, administrator of the United Nations Development Program (UNDP). "The key now is to put people first and the environment second, but also to remember that when you exhaust resources, you destroy people." With that in mind, the summiteers will wrestle with a host of difficult issues that affect both people and the environment. Among them:

• POPULATION AND HEALTH: While the number of people on earth is still rising rapidly, especially in the developing countries of Asia, the good news is that the growth rate is slowing. World population increased 48% from 1975 to 2000, compared with 64% from 1950 to 1975. As this gradual deceleration continues, the population is expected to level off eventually, perhaps at 11 billion sometime in the last half of this century.

Though it's not easy to see it from the well-fed West, a third of the world goes hungry

Economic-development and family-planning programs have helped slow the tide of people, but in some places, population growth is moderating for all the wrong reasons. In the poorest parts of the world, most notably Africa, infectious diseases such as AIDs, malaria, cholera and tuberculosis are having a Malthusian effect. Rural-land degradation is pushing people into cities, where crowded, polluted living conditions create the perfect breeding grounds for sickness. Worldwide, at least 68 million are expected to die of AIDs by 2020, including 55 million in sub-Saharan Africa. While any factor that eases population pressures may help the environment, the situation would be far less tragic if rich nations did more to help the developing world reduce birth rates and slow the spread of disease.

Efforts to provide greater access to family planning and health care have proved effective. Though women in the poorest countries still have the most children, their collective fertility rate is 50% lower than it was in 1969 and is expected to decline more by 2050. Other programs targeted at women include basic education and job training. Educated mothers not only have a stepladder out of poverty, but they also choose to have fewer babies.

Rapid development will require good health care for the young since there are more than 1 billion people ages 15 to 24. Getting programs in place to keep this youth bubble healthy could make it the most productive generation ever conceived. Says Thoraya Obaid, executive director of the U.N. Population Fund: "It's a window of opportunity to build the economy and prepare for the future."

• FOOD: Though it's not always easy to see it from the well-fed West, up to a third of the world is in danger of starving. Two billion people lack reliable access to safe, nutritious food, and 800 million of them—including 300 million children—are chronically malnourished.

Agricultural policies now in place define the very idea of unsustainable development. Just 15 cash crops such as corn, wheat and rice provide 90% of the world's food, but planting and re-planting the same crops strips fields of nutrients and makes them more vulnerable to pests. Slash-and-burn planting techniques and overreliance on pesticides further degrade the soil.

Solving the problem is difficult, mostly because of the ferocious debate over how to do it. Biotech partisans say the answer lies in genetically modified crops—foods engineered for vitamins, yield and robust growth. Environmentalists worry that fooling about with genes is a recipe for Frankensteinian disaster. There is no reason, however, that both camps can't make a contribution.

Better crop rotation and irrigation can help protect fields from exhaustion and erosion. Old-fashioned cross-breeding can yield plant strains that are heartier and more pest-resistant. But in a world that needs action fast, genetic engineering must still have a role—provided it produces suitable crops. Increasingly, those crops are being created not just by giant biotech firms but also by home-grown groups that know best what local consumers need.

The National Agricultural Research Organization of Uganda has developed corn varieties that are more resistant to disease and thrive in soil that is poor in nitrogen. Agronomists in Kenya are developing a sweet potato that wards off viruses. Also in the works are drought-tolerant, disease-defeating and vitamin-fortified forms of such crops as sorghum and cassava—hardly staples in the West, but essentials elsewhere in the world. The key, explains economist Jeffrey Sachs, head of Columbia University's Earth Institute, is not to dictate food policy from the West but to help the developing world build its own biotech infrastructure so it can produce the things it needs the most. "We can't presume that our technologies will bail out poor people in Malawi," he says. "They need their own improved varieties of sorghum and millet, not our genetically improved varieties of wheat and soybeans."

Water
In 25 years two-thirds of humanity may live in nations running short of life's elixir

---

• WATER: For a world that is 70% water, things are drying up fast. Only 2.5% of water is fresh, and only a fraction of that is accessible. Meanwhile, each of us requires about 50 quarts per day for drinking, bathing, cooking and other basic needs. At present, 1.1 billion people lack access to clean drinking water and more than 2.4 billion lack adequate sanitation. "Unless we take swift and decisive action," says U.N. Secretary-General Kofi Annan, "by 2025, two-thirds of the world's population may be living in countries that face serious water shortages."

Only 2.5% of water is fresh, and only a fraction of that is accessible

The answer is to get smart about how we use water. Agriculture accounts for about two-thirds of the fresh water consumed. A report prepared for the summit thus endorses the "more crop per drop" approach, which calls for more efficient irrigation techniques, planting of drought- and salt-tolerant crop varieties that require less water and better monitoring of growing conditions, such as soil humidity levels. Improving water-delivery systems would also help, reducing the amount that is lost en route to the people who use it.

One program winning quick support is dubbed WASH—for Water, Sanitation and Hygiene for All—a global effort that aims to provide water services and hygiene training to everyone who lacks them by 2015. Already, the U.N., 28 governments and many nongovernmental organizations (NGOs) have signed on.

Climate
Car exhaust is a major source of the heat-trapping gases that produce global warming

• ENERGY AND CLIMATE: In the U.S., people think of rural electrification as a long-ago legacy of the New Deal. In many parts of the world, it hasn't even happened yet. About 2.5 billion people have no access to modern energy services, and the power demands of developing economies are expected to grow 2.5% per year. But if those demands are met by burning fossil fuels such as oil, coal and gas, more and more carbon dioxide and other greenhouse gases will hit the atmosphere. That, scientists tell us, will promote global warming, which could lead to rising seas, fiercer storms, severe droughts and other climatic disruptions.

Of more immediate concern is the heavy air pollution caused in many places by combustion of wood and fossil fuels. A new U.N. Environment Program report warns of the effects of a haze across all southern Asia. Dubbed the "Asian brown cloud" and estimated to be 2 miles thick, it may be responsible for hundreds of thousands of deaths a year from respiratory diseases.

The better way to meet the world's energy needs is to develop cheaper, cleaner sources. Pre-Johannesburg proposals call for eliminating taxation and pricing systems that encourage oil use and replacing them with policies that provide incentives for alternative energy. In India there has been a boom in wind power because the government has made it easier for entrepreneurs to get their hands on the necessary technology and has then required the national power grid to purchase the juice that wind systems produce.

Other technologies can work their own little miracles. Micro-hydroelectric plants are already operating in numerous nations, including Kenya, Sri Lanka and Nepal. The systems divert water from streams and rivers and use it to run turbines without complex dams or catchment areas. Each plant can produce as much as 200 kilowatts—enough to electrify 200 to 500 homes and businesses—and lasts 20 years. One plant in Kenya was built by 200 villagers, all of whom own shares in the cooperative that sells the power.

The Global Village Energy Partnership, which involves the World Bank, the UNDP and various donors, wants to provide energy to 300 million people, as well as schools, hospitals and clinics in 50,000 communities worldwide over 10 years. The key will be to match the right energy source to the right users. For example, solar panels that convert sunlight into electricity might be cost-effective in remote areas, while extending the power grid might be better in Third World cities.

Biodiversity
Unless we guard wilderness, as many as half of all species could vanish in this century

• BIODIVERSITY: More than 11,000 species of animals and plants are known to be threatened with extinction, about a third of all coral reefs are expected to vanish in the next 30 years and about 36 million acres of forest are being razed annually. In his new book, *The Future of Life*, Harvard biologist Edward O. Wilson writes of his worry that unless we change our ways half of all species could disappear by the end of this century.

Once you tear out swaths of ecosystem, you harm areas you didn't want to touch

The damage being done is more than aesthetic. Many vanishing species provide humans with both food and medicine. What's more, once you start tearing out swaths of ecosystem, you upset the existing balance in ways that harm even areas you didn't intend to touch. Environmentalists have said this for decades, and now that many of them have tempered ecological absolutism with developmental realism, more people are listening.

The Equator Initiative, a public-private group, is publicizing examples of sustainable development in the equatorial belt. Among the projects already cited are one to help restore marine fisheries in Fiji and another that promotes beekeeping as a source of supplementary income in rural Kenya. The Global Conservation Trust hopes to raise $260 million to help conserve genetic material from plants for use by local agricultural programs. "When you approach sustainable development from an environmental view, the problems are global," says the U.N.'s Malloch Brown. "But from a development view, the front line is local, local, local."

If that's the message environmental groups and industry want to get out, they appear to be doing a good job of it. Increasingly, local folks act whether world political bodies do or not. California Governor Gray Davis signed a law last month requiring automakers to cut their cars' carbon emissions by 2009.

Many countries are similarly proactive. Chile is encouraging sustainable use of water and electricity; Japan is dangling financial incentives before consumers who buy environmentally sound cars; and tiny Mauritius is promoting solar cells and discouraging use of plastics and other disposables.

Business is getting right with the environment too. The Center for Environmental Leadership in Business, based in Washington, is working with auto and oil giants including Ford, Chevron, Texaco and Shell to draft guidelines for incorporating biodiversity conservation into oil and gas exploration. And the center has helped Starbucks develop purchasing guidelines that reward coffee growers whose methods have the least impact on the environment. Says Nitin Desai, secretary-general of the Johannesburg summit: "We're hoping that partnerships—involving governments, corporations, philanthropies and NGOs—will increase the credibility of the commitment to sustainable development."

Will that happen? In 1992 the big, global measures of the Rio summit seemed like the answer to what ails the world. In 2002 that illness is—in many respects—worse. But if Rio's goal was to stamp out the disease of environmental degradation, Johannesburg's appears to be subtler—and perhaps better: treating the patient a bit at a time, until the planet as a whole at last gets well.

point-counterpoint

THE
Heat
IS ON

*Conservation champion Ralph Nader squares off against
noted astrophysicist Sallie Baliunas on global warming*

Administration, corporations ignoring problem

BY RALPH NADER

Imagine a day far in the future when wind-energy technologies are able to produce 20 percent of America's electricity; automobiles get 60 miles to the gallon; and light bulbs last 10 times longer, use 75 percent less electricity and are just as bright as those used by the representatives and senators in Congress.

The truth is when it comes to energy technologies, the future has already arrived.

Today's technologies are sufficient to meet a significant percentage of our electricity needs using wind power. Toyota and Honda are selling hybrid automobiles that can go 60 miles on a gallon of gas. And almost any hardware store in the country stocks compact fluorescent light bulbs, which require a fraction of the electricity used by incandescent bulbs and more than pay for themselves. Replacing one traditional light bulb with a compact fluorescent bulb can reduce carbon dioxide emissions by about 500 pounds over the life of the bulb.

Now imagine that scientists are telling us the atmosphere is heating up and that we are likely to witness a rise in extreme weather patterns, a sea-level rise that may flood coastal and low-lying areas and submerge entire island nations, a surge in human health problems due to higher temperatures, the accelerated spread of tropical diseases and a loss of important ecosystems.

Unfortunately, that day, too, has come. There is strong scientific consensus: global warming presents a clear and present danger to health, environment and economy.

The problem of global warming demands that the United States take immediate action to reduce its emissions of greenhouse gases, foster a rapid switch to more energy-efficient technologies, dramatically step up investments in modern public transit, cut subsidies for fossil fuels and launch a program to, finally, harness the plentiful supply of solar energy.

It is now past time for the United States, the leading greenhouse gas-emitting nation, to take far-reaching steps toward the cleaner, more efficient economy that the global warming threat demands.

The Clinton administration did too little to address the global-warming problem. Unfortunately, the current Bush administration, barring a public outcry, is willing to derail even modest efforts designed to conserve energy and limit fossil-fuel emissions.

Federal policy over the past century has largely failed to promote an energy system based on safe, secure, economically affordable and environmentally benign energy sources. The tax code, budget appropriations and regulatory processes overwhelmingly have been used to subsidize with taxpayer dollars dependence on fossil fuels and nuclear power. The result: increased sickness and premature deaths; depleted family budgets; acid-rain destruction of lakes, forests and crops; oil-spill contamination; polluted rivers and loss of aquatic species; and the long-term peril of climate change and radioactive waste dumps, not to mention a dependency on foreign energy supplies.

"If America had priorities other than satisfying the oil, gas, coal and nuclear industries, we could set ourselves on a sustainable-energy course."

Resistance and Indifference. One might wonder why government officials, when presented with a host of practical solutions to address the problem of global warming, are inert.

Could the lack of action be the result of lethargy, stupidity, ineptitude, corruption or venality? Perhaps a Congress marinated in Political Action Committee campaign contributions is to blame. Or perhaps the current administration is too close to the fossil-fuel industries.

In April 2001, Greenpeace Executive Director John Passacantando wrote to the heads of the Fortune 100 corporations and asked them the following questions:

- Does your company support the ratification and entry into force of the Kyoto Protocol?
- Does your company support President Bush in his opposition to this protocol?
- Will your company support or oppose the efforts of other countries to bring the Kyoto Protocol into force without the United States?
- Does your company accept the Third Assessment Report of the Intergovernmental Panel on Climate Change as the scientific basis for action to reduce greenhouse-gas emissions, particularly carbon-dioxide emissions from the burning of coal, oil and gas?

The responses to these questions from the titans of industry were not the crisp, clear and decisive missives one might expect in corporate correspondence.

The ExxonMobil Corp. said, "Possible human-induced climate change is a long-term risk that we at ExxonMobil take very seriously."

Not to be outdone, General Motors replied, "We, as many, have raised concerns with the Kyoto Protocol, including its inadequate attention to the development and global dissemination of new technology."

General Electric noted that "global warming and the protocol present highly complex scientific, political and social questions about which there is a great deal of uncertainty and disagreement."

What is the Kyoto Protocol?

The Kyoto Protocol to the U.N. Framework Convention on Climate Change was adopted by consensus in Kyoto, Japan, in December 1997. Although the United States signed the treaty on Nov. 12, 1998, it has not been forwarded to the U.S. Senate for ratification.

The Protocol mandates that:

- Industrialized countries must reduce human-generated carbon-dioxide emissions of six key greenhouse gases by at least 5 percent below 1990 levels within the commitment period 2008 to 2012. The six gases are to be combined in a "basket," with reductions in individual gases translated into "CO_2 equivalents" that are then added up to produce a single figure.
- Each country must show "demonstrable progress" toward meeting emissions targets by 2005.
- Actual emission reductions will be much larger than 5 percent. Compared with emission levels projected for the year 2000, the richest industrialized countries would be required to reduce their collective output by a larger percentage. For example, the United States would be targeted to reduce greenhouse gases by 7 percent below 1990 levels.
- Countries will have a certain degree of flexibility in how they make and measure their emission reductions.
- Countries will pursue emissions cuts in a wide range of economic sectors.
- Both developed and developing countries agree to take measures to limit emissions and promote adaptation to future climate-change impacts; submit information on their national climate-change programs and inventories; promote technology transfer; cooperate on scientific and technical research; and promote public awareness, education and training.

—Elissa Kaupisch

I think the responses from the other corporation officials are equally noncommittal. But they do provide a snapshot of corporate America's posture on one of the most significant environmental challenges of the coming century. That snapshot tells us that too many corporate executives oppose efforts to address the causes of global warming—antiquated, inefficient and damaging technologies—or are unwilling to use any of their political or financial capital to advance policies that benefit all the would-be consumers of the planet.

Options Abound. Several alternatives to such intransigent behavior and inaction exist. But it will require an aroused and engaged citizenry with focus and stamina to break the corporate logjam that has blocked the development of an energy-independence policy.

Three decades of detailed assessments, on-the-ground results, and research and development innovations in the energy-consuming devices used in our buildings, vehicles and industries undeniably show that energy efficiency and renewable-energy technologies are superior options for society. They offer a present and future path that is economically attractive, safe and secure from large-scale and long-term risks or threats to public health, future generations, the environment and national security.

In my opinion, embarking on that path requires overcoming the power of the oil, nuclear and other conventional fuel industries to which both the Republicans and Democrats are indentured. Under the thumb of the dirty-fuel industries, Congress and the executive branch have refused to adopt even the most modest, common-sense measures. For example, when the president's Committee of Advisors on Science and Technology concluded in a 1997 report that doubling the Department of Energy's efficiency R&D funding would produce a 40-to-1 return on the investment for the nation, Congress responded by proposing deep cuts in the efficiency and renewables R&D budgets. Wasteful energy policies mean greater energy sales by energy companies.

The Clinton administration's nod to increased energy efficiency relied largely on corporate welfare. Rather than push for an increase in auto fuel-efficiency standards, the administration established the Partnership for a New Generation of Vehicles. PNGV is a $1.5 billion subsidy program for the Big Three auto companies that has done nothing to improve auto fuel efficiency. But it has served as a convenient smokescreen behind which the industry, freed from the need to competitively innovate, fended off new regulatory requirements for more efficient cars.

"Energy Innovations: A Prosperous Path to a Clean Environment," a joint study prepared by six of the nation's prominent energy and environmental research and advocacy groups, shows that a handful of simple and straightforward measures could produce a significant reduction in sulfur-dioxide emissions (the prime cause of acid rain) and nitrogen-oxide emissions (a key precursor of ground-level ozone and smog) as well as deep cuts in emissions of other damaging pollutants, including fine particles, toxic metals like mercury and hydrocarbons, and carbon dioxide emissions.

If America had priorities other than satisfying the oil, gas, coal and nuclear industries, we could set ourselves on a sustainable-energy course by:

- Ending fossil-fuel and nuclear corporate-welfare supports, including numerous special tax preferences.

- Launching a robust federal research-and-development program in sustainable renewable-energy sources, so that the practical promises of wind, solar and other forms of renewable energy are finally realized.

- Increasing long-delayed auto fuel-efficiency standards (at least to 45 miles per gallon for cars and 35 miles per gallon for light trucks, to be phased in over five years) during a transition period to zero-emissions cars. American engineers are awaiting the "Go" signal.

- Adopting stronger efficiency standards for appliances and mandatory energy-performance building codes.

- Ensuring electricity policies which promote efficient use of electricity through a range of measures, including "net metering" requirements that companies pay market prices for electricity generated by consumers and passed back to the utility, and elimination of clean air exemptions for "grandfathered" fossil-fueled facilities.

- Establishing a well-funded employee transition-assistance fund and job-retraining program for displaced coal miners, which will be easily affordable with the savings from greater energy efficiency.

As a first step, the United States should ratify the Kyoto Protocol and then progress beyond its minimal standards. Among key measures required: We must move rapidly to zero-emissions standards for cars, even as we commit to creating a satisfactory public-transit system. We must require old "grandfathered" coal plants to meet modern pollution standards. And we should place a moratorium on commercial logging in our national forests. All of these programs would likely help family pocketbooks and family health as well as reducing greenhouse gases—a win-win situation.

Also, Washington needs to follow other allied countries and end the ban on our farmers growing industrial hemp, used for fuel and paper. George Washington grew it, and former CIA director James Woolsey believes it is a national security "must" to reduce our reliance on imported oil.

Our country has more problems than it deserves and more solutions than it uses. It is time for the United States to stop letting ExxonMobil, Peabody Coal and Westinghouse shape our energy policy and for our misguided elected officials to start adopting an energy strategy based on clean renewable energy and conservation. Future generations will thank us for curbing our fossil-fuel and atomic appetites.

Ralph Nader is a consumer advocate and was the Green Party candidate for president in 2000.

Studies lack hard evidence that warming is human-induced

BY SALLIE BALIUNAS

The decision to mount America's largest military invasion, the D-Day landing on the Normandy beaches, relied on a weather forecast. Meteorologists studied decades of weather maps from the North Atlantic in order to gain forecasting acumen. Then on June 4, 1944, 5,000 ships carrying 86,000 soldiers crushed against the waters of the English Channel, while 13,000 support aircraft held for an unfavorable June 4 weather forecast. But the June 5 forecasts indicated improved conditions, so Gen. Eisenhower ordered the D-Day invasion for the next day. If that forecast hadn't been accurate, the assault troops might never have reached Normandy's beaches. Thus, modern soldiers have come to know the importance of reliable weather forecasts for technological battlefields.

Today a scientifically accurate understanding of weather and climate is essential for economies built on technology. Human use of coal, oil, natural gas and other fossil fuels has increased the concentration of carbon dioxide in the air. The belief is that this added CO_2 is causing a significant warming of the climate.

The latest report of the U.N.'s Intergovernmental Panel on Climate Change (IPCC), using several computer simulations, forecasts a human-made global-warming trend between 1.4 and 5.8 degrees C by 2100, with a middle value of about 2.8 degrees C.

To prevent the warming, the Kyoto agreement asks America to drastically cut its CO_2 emissions and energy use by about 40 percent from today's consumption, which surely will yield a worldwide economic disaster. Yet are the forecasts of human-made global warming in the century ahead reliable? Will things turn out as badly as some say? And can cutting back fossil fuel use really reduce global warming?

The answer to the first question is "not very." The second, "not likely." The third, "not much, if anything at all." To know why, we need to look at the scientific record.

Natural Causes or Not? Yes, CO_2 is a greenhouse gas, which helps keep some of the sun's energy from returning to space. The IPCC forecast of the climate's response to this small amount of extra energy comes from the encoding of present ideas about climate into sophisticated computer simulations. These simulations say that the temperature near the surface and through the first five miles of air, the troposphere, should warm. Has that happened? Compared to the previous five centuries or so, the 20th century did show a warming trend, with a globally averaged surface-temperature rise of 0.5 C.

But look deeper, and the proof of human-induced warming dissipates like so much hot air.

First, most of the warming occurred before 1940—before 80 percent of the CO_2 from human activities was added to the air. This means that the early 20th century warming must be mostly natural.

Second, the climate record of the past 1,000 years suggests this temperature rise is hardly unique. New information about historical climate change obtained from trees, glaciers, ice cores, coral and the like indicate a widespread Medieval Warm Period from about 800 to 1200 A.D. Subsequently, temperatures dropped markedly, creating a Little Ice Age that persisted nearly to the 20th century. So the 20th century's warming seems largely a natural rebound from the cold spell.

But what about the past several decades, when the CO_2 content of the air rose most dramatically?

A critical problem for those claiming human-induced warming is that the computer climate simulations predict both surface temperatures and those of the lower troposphere should rise together. Moreover, the lower troposphere should warm the most.

For more than three decades, surface temperatures actually fell slightly before starting to rise again in the late 1970s. Tropospheric temperatures showed no warming from the inception of measurements by balloon-borne instruments in 1957 until 1976. From 1976 to 1977, an upward shift occurred. But between 1979, after the advent of daily global-satellite measurements of tropospherecic temperatures, and the present, neither satellite nor balloon data show a manmade warming trend.

Proponents of human-made global warming say soot from industries has acted as an aerosol to mask a larger warming trend. But that unravels because whereas CO_2 disperses globally, aerosols tend to stick more closely to where they are released. And the southern hemisphere, which is relatively free of aerosols, actually showed a cooling trend.

The point is that the best data collected from satellites and validated by balloons to test the hypothesis of a human-induced global warming from the release of CO_2 into the atmosphere shows no meaningful trend of increasing temperatures, even as the climate models exaggerated the warmth that ought to have occurred from a build-up in CO_2.

What's Wrong with Models? Climate models are too simplistic. They must deal with more than 5 million variables, including many that are uncertain or unmeasured.

51

For example, the models lack key information about two major climate effects: water vapor and clouds. Little wonder that these models haven't reproduced the major features of present or past climate, such as the El Niño oscillations, that occur in two- to seven-year periods. They provide no proof that mankind is causing global warming to occur.

But what is causing surface temperatures to rise? A chart of surface temperatures going back more than 240 years shows a strong correlation between them and cycles of the sun's magnetism. Satellite measurements of the past two decades demonstrate the sun is a variable star, with its total energy output changing in step with periodic changes in its magnetism. This correlation suggests that changes in the sun's energy output of a few tenths of a percent over decades may explain many of the temperature changes over the centuries. Measurements made at Mount Wilson Observatory in Los Angeles of hundreds of other sun-like stars indicate the amounts of such changes are entirely possible.

America has led the scientific study of global warming with approximately $18 billion in research funding over the past decade. That research shows the threat of catastrophic warming is miniscule against the backdrop of natural change.

Evidence of any substantial human-induced warming is, at best, weak. But wouldn't such warming, if it were going on, be dangerous? Why not take precautions and cut back our use of fossil fuels?

First, the warming is likely natural, and second, warming probably poses less of a threat than cooling would. People benefited from the Medieval Warm Period, with its equable climate conditions, compared to the subse-

quent deterioration during the Little Ice Age. Fig trees grew in Koln, Germany; vineyards were found in England; and Vikings sailed the seas to colonize Iceland, Greenland and possibly Newfoundland. After the onset of the Little Ice Age, growing seasons shortened, the North Sea became stormier, and life expectancy dropped back by about 10 years due to starvation and harsh weather conditions of a colder climate.

The 20th century's warming has extended growing seasons, too. And increased CO_2 also has helped increase crop yields to feed more people. No deleterious global climate effects can be identified with energy use. Instead, vast numbers of people have been raised from poverty by the economic growth that energy use produces.

By contrast, a rash cutback in energy use, as required by the 1997 Kyoto Protocol could trigger a prolonged worldwide recession. Even economists from the Clinton administration now admit that the price tag for America would run to hundreds of billions of dollars annually. The rising energy prices needed to enforce conservation would especially hurt lower-income workers, who spend a greater proportion of their incomes on energy. And their sacrifice would accomplish little. According to the computer models global-warming alarmists rely upon, temperatures, after implementing the Kyoto Protocol, would decline less than a few tenths Celsius by the year 2100—beneath notice, given the bounds of natural climate change.

America has led the scientific study of global warming with approximately $18 billion in research funding over the past decade. That research shows the threat of catastrophic warming is miniscule against the backdrop of natural change. The best thing now would be to improve the climate simulations and better pinpoint any human effect, while readying cost-effective measures in mitigation and adaptation.

As soldiers can understand, the nation needs a more reliable climate forecast before launching an assault on global warming that could swamp the economy in energy regulations from which the world might never recover.

Sallie Baliunas is an astrophysicist and deputy director of Mount Wilson Observatory near Los Angeles.

We *Can* Build a Sustainable Economy

The keys to securing the planet's future lie in stabilizing both human population and climate. The challenges are great, but several trends look promising.

by Lester R. Brown

The world economy is growing faster than ever, but the benefits of this rapid growth have not been evenly distributed. As population has doubled since mid-century and the global economy has nearly quintupled, the demand for natural resources has grown at a phenomenal rate.

Since 1950, the need for grain has nearly tripled. Consumption of seafood has increased more than four times. Water use has tripled. Demand for beef and mutton has tripled. Firewood demand has tripled, lumber demand has more than doubled, and paper demand has gone up sixfold. The burning of fossil fuels has increased nearly fourfold, and carbon emissions have risen accordingly.

These spiraling human demands for resources are beginning to outgrow the earth's natural systems. As this happens, the global economy is damaging the foundation on which it rests.

To build an environmentally sustainable global economy, there are many obstacles, but there are also several promising trends and factors in our favor. One is that we know what an environmentally sustainable economy would look like. In a sustainable economy:

- Human births and deaths are in balance.
- Soil erosion does not exceed the natural rate of new soil formation.
- Tree cutting does not exceed tree planting.
- The fish catch does not exceed the sustainable yield of fisheries.
- The number of cattle on a range does not exceed the range's carrying capacity.
- Water pumping does not exceed aquifer recharge.
- Carbon emissions and carbon fixation are in balance.
- The number of plant and animal species lost does not exceed the rate at which new species evolve.

We know how to build an economic system that will meet our needs without jeopardizing prospects for future generations. And with some trends already headed in the right direction, we have the cornerstones on which to build such an economy.

Stabilizing Population

With population, the challenge is to complete the demographic transition, to reestablish the balance between births and deaths that characterizes a sustainable society. Since populations are rarely ever precisely stable, a stable population is defined here as one with a growth rate below 0.3%. Populations are effectively stable if they fluctuate narrowly around zero.

Thirty countries now have stable populations, including most of those in Europe plus Japan. They provide the solid base for building a world population stabilization effort. Included in the 30 are all the larger industrialized countries of Europe—France, Germany, Italy, Russia, and the United Kingdom. Collectively, these 30 countries contain 819 million people or 14% of humanity. For this goal, one-seventh of humanity is already there.

The challenge is for the countries with the remaining 86% of the world's people to reach stability. The two large nations that could make the biggest difference in this effort are China and the United States. In both, population growth is now roughly 1% per year. If the global food situation becomes desperate, both could reach stability in a decade or two if they decided it were important to do so.

The world rate of population growth, which peaked around 2% in 1970, dropped below 1.6% in 1995. Although the rate is declining, the annual addition is still close to 90 million people per year. Unless populations can be stabilized with demand below the sustainable yield of local ecosystems, these systems will be destroyed. Slowing growth may delay the eventual collapse of ecosystems, but it will not save them.

The European Union, consisting of some 15 countries and containing 360 million people, provides a model for the rest of the world of an environmentally sustainable food/population balance. At the same time that the region has reached zero population growth, movement up the food chain has come to a halt as diets have become saturated with livestock products. The result is that Europe's grain consumption has been stable for close to two decades at just under 160 million tons—a level that is within the region's carrying capacity. Indeed, there is a potential for a small but sustainable export surplus of grain that can help countries where the demand for food has surpassed the carrying capacity of their croplands.

World Fertilizer and Grainland
(Per Person, 1950-94)

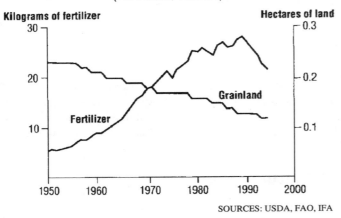

SOURCES: USDA, FAO, IFA

As other countries realize that continuing on their current population trajectory will prevent them from achieving a similar food/population balance, more and more may decide to do what China has done—launch an all-out campaign to stabilize population. Like China, other governments will have to carefully balance the reproductive rights of the current generation with the survival rights of the next generation.

Very few of the group of 30 countries with stable populations had stability as an explicit policy goal. In those that reached population stability first, such as Belgium, Germany, Sweden, and the United Kingdom, it came with rising living standards and expanding employment opportunities for women. In some of the countries where population has stabilized more recently, such as Russia and other former Soviet republics, the deep economic depression accompanying economic reform has substantially lowered birth rates, much as the Great Depression did in the United States. In addition, with the rising number of infants born with birth defects and deformities since Chernobyl, many women are simply afraid to bear children. The natural decrease of population (excluding migration) in Russia of 0.6% a year—leading to an annual population loss of 890,000—is the most rapid on record.

Not all countries are achieving population stability for the right reasons. This is true today and it may well be true in the future. As food deficits in densely populated countries expand, governments may find that there is not enough food available to import. Between fiscal year 1993 and 1996, food aid dropped from an all-time high of 15.2 million tons of grain to 7.6 million tons. This cut of exactly half in three years reflects primarily fiscal stringencies in donor countries, but also, to a lesser degree, higher grain prices in fiscal 1996. If governments fail to establish a humane balance between their people and food supplies, hunger and malnutrition may raise death rates, eventually slowing population growth.

Some developing countries are beginning to adopt social policies that will encourage smaller families. Iran, facing both land hunger and water scarcity, now limits public subsidies for housing, health care, and insurance to three children per family. In Peru, President Alberto Fujimori, who was elected overwhelmingly to his second five-year term in a predominantly Catholic country, said in his inaugural address in August 1995 that he wanted to provide better access to family-planning services for poor women. "It is only fair," he said, "to disseminate thoroughly the methods of family planning to everyone."

Stabilizing Climate

With climate, as with population, there is disagreement on the need to stabilize. Evidence that atmospheric carbon-dioxide levels are rising is clear-cut. So, too, is the greenhouse effect that these gases produce in the atmosphere. That is a matter of basic physics. What is debatable is the rate at which global temperatures will rise and what the precise local effects will be. Nonetheless, the consensus of the mainstream scientific community is that there is no alternative to reducing carbon emissions.

How would we phase out fossil fuels? There is now a highly successful "phase out" model in the case of chlorofluorocarbons (CFCs). After two British scientists discovered the "hole" in the ozone layer over Antarctica and published their findings in *Nature* in May 1985, the international community convened a conference in Montreal to draft an agreement designed to reduce CFC production sharply. Subsequent meetings in London in 1990 and Copenhagen in 1992 further advanced the goals set in Montreal. After peaking in 1988 at 1.26 million tons, the manufacture of CFCs dropped to an estimated 295,000 tons in 1994—a decline of 77% in just six years.

As public understanding of the costs associated with global warming increases, and as evidence of the effects of higher temperatures accumulates, support for reducing dependence on fossil fuels is building. At the March 1995 U.N. Climate Convention in Berlin, environmental groups were joined in lobbying for a reduction in carbon emissions by a group of 36 island communities and insurance industry representatives.

The island nations are beginning to realize that rising sea levels would, at a minimum, reduce their land area and displace people. For some low-lying island countries, it could actually threaten their survival. And the insurance industry is beginning to realize that increasing storm intensity can threaten the survival of insurance companies as well. When Hurricane Andrew tore through Florida in 1992, it took down not only thousands of buildings, but also eight insurance firms.

In September 1995, the U.S. Department of Agriculture reported a sharp drop in the estimated world grain harvest because of crop-withering heat waves in the northern tier of industrial countries. Intense late-summer heat had damaged harvests in Canada and the United States, across Europe, and in Russia. If farmers begin to see that the productivity of their land is threatened by global warming, they, too, may begin to press for a shift to renewable sources of energy.

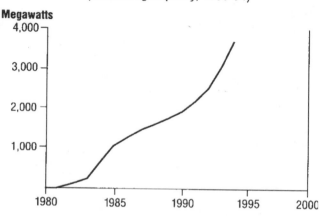

World Wind Energy
(Generating Capacity, 1980-94)

SOURCES: Gipe and Asociates, BTM Consulting

As with CFCs, there are alternatives to fossil fuels that do not alter climate. Several solar-based energy sources, including wind power, solar cells, and solar thermal power plants, are advancing rapidly in technological sophistication, resulting in steadily falling costs. The cost of photovoltaic cells has fallen precipitously over the last few decades. In some villages in developing countries where a central grid does not yet exist, it is now cheaper to install an array of photovoltaic cells than to build a centralized power plant plus the grid needed to deliver the power.

Wind power, using the new, highly efficient wind turbines to convert wind into electricity, is poised for explosive growth in the years ahead. In California, wind farms already supply enough electricity to meet the equivalent of San Francisco's residential needs.

The potential for wind energy is enormous, dwarfing that of hydropower, which provides a fifth of the world's electricity. In the United States, the harnessable wind potential in North Dakota, South Dakota, and Texas could easily meet national electricity needs. In Europe, wind power could theoretically satisfy all the continent's electricity needs. With scores of national governments planning to tap this vast resource, rapid growth in the years ahead appears inevitable.

A Bicycle Economy

Another trend to build on is the growing production of bicycles. Human mobility can be increased by investing in public transportation, bicycles, and automobiles. Of these, the first two are by far the most promising environmentally. Although China has announced plans to move toward an automobile-centered transportation system, and car production in India is expected to double by the end of the decade, there simply may not be enough land in these countries to support such a system and to meet the food needs of their expanding populations.

Against this backdrop, the creation of bicycle-friendly transportation systems, particularly in cities, shows great promise. Market forces alone have pushed bicycle production to an estimated 111 million in 1994, three times the level of automobile production. It is in the interest of societies everywhere to foster the use of bicycles and public transportation—to accelerate the growth in bicycle manufacturing while restricting that of automobiles. Not only will this help save cropland, but this technology can greatly increase human mobility without destabilizing climate. If food becomes increasingly scarce in the years ahead, as now seems likely, the land-saving, climate-stabilizing nature of bicycles will further tip the scales in their favor and away from automobiles.

The stabilization of population in some 30 countries, the stabilization of food/people balance in Europe, the reduction in CFC production, the dramatic growth in the world's wind power generating capacity, and the extraordinary growth in bicycle use are all trends for the world to build on. These cornerstones of an environmentally sustainable global economy provide glimpses of a sustainable future.

Regaining Control of Our Destiny

Avoiding catastrophe is going to take a far greater effort than is now being contemplated by the world's political leaders. We know what needs to be done, but politically we are unable to do it because of inertia and the investment of powerful interests in the status quo. Securing food supplies for the next generation depends on an all-out effort to stabilize population and climate, but we resist changing our reproductive behavior, and we refrain from converting our climate-destabilizing, fossil-fuel-based economy to a solar/hydrogen-based one.

As we move to the end of this century and beyond, food security may well come to dominate international affairs, national economic policy making, and—for much of humanity —personal concerns about survival. There is now evidence from enough countries that the old formula of substituting fertilizer for land is no longer working, so we need to search urgently for alternative formulas for humanly balancing our numbers with available food supplies.

Unfortunately, most national political leaders do not even seem to be aware of the fundamental shifts occurring in the world food economy, largely because the official projections by the World Bank and the U.N. Food and Agriculture Organization are essentially extrapolations of past trends.

If we are to understand the challenges facing us, the teams of economists responsible for world food supply-and-demand projections at these two organizations need to be replaced with an interdisciplinary team of analysts, including, for example, an agronomist, hydrologist, biologist, and meteorologist, along with an economist. Such a team could assess and incorporate into projections such things as the effect of soil erosion on land productivity, the effects of aquifer depletion on future irrigation water supplies, and the effect of increasingly intense heat waves on future harvests.

The World Bank team of economists argues that, because the past is the only guide we have to the future, simple extrapolations of past trends are the only reasonable way to make projections. But the past is also filled with a body of scientific literature on growth in finite environments, and it shows that biological growth trends typically conform to an S-shaped curve over time.

The risk of relying on these extrapolative projections is that they are essentially "no problem" projections. For example, the most recent World Bank projections, which use 1990 as a base and which were published in late 1993, are departing further and further from reality with each passing year. They show the world grain harvest climbing from 1.78 billion tons in 1990 to 1.97 billion tons in the year 2000. But instead of the projected gain of nearly 100 million tons since 1990, world grain production has not grown at all. Indeed, the 1995 harvest, at 1.69 billion tons, is 90 million tons below the 1990 harvest.

One of the most obvious needs today is for a set of country-by-country carrying-capacity assessments. Assessments using an interdisciplinary team can help provide information needed to face the new realities and formulate policies to respond to them.

Setting Priorities

The world today is faced with an enormous need for change in a period of time that is all too short. Human behavior and values, and the national priorities that reflect them, change in response to either new information or new experiences. The effort now needed to reverse the environmental degradation of the planet and ensure a sustainable future for the next generation will require mobilization on a scale comparable to World War II.

Regaining control of our destiny depends on stabilizing population as well as climate. These are both key to the achievement of a wide array of social goals ranging from the restoration of a rise in food consumption per person to protection of the diversity of plant and animal species. And neither will be easy. The first depends on a revolution in human reproductive behavior; the second, on a restructuring of the global energy system.

Serving as a catalyst for these gargantuan efforts is the knowledge that if we fail our future will spiral out of control as the acceleration of history overwhelms political institutions. It will almost guarantee a future of starvation, economic insecurity, and political instability. It will bring political conflict between societies and among ethnic and religious groups within societies. As these forces are unleashed, they will leave social disintegration in their wake.

Offsetting the dimensions of this challenge, including the opposition to change that is coming from vested interests and the momentum of trends now headed in the wrong direction, are some valuable assets. These include a well-developed global communications network, a growing body of scientific knowledge, and the possibility of using fiscal policy—a potentially powerful instrument for change—to build an environmentally sustainable economy.

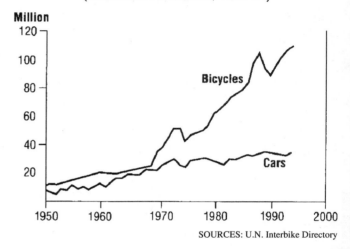

Bicycles vs. Cars
(Worldwide Production, 1950-94)

SOURCES: U.N. Interbike Directory

Policies for Progress

Satisfying the conditions of sustainability—whether it be reversing the deforestation of the planet, converting a throwaway economy into a reuse-recycle one, or stabilizing climate—will require new investment. Probably the single most useful instrument for converting an unsustainable world economy into one that is sustainable is fiscal policy. Here are a few proposals:

• **Eliminate subsidies for unsustainable activities**. At present, governments subsidize many of the very activities that threaten the sustainability of the economy. They support fishing fleets to the extent of some $54 billion a year, for example, even though existing fishing capacity already greatly exceeds the sustainable yield of oceanic fisheries. In Germany, coal production is subsidized even though the country's scientific community has been outspoken in its calls for reducing carbon emissions.

• **Institute a carbon tax**. With alternative sources of energy such as wind power, photovoltaics, and solar thermal power plants becoming competitive or nearly so, a carbon tax that would reflect the cost to society of burning fossil fuels—the costs, that is, of air pollution, acid rain, and global warming—could quickly tip the scales away from further investment in fossil fuel production to investment in wind and solar energy.

Today's fossil-fuel-based energy economy can be replaced with a solar/hydrogen economy that can meet all the energy needs of a modern industrial society without causing disruptive temperature rises.

• **Replace income taxes with environmental taxes**. Income taxes discourage work and savings, which are both positive activities that should be encouraged. Taxing environmentally destructive activities instead would help steer the global economy in an environmentally sustainable direction. Among the activities to be taxed are the use of pesticides, the generation of toxic wastes, the use of virgin raw materials, the conversion of cropland to nonfarm uses, and carbon emissions.

The time may have come also to limit tax deductions for children to two per couple: It may not make sense to subsidize childbearing beyond replacement level when the most pressing need facing humanity is to stabilize population.

The challenge for humanity is a profound one. We have the information, the technology, and the knowledge of what needs to be done. The question is, Can we do it? Can a species that is capable of formulating a theory that explains the birth of the universe now implement a strategy to build an environmentally sustainable economic system?

About the Author Lester R. Brown is president of the Worldwatch Institute, 1776 Massachusetts Avenue, N.W., Washington, D.C. 20036. Telephone 202/452-1999; fax 202/296-7365.

From *The Futurist*, July/August 1996, pp. 8–12. © 1996 by The World Future Society, Bethesda, MD; http://www.wfs.org/wfs. Reprinted by permission.

UNIT 4
Political Economy

Unit Selections

Key Points to Consider

- Are those who argue that there is in fact a process of globalization overly optimistic? Why or why not?

- What are some of the impediments to a truly global political economy?

- How are the political economies of traditional societies different from those of the consumer-oriented societies?

- What are some of the barriers that make it difficult for non-industrial countries to develop?

- How are China and other emerging countries trying to alter their ways of doing business in order to meet the challenges of globalization? Are they likely to succeed?

- What economic challenges do countries like Japan and the United States face in the years to come?

- What is the nature of the debate surrounding the practices of international organizations like the International Monetary Fund and World Trade Organization?

 **Links: www.dushkin.com/online/**
These sites are annotated in the World Wide Web pages.

Belfer Center for Science and International Affairs (BCSIA)
http://ksgwww.harvard.edu/csia/

Communications for a Sustainable Future
http://csf.colorado.edu

U.S. Agency for International Development
http://www.info.usaid.gov

Virtual Seminar in Global Political Economy/Global Cities & Social Movements
http://csf.colorado.edu/gpe/gpe95b/resources.html

World Bank
http://www.worldbank.org

A defining characteristic of the twentieth century's social history was the contest between two dramatically opposing views about how economic systems should be organized. The focus of the debate was on what role government should play in the management of a country's economy. For some the dominant capitalist economic system appeared to be organized primarily for the benefit of the few. From their perspective, the masses were trapped in poverty, supplying cheap labor to further enrich the wealthy. These critics argued that the capitalist system could be changed only by gaining control of the political system and radically changing the ownership of the means of production. In striking contrast to this perspective, others argued that the best way to create wealth and eliminate poverty was through the profit motive, which encouraged entrepreneurs to innovate. An open and competitive marketplace that minimized government interference was the best system for making decisions about production, wages, and the distribution of goods and services.

Both abstract theorizing and very pragmatic and often violent conflict characterized the debate between socialism/communism and capitalism. The Russian and Chinese revolutions overthrew the old social order and created radical changes in the political and economic systems in these two important countries. The political structures that were created to support new systems of agricultural and industrial production (along with the centralized planning of virtually all aspects of economic activity) eliminated most private ownership of property. These two revolutions were, in short, unparalleled experiments in social engineering.

The collapse of the Soviet Union and the dramatic reforms that have taken place in China have recast the debate about how to best structure contemporary economic systems. Some believe that with the end of communism and the resulting participation of hundreds of millions of new consumers in the global market, an unprecedented era has been entered. Many have noted that this process of "globalization" is being accelerated by the revolution in communication and computer technologies. Proponents of this view argue that a new global economy is emerging that will ultimately eliminate national economic systems.

Others are less optimistic about the prospects of globalization. They argue that the creation of a single economic system where there are no boundaries to impede the flow of capital and goods and services does not mean a closing of the gap between the world's rich and poor. Rather, they argue that giant corporations will have fewer legal constraints on their behavior, and this will lead to greater exploitation of workers and the accelerated destruction of the environment.

The use of the term *political economy* for the title of this unit is recognition that economic and political systems are not separate. All economic systems have some type of marketplace where goods and services are bought and sold. Government (either national or international) regulates these transactions to some degree; that is, government sets the rules that regulate the marketplace.

One of the most important concepts in assessing the contemporary political economy is "development." For the purposes of this unit, the term *development* is defined as an improvement in the basic aspects of life: lower infant mortality rates, longer life expectancy, lower disease rates, higher rates of literacy, healthier diets, and improved sanitation. Judged by these standards, some countries are more "developed" than others. A fundamental question that a thoughtful reader must consider is whether globalization is resulting in increased development not only for a few people but also for all of those participating in the global political economy.

The unit is organized into two subsections. The first is a general discussion of the concept of globalization. How do various experts define the term, and what are their differing perspectives on it? For example, is the idea of a global economy merely wishful thinking by those who sit on top of the power hierarchy, self-deluded into believing that globalization is an inexorable force that will evolve in its own way, following its own rules? Or will there continue to be the traditional tensions of power politics, that is, between the powerful and those who are either ascending or descending in power?

The second subsection is a selection of case studies that focus on specific countries or economic sectors (e.g., international media). These case studies have been selected to challenge the reader to develop his or her own conclusions about the positive and negative consequences of the globalization process. Does the contemporary global political economy result in an age-old system of winners and losers, or can everyone positively benefit from its system of wealth creation and distribution?

The Complexities and Contradictions of Globalization

Globalization, we are told, is what every business should be pursuing, and what every nation should welcome. But what, exactly, is it? James Rosenau offers a nuanced understanding of a process that is much more real, and transforming, than the language of the marketplace expresses.

JAMES N. ROSENAU

The mall at Singapore's airport has a food court with 15 food outlets, all but one of which offering menus that cater to local tastes; the lone standout, McDonald's, is also the only one crowded with customers. In New York City, experts in *feng shui*, an ancient Chinese craft aimed at harmonizing the placement of man-made structures in nature, are sought after by real estate developers in order to attract a growing influx of Asian buyers who would not be interested in purchasing buildings unless their structures were properly harmonized.

Most people confronted with these examples would probably not be surprised by them. They might even view them as commonplace features of day-to-day life late in the twentieth century, instances in which local practices have spread to new and distant sites. In the first case the spread is from West to East and in the second it is from East to West, but both share a process in which practices spread and become established in profoundly different cultures. And what immediately comes to mind when contemplating this process? The answer can be summed up in one word: globalization, a label that is presently in vogue to account for peoples, activities, norms, ideas, goods, services, and currencies that are decreasingly confined to a particular geographic space and its local and established practices.

Indeed, some might contend that "globalization" is the latest buzzword to which observers resort when things seem different and they cannot otherwise readily account for them. That is why, it is reasoned, a great variety of activities are labeled as globalization, with the result that no widely accepted formulation of the concept has evolved. Different observers use it to describe different phenomena, and often there is little overlap among the various usages. Even worse, the elusiveness of the concept of globalization is seen as underlying the use of a variety of other, similar terms—world society, interdependence, centralizing tendencies, world system, globalism, universalism, internationalization, globality—that come into play when efforts are made to grasp why public affairs today seem significantly different from those of the past.

Such reasoning is misleading. The proliferation of diverse and loose definitions of globalization as well as the readiness to use a variety of seemingly comparable labels are not so much a reflection of evasive confusion as they are an early stage in a profound ontological shift, a restless search for new ways of understanding unfamiliar phenomena. The lack of precise formulations may suggest the presence of buzzwords for the inexplicable, but a more convincing interpretation is that such words are voiced in so many different contexts because of a shared sense that the human condition is presently undergoing profound transformations in all of its aspects.

WHAT IS GLOBALIZATION?

Let us first make clear where globalization fits among the many buzzwords that indicate something new in world affairs that is moving important activities and concerns beyond the national seats of power that have long served as the foundations of economic, political, and social life. While all the buzzwords seem to cluster around the same dimension of the present human condition, useful distinctions can be drawn among them.

Most notably, if it is presumed that the prime characteristic of this dimension is change—a transformation of practices and norms—then the term "globalization" seems appropriate to denote the "something" that is changing humankind's preoccupation with territoriality and the traditional arrangements of the state system. It is a term that directly implies change, and thus differentiates the phenomenon as a process rather than as a prevailing condition or a desirable end state.

Conceived as an underlying process, in other words, globalization is not the same as globalism, which points to aspirations for a state of affairs where values are shared by or pertinent to all the world's more than 5 billion people, their environment, and their role as citizens, consumers, or producers with an interest in collective action to solve common problems. And it can also be distinguished from universalism, which refers to those values that embrace all of humanity (such as the values that science or religion draws on), at any time or place. Nor is it coterminous with complex interdependence, which signifies structures that link people and communities in various parts of the world.

Although related to these other concepts, the idea of globalization developed here is narrower in scope. It refers neither to values nor to structures, but to sequences that unfold either in the mind or in behavior, to processes that evolve as people and organizations go about their daily tasks and seek to realize their particular goals. What distinguishes globalizing processes is that they are not hindered or prevented by territorial or jurisdictional barriers. As indicated by the two examples presented at the outset, such processes can readily spread in many directions across national boundaries, and are capable of reaching into any community anywhere in the world. They consist of all those forces that impel individuals, groups, and institutions to engage in similar forms of behavior or to participate in more encompassing and coherent processes, organizations, or systems.

Contrariwise, localization derives from all those pressures that lead individuals, groups, and institutions to narrow their horizons, participate in dissimilar forms of behavior, and withdraw to less encompassing processes, organizations, or systems. In other words, any technological, psychological, social, economic, or political developments that foster the expansion of interests and practices beyond established boundaries are both sources and expressions of the processes of globalization, just as any developments in these realms that limit or reduce interests are both sources and expressions of localizing processes.

Note that the processes of globalization are conceived as only capable of being worldwide in scale. In fact, the activities of no group, government, society, or company have never been planetary in magnitude, and few cascading sequences actually encircle and encompass the entire globe. Televised events such as civil wars and famines in Africa or protests against governments in Eastern Europe may sustain a spread that is worldwide in scope, but such a scope is not viewed as a prerequisite of globalizing dynamics. As long as it has the potential of an unlimited spread that can readily transgress national jurisdictions, any interaction sequence is considered to reflect the operation of globalization.

Obviously, the differences between globalizing and localizing forces give rise to contrary conceptions of territoriality. Globalization is rendering boundaries and identity with the land less salient while localization, being driven by pressures to narrow and withdraw, is highlighting borders and intensifying the deep attachments to land that can dominate emotion and reasoning.

In short, globalization is boundary-broadening and localization is boundary-heightening. The former allows people, goods, information, norms, practices, and institutions to move about oblivious to despite boundaries. The boundary-heightening processes of localization are designed to inhibit or prevent the movement of people, goods, information, norms, practices, and institutions. Efforts along this line, however, can be only partially successful. Community and state boundaries can be heightened to a considerable extent, but they cannot be rendered impervious. Authoritarian governments try to make them so, but their policies are bound to be undermined in a shrinking world with increasingly interdependent economies and communications technologies that are not easily monitored. Thus it is hardly surprising that some of the world's most durable tensions flow from the fact that no geographic borders can be made so airtight to prevent the infiltration of ideas and goods. Stated more emphatically, some globalizing dynamics are bound, at least in the long run, to prevail.

The boundary-expanding dynamics of globalization have become highly salient precisely because recent decades have witnessed a mushrooming of the facilities, interests, and markets through which a potential for worldwide spread can be realized. Likewise, the boundary-contracting dynamics of localization have also become increasingly significant, not least because some people and cultures feel threatened by the incursions of globalization. Their jobs, their icons, their belief systems, and their communities seem at risk as the boundaries that have sealed them off from the outside world in the past no longer assure protection. And there is, of course, a basis of truth in these fears. Globalization does intrude; its processes do shift jobs elsewhere; its norms do undermine traditional mores. Responses to these threats can vary considerably. At one extreme are adaptations that accept the boundary-broadening processes and make the best of them by integrating them into local customs and practices. At the other extreme are responses intended to ward off the globalizing processes by resort to ideological purities, closed borders, and economic isolation.

THE DYNAMICS OF FRAGMEGRATION

The core of world affairs today thus consists of tensions between the dynamics of globalization and localization. Moreover, the two sets of dynamics are causally linked, almost as if every increment of globalization gives rise to an increment of localization, and vice versa. To account for these tensions I have long used the term "fragmegration," an awkward and perhaps even grating label that has the virtue of capturing the pervasive interactions between the fragmenting forces of localization and the integrative forces of globalization.1 One can readily observe the unfolding of fragmegrative dynamics in the struggle of the European Union to cope with proposals for monetary unification or in the electoral campaigns and successes of Jean-Marie Le Pen in France, Patrick Buchanan in the United States, and Pauline Hanson in Australia—to mention only three examples.

It is important to keep in mind that fragmegration is not a single dynamic. Both globalization and localization are clusters of forces that, as they interact in different ways and through different channels, contribute to more encompassing processes in the case of globalization and to less encompassing processes in the case of localization. These various dynamics, moreover, operate in all realms of human activity, from the cultural and social to the economic and political.

In the political realm, globalizing dynamics underlie any developments that facilitate the expansion of authority, policies, and interests beyond existing socially constructed territorial boundaries, whereas the politics of localization involves any trends in which the scope of authority and policies undergoes contraction and reverts to concerns, issues, groups, and institutions that are less extensive than the prevailing socially constructed territorial boundaries. In the economic realm, globalization encompasses the expansion of production, trade, and investments beyond their prior locales, while localizing dynamics are at work when the activities of producers and consumers are constricted to narrower boundaries. In the social and cultural realms, globalization operates to extend ideas, norms, and practices beyond the settings in which they originated, while localization highlights or compresses the original settings and thereby inhibits the inroad of new ideas, norms, and practices.

It must be stressed that the dynamics unfolding in all these realms are long-term processes. They involve fundamental human needs and thus span all of human history. Globalizing dynamics derive from peoples' need to enlarge the scope of their self-created orders so as to increase the goods, services, and ideas available for their well-being. The agricultural revolution, followed by the industrial and postindustrial transformations, are among the major sources that have sustained globalization. Yet even as these forces have been operating, so have contrary tendencies toward contraction been continuously at work. Localizing dynamics derive from people's need for the psychic comforts of close-at-hand, reliable support—for the family and neighborhood, for local cultural practices, for a sense of "us" that is distinguished from "them." Put differently, globalizing dynamics have long fostered large-scale order, whereas localizing dynamics have long created pressure for small-scale order. Fragmegration, in short, has always been an integral part of the human condition.

GLOBALIZATION'S EVENTUAL PREDOMINANCE

Notwithstanding the complexities inherent in the emergent structures of world affairs, observers have not hesitated to anticipate what lies beyond fragmegration as global history unfolds. All agree that while the contest between globalizing and localizing dynamics is bound to be marked by fluctuating surges in both directions, the underlying tendency is for the former to prevail over the latter. Eventually, that is, the dynamics of globalization are expected to serve as the bases around which the course of events is organized.

Consensus along these lines breaks down, however, over whether the predominance of globalization is likely to have desirable or noxious consequences. Those who welcome globalizing processes stress the power of economic variables. In this view the globalization of national economies through the diffusion of technology and consumer products, the rapid transfer of financial resources, and the efforts of transnational companies to extend their market shares is seen as so forceful and durable as to withstand and eventually surmount any and all pressures toward fragmentation. This line acknowledges that the diffusion that sustains the processes of globalization is a centuries-old dynamic, but the difference is that the present era has achieved a level of economic development in which it is possible for innovations occurring in any sector of any country's economy to be instantaneously transferred to and adapted in any other country or sector. As a consequence,

> when this process of diffusion collides with cultural or political protectionism, it is culture and protectionism that wind up in the shop for repairs. Innovation accelerates. Productivity increases. Standards of living improve. There are setbacks, of course. The newspaper headlines are full of them. But we believe that the time required to override these setbacks has shortened dramatically in the developed world. Indeed, recent experience suggests that, in most cases, economic factors prevail in less than a generation....
>
> Thus understood, globalization—the spread of economic innovations around the world and the political and cultural adjustments that accompany this diffusion—cannot be stopped.... As history teaches, the political organizations and ideologies that yield superior economic performance survive, flourish, and replace those that are less productive.[2]

While it is surely the case that robust economic incentives sustain and quicken the processes of globalization, this line of theorizing nevertheless suffers from not allowing for its own negation. The theory offers no alternative interpretations as to how the interaction of economic, political, and social dynamics will play out. One cannot demonstrate the falsity—if falsity it is—of the theory because any contrary evidence is seen merely as "setbacks," as expectable but temporary deviations from the predicted course. The day may come, of course, when event so perfectly conform to the predicted patterns of globalization that one is inclined to conclude that the theory has been affirmed. But in the absence of alternative scenarios, the theory offers little guidance as to how to interpret intervening events, especially those that highlight the tendencies toward fragmentation. Viewed in this way, it is less a theory and more an article of faith to which one can cling.

Other observers are much less sanguine about the future development of fragmegration. They highlight a litany of noxious consequences that they see as following from the eventual predominance of globalization: "its economism; its economic reductionism; its technological determinism; its political cynicism, defeatism, and immobilism; its de-socialization of the subject and resocialization of risk; its teleological subtext of inexorable global 'logic' driven exclusively by capital accumulation and the market; and its ritual exclusion of factors, causes, or goals other than capital accumulation and the market from the priority of values to be pursued by social action."[3]

Still another approach, allowing for either desirable or noxious outcomes, has been developed by Michael Zurn. He identifies a mismatch between the rapid extension of boundary-crossing activities and the scope of effective governance. Consequently, states are undergoing what is labeled "uneven denationalization," a primary process in which "the rise of international governance is still remarkable, but not accompanied by mechanisms for... democratic control; people, in addition, become alienated from the remote political process.... The democratic state in the Western world is confronted with a situation in which it is undermined by the process of globalization and overarched by the rise of international institutions."[4]

> *There is no inherent contradiction between localizing and globalizing tendencies.*

While readily acknowledging the difficulties of anticipating where the process of uneven denationalization is driving the world, Zurn is able to derive two scenarios that may unfold: "Whereas the pessimistic scenario points to instances of fragmentation and emphasizes the disruption caused by the transition, the optimistic scenario predicts, at least in the long run, the triumph of centralization." The latter scenario rests on the presumption that the increased interdependence of societies will propel them to develop ever more effective democratic controls over the very complex arrangements on which international institutions must be founded.

UNEVEN FRAGMEGRATION

My own approach to theorizing about the fragmegrative process builds on these other perspectives and a key presumption of my own—that there is no inherent contradiction between localizing and globalizing tendencies—to develop an overall hypothesis that anticipates fragmegrative outcomes and that allows for its own negation: the more pervasive globalizing tendencies become, the less resistant localizing reactions will be to further globalization. In other words, globalization and localization will coexist, but the former will continue to set the context for the latter. Since the degree of coexistence will vary from situation to situation (depending on the salience of the global economy and the extent to which ethnic and other noneconomic

factors actively contribute to localization), I refer, borrowing from Zurn, to the processes depicted by the hypothesis as uneven fragmegration. The hypothesis allows for continuing pockets of antagonism between globalizing and localizing tendencies even as increasingly (but unevenly) the two accommodate each other. It does not deny the pessimistic scenario wherein fragmentation disrupts globalizing tendencies; rather it treats fragmentation as more and more confined to particular situations that may eventually be led by the opportunities and requirements of greater interdependence to conform to globalization.

For globalizing and localizing tendencies to accommodate each other, individuals have to come to appreciate that they can achieve psychic comfort in collectivities through multiple memberships and multiple loyalties, that they can advance both local and global values without either detracting from the other. The hypothesis of uneven fragmegration anticipates a growing appreciation along these lines because the contrary premise, that psychic comfort can only be realized by having a highest loyalty, is becoming increasingly antiquated. To be sure, people have long been accustomed to presuming that, in order to derive the psychic comfort they need through collective identities, they had to have a hierarchy of loyalties and that, consequently, they had to have a highest loyalty that could only be attached to a single collectivity. Such reasoning, however, is a legacy of the state system, of centuries of crises that made people feel they had to place nation-state loyalties above all others. It is a logic that long served to reinforce the predominance of the state as the "natural" unit of political organization and that probably reached new heights during the intense years of the cold war.

But if it is the case, as the foregoing analysis stresses, that conceptions of territoriality are in flux and that the failure of states to solve pressing problems has led to a decline in their capabilities and a loss of legitimacy, it follows that the notion that people must have a "highest loyalty" will also decline and give way to the development of multiple loyalties and an understanding that local, national, and transnational affiliations need not be mutually exclusive. For the reality is that human affairs are organized at all these levels for good reasons; people have needs that can only be filled by close-at-hand organizations and other needs that are best served by distant entities at the national or transnational level.

In addition, not only is an appreciation of the reality that allows for multiple loyalties and memberships likely to grow as the effectiveness of states and the salience of national loyalties diminish, but it also seems likely to widen as the benefits of the global economy expand and people become increasingly aware of the extent to which their well-being is dependent on events and trends elsewhere in the world. At the same time, the distant economic processes serving their needs are impersonal and hardly capable of advancing the need to share with others in a collective affiliation. This need was long served by the nation-state, but with fragmegrative dynamics having undermined the national level as a source of psychic comfort and with transnational entities seeming too distant to provide the psychic benefits of affiliation, the satisfactions to be gained through more close-at-hand affiliations are likely to seem ever more attractive.

THE STAKES

It seems clear that fragmegration has become an enduring feature of global life; it is also evident that globalization is not merely a buzzword, that it encompasses pervasive complexities and contradictions that have the potential both to enlarge and to degrade our humanity. In order to ensure that the enlargement is more prevalent than the degradation, it is important that people and their institutions become accustomed to the multiple dimensions and nuances as our world undergoes profound and enduring transformations. To deny the complexities and contradictions in order to cling to a singular conception of what globalization involves is to risk the many dangers that accompany oversimplification.

NOTES

1. For an extensive discussion of the dynamics of fragmegration, see James N. Rosenau, *Along the Domestic-Foreign Frontier: Exploring Governance in a Turbulent World* (Cambridge: Cambridge University Press, 1997), ch. 6.
2. William W. Lewis and Marvin Harris, "Why Globalization Must Prevail," *The McKinsey Quarterly*, no. 2 (1992), p. 115.
3. Barry K. Gills, "Editorial: 'Globalization' and the 'Politics of Resistance,'" *New Political Economy*, vol. 2 (March 1997), p. 12.
4. Michael Zurn, "What Has Changed in Europe? The Challenge of Globalization and Individualization," paper presented at a meeting on What Has Changed? Competing Perspectives on World Order (Copenhagen, May 14–16, 1993), p. 40.

JAMES N. ROSENAU *is University Professor of International Affairs at George Washington University. His latest book is* Along the Domestic-Foreign Frontier: Exploring Governance in a Turbulent World *(Cambridge: Cambridge University Press, 1997). This article draws on the author's "New Dimensions of Security: The Interaction of Globalizing and Localizing Dynamics,"* Security Dialogue, *September 1994, and "The Dynamics of Globalization: Toward an Operational Formulation,"* Security Dialogue, *September 1996).*

Dueling Globalizations:

A DEBATE BETWEEN THOMAS L. FRIEDMAN AND IGNACIO RAMONET

DOSCAPITAL

by Thomas L. Friedman

If there can be a statute of limitations on crimes, then surely there must be a statute of limitations on foreign-policy clichés. With that in mind, I hereby declare the "post–Cold War world" over.

For the last ten years, we have talked about this "post–Cold War world." That is, we have defined the world by what it wasn't because we didn't know what it was. But a new international system has now clearly replaced the Cold War: globalization. That's right, globalization—the integration of markets, finance, and technologies in a way that is shrinking the world from a size medium to a size small and enabling each of us to reach around the world farther, faster, and cheaper than ever before. It's not just an economic trend, and it's not just some fad. Like all previous international systems, it is directly or indirectly shaping the domestic politics, economic policies, and foreign relations of virtually every country.

As an international system, the Cold War had its own structure of power: the balance between the United States and the USSR, including their respective allies. The Cold War had its own rules: In foreign affairs, neither superpower would encroach on the other's core sphere of influence, while in economics, underdeveloped countries would focus on nurturing their own national industries, developing countries on export-led growth, communist countries on autarky, and Western economies on regulated trade. The Cold War had its own dominant ideas: the clash between communism and capitalism, as well as détente, nonalignment, and perestroika. The Cold War had its own demographic trends: The movement of peoples from East to West was largely frozen by the Iron Curtain; the movement from South to North was a more steady flow. The Cold War had its own defining technologies: Nuclear weapons and the Second Industrial Revolution were dominant, but for many developing countries, the hammer and sickle were still relevant tools. Finally, the Cold War had its own defining anxiety: nuclear annihilation. When taken all together, this Cold War system didn't shape everything, but it shaped many things.

Today's globalization system has some very different attributes, rules, incentives, and characteristics, but it is equally influential. The Cold War system was characterized by one overarching feature: division. The world was chopped up, and both threats and opportunities tended to grow out of whom you were divided from. Appropriately, that Cold War system was symbolized by a single image: the Wall. The globalization system also has one overarching characteristic: integration. Today, both the threats and opportunities facing a country increasingly grow from whom it is connected to. This system is also captured by a single symbol: the World Wide Web. So in the broadest sense, we have gone from a system built around walls to a system increasingly built around networks.

Once a country makes the leap into the system of globalization, its élite begin to internalize this perspective of integration and try to locate themselves within a global context. I was visiting Amman, Jordan, in the summer of 1998 when I met my friend, Rami Khouri, the country's leading political columnist, for coffee at the Hotel Inter-Continental. We sat down, and I asked him what was new. The first thing he said to me was "Jordan was just added to CNN's worldwide weather highlights." What Rami was saying was that it is important for Jordan to know that those institutions that think globally believe it is now worth knowing what the weather is like in Amman. It makes Jordanians feel more important and holds out the hope that they will profit by having more tourists or global investors visiting. The day after seeing Rami I happened to interview Jacob Frenkel, governor of the Bank of Israel and a University of Chicago-trained economist. He remarked to me: "Before, when we talked about macroeconomics, we started by looking at the local markets, local financial system, and the interrelationship between them, and then, as an afterthought, we looked at the international economy. There was a feeling that what we do is primarily our own business and then there are some outlets where we will sell abroad. Now, we reverse the perspective. Let's not ask what markets we should export to after having decided what to produce; rather, let's first study the global framework within which we operate and then decide what to produce. It changes your whole perspective."

Integration has been driven in large part by globalization's defining technologies: computerization, miniaturization, digitization, satellite communications, fiber optics, and the Internet. And that integration, in turn, has led to many other differences between the Cold War and globalization systems.

Unlike the Cold War system, globalization has its own dominant culture, which is why integration tends to be homogenizing. In previous eras, cultural homogenization happened on a regional scale—the Romanization of Western Europe and the Mediterranean world, the Islamization of Central Asia, the Middle East, North Africa, and Spain by the Arabs, or the Russification of Eastern and Central Europe, and parts of Eurasia, under the Soviets. Culturally speaking, globalization is largely the spread (for better and for worse) of Americanization—from Big Macs and iMacs to Mickey Mouse.

Whereas the defining measurement of the Cold War was weight, particularly the throw-weight of missiles, the defining measurement of the globalization system is speed—the speed of commerce, travel, communication, and innovation. The Cold War was about Einstein's mass-energy equation, $e=mc^2$. Globalization is about Moore's Law, which states that the performance power of microprocessors will double every 18 months. The defining document of the Cold War system was "the treaty." The defining document of the globalization system is "the deal." If the defining anxiety of the Cold War was fear of annihilation from an enemy you knew all too well in a world struggle that was fixed and stable, the defining anxiety in globalization is fear of rapid change from an enemy you cannot see, touch, or feel—a sense that your job, community, or workplace can be changed at any moment by anonymous economic and technological forces that are anything but stable.

If the defining economists of the Cold War system were Karl Marx and John Maynard Keynes, each of whom wanted to tame capitalism, the defining economists of the globalization system are Joseph Schumpeter and Intel chairman Andy Grove, who prefer to unleash capitalism. Schumpeter, a former Austrian minister of finance and Harvard University professor, expressed the view in his classic work *Capitalism, Socialism, and Democracy* (1942) that the essence of capitalism is the process of "creative destruction"—the perpetual cycle of destroying old and less efficient products or services and replacing them with new, more efficient ones. Grove took Schumpeter's insight that only the paranoid survive for the title of his book about life in Silicon Valley and made it in many ways the business model of globalization capitalism. Grove helped popularize the view that dramatic, industry-transforming innovations are taking place today faster and faster. Thanks to these technological breakthroughs, the speed at which your latest invention can be made obsolete or turned into a commodity is now lightening quick. Therefore, only the paranoid will survive—only those who constantly look over their shoulders to see who is creating something new that could destroy them and then do what they must to stay one step ahead. There will be fewer and fewer walls to protect us.

If the Cold War were a sport, it would be sumo wrestling, says Johns Hopkins University professor Michael Mandelbaum. "It would be two big fat guys in a ring, with all sorts of posturing and rituals and stomping of feet, but actually very little contact until the end of the match, when there is a brief moment of shoving and the loser gets pushed out of the ring, but nobody gets killed." By contrast, if globalization were a sport, it would be the 100-meter dash, over and over and over. No matter how many times you win, you have to race again the next day. And if you lose by just one-hundredth of a second, it can be as if you lost by an hour.

Last, and most important, globalization has its own defining structure of power, which is much more complex than the Cold War structure. The Cold War system was built exclusively around nation-states, and it was balanced at the center by two superpowers. The globalization system, by contrast, is built around three balances, which overlap and affect one another.

The first is the traditional balance between nation-states. In the globalization system, this balance still matters. It can still explain a lot of the news you read on the front page of the paper, be it the containment of Iraq in

A Tale of Two Systems	Cold War	Globalization
"The Cold War had its own dominant ideas: the clash between communism and capitalism.... The driving idea behind globalization is free-market capitalism."	In 1961, dressed in military fatigues, Cuban president Fidel Castro made his famous declaration: "I shall be a Marxist-Leninist for the rest of my life."	This January, Castro donned a business suit for a conference on globalization in Havana. Financier George Soros and conservative economist Milton Friedman were invited.
	In February 1972, President Richard Nixon traveled to China to discuss a strategic alliance between the two countries against the USSR.	In April 1999, Chinese premier Zhu Rongji came to Washington to discuss China's admission to the World Trade Organization.
"These countries that are most willing to let capitalism quickly destroy inefficient companies, so that money can be freed up and directed to more innovative ones, will thrive in the era of globalization. Those which rely on governments to protect them from such creative destruction will fall behind."	Many countries raised trade barriers and tried import substitution industrialization, nationalization, price controls, and interventionist policies.	Economic development relies on private-sector ownership, transparency and accountability, as well as investments in human capital and social infrastructure.
	The International Monetary Fund (IMF) and the World Bank were always present but rarely heeded.	The IMF plays a critical role, but must be enmeshed in a web of other organizations that support social welfare and the environment while promoting economic growth.
	Result: Hyperinflation, overwhelming external debt, corruption, and inefficient industries ruled the day. Only 8 percent of countries had liberal capital regimes in 1975 and foreign direct investment was at a low of $23 billion.	**Result**: Foreign direct investment increased five-fold between 1990 and 1997, jumping into $644 billion, and the number of countries with liberal regimes tripled to 28 percent.
"The balance between individuals and nation-states [has changed].... So you have today not only a superpower, not only Supermarkets, but... Super-empowered individuals."	In 1956, there were 973 international nongovernmental organizations (NGOs) in the world.	In 1996, there were 5,472 international NGOs in the world.
	In 1972, the total volume of world trade was only a fraction larger than the gross national product of the USSR.	The estimated annual revenue of transnational organized crime as of 1997, $750 billion, is larger than the gross domestic product of Russia.
	In 1970, there were only 7,000 transnational corporations (TNCs) in the world.	By 1994 the number of TNCs grew to 37,000 parent companies with 200,000 affiliates worldwide—controlling 33 percent of the world's productive assets.

SOURCE: Quotes taken from *The Lexus and the Olive Tree*, by Thomas Friedman (New York: Farrar, Straus, and Giroux, 1999).

the Middle East or the expansion of NATO against Russia in Central Europe.

The second critical balance is between nation-states and global markets. These global markets are made up of millions of investors moving money around the world with the click of a mouse. I call them the "Electronic herd." They gather in key global financial centers, such as Frankfurt, Hong Kong, London, and New York—the "supermarkets." The United States can destroy you by dropping bombs and the supermarkets can destroy you by downgrading your bonds. Who ousted President Suharto in Indonesia? It was not another superpower, it was the supermarkets.

The third balance in the globalization system—the one that is really the newest of all—is the balance between individuals and nation-states. Because globalization has brought down many of the walls that limited the movement and reach of people, and because it has simultaneously wired the world into networks, it gives more direct power to individuals than at any time in history. So we have today not only a superpower, not only supermarkets, but also super-empowered individuals. Some of these super-empowered individuals are quite angry, some of them quite constructive—but all are now able to act directly on the world stage without the traditional mediation of governments or even corporations.

Jody Williams won the Nobel Peace Prize in 1997 for her contribution to the International Campaign to Ban Landmines. She managed to build an international coalition in favor of a landmine ban without much government help and in the face of opposition from the major powers. What did she say was her secret weapon for organizing 1,000 different human rights and arms control groups on six continents? "E-mail."

By contrast, Ramzi Ahmed Yousef, the mastermind of the February 26, 1993, World Trade Center bombing in New York, is the quintessential "super-empowered angry man." Think about him for a minute. What was his program? What was his ideology? After all, he tried to blow up two of the tallest buildings in America. Did he want an Islamic state in Brooklyn? Did he want a Palestinian state in New Jersey? No. He just wanted to blow up two of the tallest buildings in America. He told the Federal District Court in Manhattan that his goal was to set off an explosion that would cause one World Trade Center tower to fall onto the other and kill 250,000 civilians. Yousef's message was that he had no message, other than to rip up the message coming from the all-powerful America to his society. Globalization (and Americanization) had gotten in his face and, at the same time, had empowered him as an individual to do something about it. A big part of the U.S. government's conspiracy case against Yousef (besides trying to blow up the World Trade Center in 1993, he planned to blow up a dozen American airliners in Asia in January 1995) relied on files found in the off-white Toshiba laptop computer that Philippine police say Yousef abandoned as he fled his Manila apartment in January 1995, shortly before his arrest. When investigators got hold of Yousef's laptop and broke into its files, they found flight schedules, projected detonation times, and sample identification documents bearing photographs of some of his co-conspirators. I loved that—Ramzi Yousef kept all his plots on the C drive of his Toshiba laptop! One should have no illusions, though. The super-empowered angry men are out there, and they present the most immediate threat today to the United States and the stability of the new globalization system. It's not because Ramzi Yousef can ever be a superpower. It's because in today's world, so many people can be Ramzi Yousef.

So, we are no longer in some messy, incoherent "post–Cold War world." We are in a new international system, defined by globalization, with its own moving parts and characteristics. We are still a long way from fully understanding how this system is going to work. Indeed, if this were the Cold War, the year would be about 1946. That is, we understand as much about how this new system is going to work as we understood about how the Cold War would work in the year Churchill gave his "Iron Curtain" speech.

Nevertheless, it's time we recognize that there is a new system emerging, start trying to analyze events within it, and give it its own name. I will start the bidding. I propose that we call it "DOScapital."

THOMAS L. FRIEDMAN *is a foreign affairs columnist for the* New York Times *and author of* The Lexus and the Olive Tree *(New York: Farrar, Straus, and Giroux, 1999).*

A NEW TOTALITARIANISM

by Ignacio Ramonet

We have known for at least ten years that globalization is the dominant phenomenon of this century. No one has been waiting around for Thomas Friedman to discover this fact. Since the end of the 1980s, dozens of authors have identified, described, and analyzed globalization inside and out. What is new in Friedman's work—and debatable—is the dichotomy he establishes between globalization and the Cold War: He presents them as opposing, interchangeable "systems." His constant repetition of this gross oversimplification reaches the height of annoyance.

Just because the Cold War and globalization are dominant phenomena in their times does not mean that they are both systems. A system is a set of practices and institutions that provides the world with a practical and theoretical framework. By this fight, the Cold War never constituted a system—Friedman makes a gross error by suggesting otherwise. The term "Cold War," coined by the media, is shorthand for a period of contemporary history (1946–89) characterized by the predominance of geopolitical and geostrategic concerns. However, it does not explain a vast number of unrelated events that also shaped that era: the expansion of multinational corporations, the development of air transportation, the worldwide extension of the United Nations, the decolonization of Africa, apartheid in South Africa, the advancement of environmentalism, or the development of computers and high-tech industries such as genetic engineering. And the list goes on.

Furthermore, tension between the West and the Soviet Union, contrary to Friedman's ideas, dates from before the Cold War. In fact, that very tension was formative in shaping the way democratic states understood Italian fascism in the 1920s, Japanese militarism in the 1930s, German rearmament after the rise of Adolf Hitler in 1933, and the Spanish Civil War between 1936 and 1939.

Friedman is right, however, to argue that globalization has a systemic bent. Step by step, this two-headed monster of technology and finance throws everything into confusion. Friedman, by contrast, tells a tale of globalization fit for Walt Disney. But the chaos that seems to delight our author so much is hardly good for the whole of humanity.

Friedman notes, and rightly so, that everything is now interdependent and that, at the same time, everything is in conflict. He also observes that globalization embodies (or infects) every trend and phenomenon at work in the world today—whether political, economic, social, cultural, or ecological. But he forgets to remark that there are groups from every nationality, religion, and ethnicity that vigorously oppose the idea of global unification and homogenization.

Furthermore, our author appears incapable of observing that globalization imposes the force of two powerful and contradictory dynamics on the world: fusion and fission. On the one hand, many states seek out alliances. They pursue fusion with others to build institutions, especially economic ones, that provide strength—or safety—in numbers. Like the European Union, groups of countries in Asia, Eastern Europe, North Africa, North America, and South America are signing free-trade agreements and reducing tariff barriers to stimulate commerce, as well as reinforcing political and security alliances.

But set against the backdrop of this integration, several multinational communities are falling victim to fission, cracking or imploding into fragments before the astounded eyes of their neighbors. When the three federal states of the Eastern bloc—Czechoslovakia, the USSR, and Yugoslavia—broke apart, they gave birth to some 22 independent states! A veritable sixth continent!

The political consequences have been ghastly. Almost everywhere, the fractures provoked by globalization have reopened old wounds. Borders are increasingly contested, and pockets of minorities give rise to dreams of annexation, secession, and ethnic cleansing. In the Balkans and the Caucasus, these tensions unleashed wars (in Abkhazia, Bosnia, Croatia, Kosovo, Moldova, Nagorno-Karabakh, Slovenia, and South Ossetia).

The social consequences have been no kinder. In the 1980s, accelerating globalization went hand in hand with the relentless ultraliberalism of British prime minister Margaret Thatcher and U.S. president Ronald Reagan. Quickly, globalization became associated with increased inequality, hikes in unemployment, deindustrialization, and deteriorated public services and goods.

Now, accidents, uncertainty, and chaos have become the parameters by which we measure the intensity of globalization. If we sized up our globalizing world today, what would we find? Poverty, illiteracy, violence, and illness are on the rise. The richest fifth of the world's population owns 80 percent of the world's resources, while the poorest fifth owns barely .5 percent. Out of a global population of 5.9 billion, barely 500 million people live comfortably, while 4.5 billion remain in need. Even in the European Union, there are 16 million people unemployed and 50 million living in poverty. And the combined fortune of the 358 richest people in the world (billionaires, in dollars) equals more than the annual revenue of 45 percent of the poorest in the world, or 2.6 billion people. That, it seems, is the brave new world of globalization.

> *Dazzled by the glimmer of fast profits, the champions of globalization are incapable of taking stock of the future.*

Beware of Dogma

Globalization has little to do with people or progress and everything to do with money. Dazzled by the glimmer of fast profits, the champions of globalization are incapable of taking stock of the future, anticipating the needs of humanity and the environment, planning for the expansion of cities, or slowly reducing inequalities and healing social fractures.

According to Friedman, all of these problems will be resolved by the "invisible hand of the market" and by macroeconomic growth—so goes the strange and insidious logic of what we in France call the *pensée unique*. The *pensée unique*, or "single thought," represents the interests of a group of economic forces—in particular, free-flowing international capital. The arrogance of the *pensée unique* has reached such an extreme that one can, without exaggerating, call it modern dogmatism. Like a cancer, this vicious doctrine imperceptibly surrounds any rebellious logic, then inhibits it, disturbs it, paralyzes it, and finally kills it. This doctrine, this *pensée unique*, is the only ideology authorized by the invisible and omnipresent opinion police.

The *pensée unique* was born in 1944, at the time of the Bretton Woods Agreement. The doctrine sprang from the world's large economic and monetary institutions—the Banque de France, Bundesbank, European Commission, International Monetary Fund, Organisation for Economic Cooperation and Development, World Bank, and World Trade Organization—which tap their deep coffers to enlist research centers, universities, and foundations around the planet to spread the good word.

Almost everywhere, university economics departments, journalists (such as Friedman), writers, and political leaders take up the principal commandments of these new tablets of law and, through the mass media, repeat them until they are blue in the face. Their dogma is echoed dutifully by the mouthpieces of economic information and notably by the "bibles" of investors and stockbrokers—the *Economist, Far Eastern Economic Review,* Reuters, and *Wall Street Journal,* for starters—which are often owned by large industrial or financial groups. And of course, in our media-mad society, repetition is as good as proof.

So what are we told to believe? The most basic principle is so strong that even a Marxist, caught offguard, would agree: The economic prevails over the political. Or as the writer Alain Minc put it, "Capitalism cannot collapse, it is the natural state of society. Democracy is not the natural state of society. The market, yes." Only an economy disencumbered of social speed bumps and other "inefficiencies" can steer clear of regression and crisis.

The remaining key commandments of the *pensée unique* build upon the first. For instance, the market's "invisible hand corrects the unevenness and malfunctions of capitalism" and, in particular, financial markets, whose "signals orient and determine the general movement of the economy." Competition and competitiveness "stimulate and develop businesses, bringing them permanent and beneficial modernization." Free trade without barriers is "a factor of the uninterrupted development of commerce and therefore of societies." Globalization of manufactured production and especially financial flows should be encouraged at all costs. The international division of labor "moderates labor demands and lowers labor costs." A strong currency is a must, as is deregulation and privatization at every turn. There is always "less of the state" and a constant bias toward the interests of capital to the detriment of the interests of labor, not to mention a callous indifference to ecological costs. The constant repetition of this catechism in the media by almost all political decision makers, Right and Left alike (think of British and German prime ministers Tony Blair and Gerhard Schroder's "Third Way" and "New Middle"), gives it such an intimidating power that it snuffs out every tentative free thought.

Magnates and Misfits

Globalization rests upon two pillars, or paradigms, which influence the way globalizers such as Friedman think. The first pillar is communication. It has tended to replace, little by little, a major driver of the last two centuries: progress.

From schools to businesses, from families and law to government, there is now one command: Communicate.

The second pillar is the market. It replaces social cohesion, the idea that a democratic society must function like a clock. In a clock, no piece is unnecessary and all pieces are unified. From this eighteenth-century mechanical metaphor, we can derive a modern economic and financial version. From now on, everything must operate according to the criteria of the "master market." Which of our new values are most fundamental? Windfall profits, efficiency, and competitiveness.

In this market-driven, interconnected world, only the strongest survive. Life is a fight, a jungle. Economic and social Darwinism, with its constant calls for competition, natural selection, and adaptation, forces itself on everyone and everything. In this new social order, individuals are divided into "solvent" or "nonsolvent"—i.e., apt to integrate into the market or not. The market offers protection to the solvents only. In this new order, where human solidarity is no longer an imperative, the rest are misfits and outcasts.

Thanks to globalization, only activities possessing four principal attributes thrive—those that are planetary, permanent, immediate, and immaterial in nature. These four characteristics recall the four principal attributes of God Himself. And in truth, globalization is set up to be a kind of modern divine critic, requiring submission, faith, worship, and new rites. The market dictates the Truth, the Beautiful, the Good, and the Just. The "laws" of the market have become a new stone tablet to revere.

Friedman warns us that straying from these laws will bring us to ruin and decay. Thus, like other propagandists of the New Faith, Friedman attempts to convince us that there is one way, and one way alone—the ultraliberal way—to manage economic affairs and, as a consequence, political affairs. For Friedman, the political is in effect the economic, the economic is finance, and finances are markets. The Bolsheviks said, "All power to the Soviets!" Supporters of globalization, such as Friedman, demand, "All power to the market!" The assertion is so peremptory that globalization has become, with its dogma and high priests, a kind of new totalitarianism.

IGNACIO RAMONET *is editor of* Le Monde diplomatique.

DOSCAPITAL 2.0

by Thomas L. Friedman

Ignacio Ramonet makes several points in his provocative and impassioned anti-globalization screed. Let me try to respond to what I see as the main ones.

Ramonet argues that the Cold War was not an international system. I simply disagree. To say that the Cold War was not an international system because it could not explain everything that happened during the years 1946 to 1989—such as aerial transport or apartheid—is simply wrong. An international system doesn't explain everything that happens in a particular era. It is, though, a dominant set of ideas, power structures, economic patterns, and rules that shape the domestic politics and international relations of more countries in more places than anything else.

Diplomacy then: Soviet premier Nikita Khrushev and U.S. vice president Richard Nixon argue over the merits of capitalism in 1959's "Kitchen Debate"...

Not only was the Cold War such an international system, but France had a very comfortable, unique, and, at times, constructive niche in that system, bridging the two superpower camps. Now that this old order is gone, it is obvious France is looking for a new, singular, and equally comfortable niche in today's system of globalization. Just as in the Cold War, France, like every other country, will have to define itself in relation to this new system. The obsession with globalization in the pages of *Le Monde diplomatique* is eloquent testimony to the fact that this search is alive and well in France.

Ramonet says that I "forget to remark that there are groups from every nationality, religion, ethnicity, etc., who vigorously oppose… globalization." In my book *The Lexus and the Olive Tree*, however, I have five separate chapters dealing with different aspects of that backlash. The penultimate chapter, in fact, lays out why I believe that globalization is not irreversible and identifies the five major threats to it: Globalization may be "just too hard" for too many people; it may be "just too connected" so that small numbers of people can disrupt the whole wired world today; it may be "just too intrusive" into people's lives; it may be "just too unfair to too many people"; and lastly, it may be "just too dehumanizing." My approach could hardly be called the Walt Disney version of globalization.

Frankly, I can and do make a much stronger case for the downsides of globalization than Ramonet does. I know that globalization is hardly all good, but unlike Ramonet I am not utterly blind to the new opportunities it creates for people—and I am not just talking about the wealthy few. Ask the high-tech workers in Bangalore, India, or Taiwan, or the Bordeaux region of France, or Finland, or coastal China, or Idaho what they think of the opportunities created by globalization. They are huge beneficiaries of the very market forces that Ramonet decries. Don't they count? What about all the human rights and environmental nongovernmental organizations that have been empowered by the Internet and globalization? Don't they count? Or do only French truck drivers count?

Ramonet says I am "incapable of observing that globalization imposes the force of two powerful contradictory dynamics on the world: fusion and fission." Say what? Why does he think I called my book *The Lexus and the Olive Tree?* It is all about the interaction between what is old and inbred—the quest for community, nation, family, tribe, identity, and one's own olive tree—and the economic pressures of globalization that these aspirations must interact with today, represented by the Lexus. These age-old passions are bumping up against, being squashed by, ripping through, or simply learning to live in balance with globalization.

What Ramonet can accuse me of is a belief that for the moment, the globalization system has been dominating the olive-tree impulses in most places. Many critics have pointed out that my observation that no two countries have ever fought a war against each other while they both had a McDonald's was totally disproved by the war in Kosovo. This is utter nonsense. Kosovo was only a temporary exception that in the end proved my rule. Why did airpower work to bring the Balkan war to a close after only 78 days? Because NATO bombed the Serbian tanks and troops out of Kosovo? No way. Airpower alone worked because NATO bombed the electricity stations, water system, bridges, and economic infrastructure in Belgrade—a modern European city, a majority of whose citizens wanted to be integrated with Europe and the globalization system. The war was won on the power grids of Belgrade, not in the trenches of Kosovo. One of the first things to be reopened in Belgrade was the McDonald's. It turns out in the end the Serbs wanted to wait in line for burgers, not for Kosovo.

The wretched of the earth want to go to Disneyworld, not to barricades. They want the Magic Kingdom, not Les Misérables. Just ask them.

Ramonet falls into a trap that often ensnares French intellectuals, and others, who rail against globalization. They assume that the rest of the world hates it as much as they do, and so they are always surprised in the end when the so-called little people are ready to stick with it. My dear Mr. Ramonet, with all due respect to you and Franz Fanon, the fact is the wretched of the earth want to go to Disneyworld, not to the barricades. They want the Magic Kingdom, not *Les Misérables.* Just ask them.

Finally, Ramonet says that I believe all the problems of globalization will be solved by the "invisible hand of the market." I have no idea where these quotation marks came from, let alone the thought. It certainly is not from anything I have written. The whole last chapter of my book lays out in broad strokes what I believe governments—the American government in particular—must do to "democratize" globalization, both economically and politically. Do I believe that market forces and the Electronic Herd are very powerful today and can, at times, rival governments? Absolutely. But do I believe that market forces will solve everything? Absolutely not. Ramonet, who clearly doesn't know a hedge fund from a hedge hog, demonizes markets to an absurd degree. He may think governments are powerless against such monsters, but I do not.

I appreciate the passion of Ramonet's argument, but he confuses my analysis for advocacy. My book is not a tract for or against globalization, and any careful reader will see that. It is a book of reporting about the world we now live in and the dominant international system that is shaping it—a system driven largely by forces of technology that I did not start and cannot stop. Ramonet treats globalization as a choice, and he implicitly wants us to choose something different. That is his politics. I view globalization as a reality, and I want us first to understand that reality and then, by understanding it, figure out how we can get the best out of it and cushion the worst. That is my politics.

. . . and now: Microsoft boss Bill Gates gives Russia's former first deputy premier Anatoly Chubais a crash course on the new economy in Moscow, 1997.

Let me share a secret with Ramonet. I am actually rooting for France. I hope that it can preserve all that is good and unique in its culture and way of life from the brutalizing, homogenizing forces of globalization. There is certainly room for a different path between the United States and North Korea, and good luck to France in finding it. But the readers of *Le Monde diplomatique* will get a lot better idea of how to find that middle path by reading my book than by reading Ramonet's critique.

Unfortunately, his readers will have to read *The Lexus and the Olive Tree* in a language other than French. The book is coming out in Arabic, Chinese, German, Japanese, and Spanish. There is only one major country where my American publisher could not find a local publisher to print it: France.

LET THEM EAT BIG MACS

by Ignacio Ramonet

It is truly touching when Thomas Friedman says, "The wretched of the earth want to go to Disneyworld, not to the barricades." Such a sentence deserves a place in posterity alongside Queen Marie-Antoinette's declaration in 1789, when she learned that the people of Paris were revolting and demanding bread: "Let them eat cake!"

My dear Mr. Friedman, do reread the 1999 *Human Development Report* from the United Nations Development Programme. It confirms that 1.3 billion people (or one-quarter of humanity) live on less than one dollar a day. Going to Disneyworld would probably not displease them, but I suspect they would prefer, first off, to eat well, to have a decent home and decent clothes, to be better ed-

ucated, and to have a job. To obtain these basic needs, millions of people around the world (their numbers grow more numerous each day) are without a doubt ready to erect barricades and resort to violence.

I deplore this kind of solution as much as Friedman does. But if we are wise, it should never come to that. Rather, why not allocate a miniscule part of the world's wealth to the "wretched of the earth"? If we assigned just 1 percent of this wealth for 20 years to the development of the most unhappy of our human brothers, extreme misery might disappear, and with it, risks of endemic violence.

But globalization is deaf and blind to such considerations—and Friedman knows it. On the contrary, it worsens differences and divides and polarizes societies. In 1960, before globalization, the most fortunate 20 percent of the planet's population were 30 times richer than the poorest 20 percent. In 1997, at the height of globalization, the most fortunate were 74 times richer than the world's poorest! And this gap grows each day. Today, if you add up the gross national products of all the world's underdeveloped countries (with their 600 million inhabitants) they still will not equal the total wealth of the three richest people in the world. I am sure, my dear Mr. Friedman, that those 600 million people have only one thing on their minds: Disneyworld!

It is true that there is more to globalization than just the downsides, but how can we overlook the fact that during the last 15 years of globalization, per capita income has decreased in more than 80 countries, or in almost half the states of the world? Or that since the fall of communism, when the West supposedly arranged an economic miracle cure for the former Soviet Union—more or less, as Friedman would put it, new McDonald's restaurants—more than 150 million ex-Soviets (out of a population of approximately 290 million) have fallen into poverty?

If you would agree to come down out of the clouds, my dear Mr. Friedman, you could perhaps understand that globalization is a symptom of the end of a cycle. It is not only the end of the industrial era (with today's new technology), not only the end of the first capitalist revolution (with the financial revolution), but also the end of an intellectual cycle—the one driven by reason, as the philosophers of the eighteenth century defined it. Reason gave birth to modern politics and sparked the American and French Revolutions. But almost all that modern reason constructed—the state, society, industry, nationalism, socialism—has been profoundly changed. In terms of political philosophy, this transformation captures the enormous significance of globalization. Since ancient times, humanity has known two great organizing principles: the gods, and then reason. From here on out, the market succeeds them both.

Now the triumph of the market and the irresistible expansion of globalization cause me to fear an inevitable showdown between capitalism and democracy. Capitalism inexorably leads to the concentration of wealth and economic power in the hands of a small group. And this in turn leads to a fundamental question: How much redistribution will it take to make the domination of the rich minority acceptable to the majority of the world's popu-

lation? The problem, my dear Mr. Friedman, is that the market is incapable of responding. All over the world, globalization is destroying the welfare state.

What can we do? How do we keep half of humanity from revolting and choosing violence? I know your response, dear Mr. Friedman: Give them all Big Macs and send them to Disneyworld!

WANT TO KNOW MORE?

An insightful overview of the social transformations that globalization has ushered in can be found in Malcolm Waters' *Globalization* (New York: Routledge, 1995). In *Capitalism, Socialism, and Democracy* (London: Harper, 1942), Joseph Schumpeter argues that only innovation can compensate for the destructive forces of the market. Benjamin Barber looks at culture clash in his book *Jihad versus McWorld* (New York: Times Books, 1995). William Greider argues for more managed globalization in *One World Ready or Not: The Manic Logic of Global Capitalism* (New York: Simon & Schuster, 1997). In his book, *The Post-Corporate World: Life after Capitalism* (San Francisco: Berrett-Koehler, 1999), David Korten stipulates that corporate capitalism could unravel the cohesion of society. Robert Reich considers how international labor markets will react to a shrinking world in *The Work of Nations: Preparing Ourselves for the 21st Century* (New York: Alfred A. Knopf, 1991). For a view on how information technology has changed the world economy, see Frances Cairncross' *The Death of Distance* (Cambridge: Harvard Business School Press, 1997). For a provocative advocate of Americanization, see David Rothkopf's **"In Praise of Cultural Imperialism"** (FOREIGN POLICY, Summer 1997). Refraining from taking sides, Dani Rodrik reexamines some of the faulty assumptions made on both sides of the globalization debate in *"Sense and Nonsense in the Globalization Debate"* (FOREIGN POLICY, Summer 1997). Ignacio Ramonet's wide-ranging commentary can be found in back issues of *Le Monde diplomatique*, archived online. Rigorous critiques of Thomas Friedman's new book, *The Lexus and the Olive Tree* (New York: Farrar Straus and Giroux, 1999) can be found in the *New Yorker* (May 10, 1999), *Nation* (June 14, 1999), *Financial Times* (May 15, 1999) and *New Statesman* (July 5,1999).

For links to relevant Web sites, as well as a comprehensive index of related FOREIGN POLICY articles, access **www.foreignpolicy.com**.

Will Globalization Go Bankrupt?

Global integration is driven not by politics or the Internet or the World Trade Organization or even—believe it or not—McDonald's. No, throughout history, globalization has been driven primarily by monetary expansions. Credit booms spark periods of economic integration, while credit contractions quickly squelch them. Is today's world on the verge of another globalization bust?

By Michael Pettis

Only the young generation which has had a college education is capable of comprehending the exigencies of the times," wrote Alphonse, a third-generation Rothschild, in a letter to a family member in 1865. At the time the world was in the midst of a technological boom that seemed to be changing the globe beyond recognition, and certainly beyond the ability of his elders to understand. As part of that boom, capital flowed into remote corners of the earth, dragging isolated societies into modernity. Progress seemed unstoppable.

Eight years later, however markets around the world collapsed. Suddenly, investors turned away from foreign adventures and new technologies. In the depression that ensued, many of the changes eagerly embraced by the educated young—free markets, deregulated banks, immigration—seemed too painful to continue. The process of globalization, it seems, was neither inevitable nor irreversible.

What today we call economic globalization—a combination of rapid technological progress, large-scale capital flows, and burgeoning international trade—has happened many times before in the last 200 years. During each of these periods (including our own), engineers and entrepreneurs became folk heroes and made vast fortunes while transforming the world around them. They exploited scientific advances, applied a suc-

cession of innovations to older discoveries, and spread the commercial application of these technologies throughout the developed world [see box]. Communications and transportation were usually among the most affected areas, with each technological surge causing the globe to "shrink" further.

But in spite of the enthusiasm for science that accompanied each wave of globalization, as a historical rule it was primarily commerce and finance that drove globalization, not science or technology, and certainly not politics or culture. It is no accident that each of the major periods of technological progress coincided with an era of financial market expansion and vast growth in international commerce. Specifically, a sudden expansion of financial liquidity in the world's leading banking centers—whether an increase in British gold reserves in the 1820s or the massive transformation in the 1980s of illiquid mortgage loans into very liquid mortgage securities, or some other structural change in the financial markets—has been the catalyst behind every period of globalization.

If liquidity expansions historically have pushed global integration forward, subsequent liquidity contractions have brought globalization to an unexpected halt. Easy money had allowed investors to earn fortunes for their willingness to take risks, and the wealth generated by rising asset values and new investments made the liberal ideology behind the rapid market expansion seem unassailable. When conditions changed, however, the outflow of money from the financial centers was reversed. Investors rushed to pull their money out of risky ventures and into safer assets. Banks tightened up their lending requirements and refused to make new loans. Asset values collapsed. The costs of globalization, in the form of social disruption, rising income inequality, and domination by foreign elites, became unacceptable. The political and intellectual underpinnings of globalization, which had once seemed so secure, were exposed as fragile, and the popular counterattack against the logic of globalization grew irresistible.

THE BIG BANG

The process through which monetary expansions lead to economic globalization has remained consistent over the last two centuries. Typically, every few decades, a large shift in income, money supply, saving patterns, or the structure of financial markets results in a major liquidity expansion in the rich-country financial centers. The initial expansion can take a variety of forms. In England, for example, the development of joint-stock banking (limited liability corporations that issued currency) in the 1820s and 1830s—and later during the 1860s and 1870s—produced a rapid expansion of money, deposits, and bank credit, which quickly spilled over into speculative investing and international lending. Other monetary expansions were sparked by large increases in U.S. gold reserves in the early 1920s, or by major capital recyclings, such as the massive French indemnity payment after the Franco-Prussian War of 1870, the petrodollar recycling of the 1970s, or the recycling of Japan's huge trade surplus in the 1980s and 1990s. Monetary expansions also can result from the conversion of assets into more liquid instruments, such as with the explosion in U.S. speculative real-estate lending in the 1830s or the creation of the mortgage securities market in the 1980s.

The expansion initially causes local stock markets to boom and real interest rates to drop. Investors, hungry for high yields, pour money into new, nontraditional investments, including ventures aimed at exploiting emerging technologies. Financing becomes available for risky new projects such as railways, telegraph cables, textile looms, fiber optics, or personal computers, and the strong business climate that usually accompanies the liquidity expansion quickly makes these investments profitable. In turn, these new technologies enhance productivity and slash transportation costs, thus speeding up economic growth and boosting business profits. The cycle is self-reinforcing: Success breeds success, and soon the impact of rapidly expanding transportation and communication technology begins to cause a noticeable impact on social behavior, which adapts to these new technologies.

But it is not just new technology ventures that attract risk capital. Financing also begins flowing to the "peripheral" economies around the world, which, because of their small size, are quick to respond. These countries then begin to experience currency strength and real economic growth, which only reinforce the initial investment decision. As more money flows in, local markets begin to grow. As a consequence of the sudden growth in both asset values and gross domestic product, political leaders in developing countries often move to reform government policies in these countries—whether reform consists of expelling a backward Spanish monarch in the 1820s, expanding railroad transportation across the Andes in the 1860s, transforming the professionalism of the Mexican bureaucracy in the 1890s, deregulating markets in the 1920s, or privatizing bloated state-owned firms in the 1990s. By providing the government with the resources needed to overcome the resistance of local elites, capital inflows enable economic-policy reforms.

This relationship between capital and reform is frequently misunderstood: Capital inflows do not simply respond to successful economic reforms, as is commonly thought; rather, they create the conditions for reforms to take place. They permit easy financing of fiscal deficits, provide industrialists who might oppose free trade with low-cost capital, build new infrastructure, and generate so much asset-based wealth as to mollify most members of the economic and political elite who might ordinarily oppose the reforms. Policymakers tend to design such reforms to appeal to foreign investors, since policies that encourage foreign investment seem to be quickly and richly rewarded during periods of liquidity. In reality, however, capital is just as likely to flow into countries that have failed to introduce reforms. It is not a coincidence that the most famous "money doctors"—Western-trained thinkers like French economist Jean-Gustave Courcelle-Seneuil in the 1860s, financial historian Charles Conant in the 1890s, and Princeton University economist Edwin Kemmerer in the 1920s, under whose influence many developing countries undertook major liberal reforms—all exerted their maximum influence during these periods. During the 1990s, their modern counterparts advised Argentina on its currency board, brought "shock therapy" to Russia, convinced China of the benefits of membership in the World Trade Organization, and everywhere spread the ideology of free trade.

Globalization takes place largely because sudden monetary expansions encourage investors to embrace new risks.

The pattern is clear: Globalization is primarily a monetary phenomenon in which expanding liquidity induces investors to take more risks. This greater risk appetite translates into the financing of new technologies and investment in less developed markets. The combination of the two causes a "shrinking" of the globe as communications and transportation technologies improve and investment capital flows to every part of the globe. Foreign trade, made easier by the technological advances, expands to accommodate these flows. Globalization takes place, in other words, largely because investors are suddenly eager to embrace risk.

THE BIG CRUNCH

As is often forgotten during credit and investment booms, however, monetary conditions contract as well as expand. In fact, the contraction is usually the inevitable outcome of the very conditions that prompted the expansion. In times of growth, financial institutions often overextend themselves, creating distortions in financial markets and leaving themselves vulnerable to external shocks that can force a sudden retrenchment in credit and investment. In a period of rising asset prices, for example, it is often easy for even weak borrowers to obtain collateral-based loans, which of course increases the risk to the banking system of a fall in the value of the collateral. For example, property loans in the 1980s dominated and ultimately brought down the Japanese banking system. As was evident in Japan, if the financial structure has become sufficiently fragile, a retrenchment can lead to a collapse that quickly spreads throughout the economy.

Since globalization is mainly a monetary phenomenon, and since monetary conditions eventually must contract, then the process of globalization can stop and even reverse itself. Historically, such reversals have proved extraordinarily disruptive. In each of the globalization periods before the 1990s, monetary contractions usually occurred when bankers and financial authorities began to pull back from market excesses. If liquidity contracts—in the context of a perilously overextended financial system—the likelihood of bank defaults and stock market instability is high. In 1837, for example, the U.S. and British banking systems, overdependent on real estate and commodity loans, collapsed in a series of crashes that left Europe's financial sector in tatters and the United States in the midst of bank failures and state government defaults.

The same process occurred a few decades later. Alphonse Rothschild's globalizing cycle of the 1860s ended with the stock market crashes that began in Vienna in May 1873 and spread around the world during the next four months, leading, among other things, to the closing of the New York Stock Exchange (NYSE) that September amid the near-collapse of American railway securities. Conditions were so bad that the rest of the decade after 1873 was popularly referred to in the United States as the Great Depression. Nearly 60 years later, that name was reassigned to a similar episode—the one that ended the Roaring Twenties and began with the near-breakdown of the U.S. banking system in 1930–31. The expansion of the 1960s was somewhat different in that it began to unravel during the early and mid-1970s when, thanks partly to the OPEC oil price hikes and subsequent petrodollar recycling, a second liquidity boom occurred, and lending to sovereign borrowers in the developing world continued through the end of the decade. However, the cycle finally broke down altogether when rising interest rates and contracting money engineered by then Federal Reserve Chairman Paul Volcker helped precipitate the Third World debt crisis of the 1980s. Indeed, with the exception of the globalization period of the early 1900s, which ended with the advent of World War I, each of these eras of international integration concluded with sharp monetary contractions that led to a banking system collapse or retrenchment, declining asset

values, and a sharp reduction in both investor risk appetite and international lending.

Following most such market crashes, the public comes to see prevalent financial market practices as more sinister, and criticism of the excesses of bankers becomes a popular sport among politicians and the press in the advanced economies. Once capital stops flowing into the less developed, capital-hungry countries, the domestic consensus in favor of economic reform and international integration begins to disintegrate. When capital inflows no longer suffice to cover the short-term costs to the local elites and middle classes of increased international integration-including psychic costs such as feelings of wounded national pride-support for globalization quickly wanes. Populist movements, never completely dormant, become reinvigorated. Countries turn inward. Arguments in favor of protectionism suddenly start to sound appealing. Investment flows quickly become capital flight.

> **Following market crashes, the public comes to see financial markets as more sinister, and criticism of bankers becomes a popular sport among politicians and the press.**

This pattern emerged in the aftermath of the 1830s crash, when confidence in free markets nose-dived and the subsequent populist and nationalist backlash endured until the failure of the muchdreaded European liberal uprisings of 1848, which saw the earliest stirrings of communism and the publication of the *Communist Manifesto*. Later, in the 1870s, the economic depression that followed the mass bank closings in Europe, the United States, and Latin America was accompanied by an upsurge of political radicalism and populist outrage, along with bouts of protectionism throughout Europe and the United States by the end of the decade. Similarly, the Great Depression of the 1930s also fostered political instability and a popular revulsion toward the excesses of financial capitalism, culminating in burgeoning left-wing movements, the passage of anti-bank legislation, and even the jailing of the president of the NYSE.

PROFITS OF DOOM

Will these patterns manifest themselves again? Indeed, a new global monetary contraction already may be under way. In each of the previous contractions, stock markets fell, led by the collapse of the once-high-flying technology sector; lending to emerging markets dried up, bringing with it a series of sovereign defaults; and investors clamored for safety and security. Consider the crash of 1873, a typical case: Then, the equivalent of today's high-tech sector was the market for railway stocks and bonds, and the previous decade had seen a rush of new stock and bond offerings that reached near-manic proportions in the early 1870s. The period also saw rapid growth of lending to

Latin America, southern and eastern Europe, and the Middle East. Wall Street veterans had expressed nervousness about market excesses for years leading up to the crash, but the exuberance of investors who believed in the infinite promise of the railroads, at home and abroad, coupled with the rising prominence of bull-market speculators like Jay Gould and Diamond Jim Brady, swept them aside. When the market collapsed in 1873, railway securities were the worst hit, with many companies going bankrupt and closing their doors. Major borrowers from the developing world were unable to find new financing, and a series of defaults spread from the Middle East to Latin America in a matter of months. In the United States, the Congress and press became furious with the actions of stock market speculators and pursued financial scandals all the way to President Ulysses S. Grant's cabinet. Even Grant's brother-in-law was accused of being in cahoots with a notorious group that attempted a brutal gold squeeze.

Today, we see many of the same things. The technology sector is in shambles, and popular sentiment has turned strongly against many of the Wall Street heroes who profited most from the boom. Lending to emerging markets has all but dried up. As of this writing, the most sophisticated analysts predict that a debt default in Argentina is almost certain—and would unleash a series of other sovereign defaults in Latin America and around the world. The yield differences between risky assets and the safest and most liquid assets are at historical highs. In short, investors seem far more reluctant to take on risk than they were just a few years ago.

This lower risk tolerance does not bode well for poor nations. Historically, many developing countries only seem to experience economic growth during periods of heavy capital inflow, which in turn tend to last only as long as the liquidity-inspired asset booms in rich-country financial markets. Will the international consensus that supports globalization last when capital stops flowing? The outlook is not very positive. While there is still broad support in many circles for free trade, economic liberalization, technological advances, and free capital flows—even when the social and psychic costs are acknowledged—we already are witnessing a strong political reaction against globalization. This backlash is evident in the return of populist movements in Latin America; street clashes in Seattle, Prague, and Quebec; and the growing disenchantment in some quarters with the disruptions and uncertainties that follow in the wake of globalization.

The leaders now gathered in opposition to globalization—from President Hugo Chávez in Venezuela to Malaysian Prime Minister Mahathir bin Mohamad to anti-trade activist Lori Wallach in the United States—should not be dismissed too easily, no matter how dubious or fragile some of their arguments may seem. The logic of their arguments may not win the day, but rather a global monetary contraction may reverse the political consensus that was necessary to support the broad and sometimes disruptive social changes that accompany globalization. When that occurs, policy debates will be influenced by the less emotional and more thoughtful attacks on globalization by the likes of Robert Wade, a professor of political economy at the London School of Economics, who argues forcefully that glo-

balization has actually resulted in greater global income inequality and worse conditions for the poor.

Investing in the Future

In past periods of monetary expansion and globalization, Western societies have experienced the rapid development and commercial application of new technologies.

1822–37: expansion of canal building, first railway boom, application of steam power to the manufacturing process, advances in machine tool design, invention of McCormick's reaper, first gas-lighting enterprises, and development of the telegraph

1851–73: advances in mining, second railway boom, developments in shipping, and rapid growth in the number of corporations in continental Europe

1881–1914: explosive productivity growth in Europe and the United States, improvement in steel production and heavy chemical manufacturing, first power station, spread of electricity, development of the internal combustion engine, another railway boom, innovation in newspaper practice and technology, and developments in canning and refrigeration

1922–30: commercialization of automobiles and aircraft, new forms of mass media, rising popularity of cinema and radio broadcasting, spread of artificial fibers and plastics, widespread use of electricity in U.S. factories, the creation and sale of a variety of new electric appliances, and expanded telephone ownership

1960–73: development and application of transistor technology, advances in commercial flying and shipping, and the spread of telecommunications and software

1985–present: ubiquity of information processing, explosion in computer memory, advances in biotechnology and medical technologies, and commercial application of the Internet

---

If a global liquidity contraction is under way, antiglobalization arguments will resonate more strongly as many of the warnings about the greed of Wall Street and the dangers of liberal reform will seem to come true. Supposedly irreversible trends will suddenly reverse themselves. Further attempts to deepen economic reform, spread free trade, and increase capital and labor mobility may face political opposition that will be very difficult to overcome, particularly since bankers, the most committed supporters of globalization, may lose much of their prestige and become the target of populist attacks following a serious stock market decline. Because bankers are so identified with globalization, any criticism of Wall Street will also implicitly be a criticism of globalizing markets.

Financiers, after all, were not the popular heroes in the 1930s that they were during the 1920s, and current events seem to mirror past backlashes. Already the U.S. Securities and Exchange Commission, which was created during the Great Depression of the 1930s, is investigating the role of bankers and analysts in misleading the public on the market excesses of the 1990s. In June 2001, the industry's lobby group, the Securities Industry Association, proposed a voluntary code, euphemized as "a compilation of best practices... to ensure the ongoing integrity of securities research and analysis," largely to head off an expansion of external regulation. Increasingly, experts bewail the conflicts of interest inherent among the mega-banks that dominate U.S. and global finance.

Globalization itself always will wax and wane with global liquidity. For those committed to further international integration within a liberal economic framework, the successes of the recent past should not breed complacency since the conditions will change and the mandate for liberal expansion will wither. For those who seek to reverse the socioeconomic changes that globalization has wrought, the future may bring far more progress than they hoped. If global liquidity contracts and if markets around the world pull back, our imaginations will once again turn to the increasingly visible costs of globalization and away from the potential for all peoples to prosper. The reaction against globalization will suddenly seem unstoppable.

Michael Pettis is an investment banker and professor of finance at Columbia University. He is author of The Volatility Machine: Emerging Economies and the Threat of Financial Collapse *(New York: Oxford University Press, 2001).*

Want to Know More?

Charles P. Kindleberger's *A Financial History of Western Europe*, 2nd ed. (New York: Oxford University Press, 1993) is probably the best single volume for anyone interested in understanding the financial history of globalization. For a discussion of how structural changes in financial systems can lead to overextension and banking crises, consult Hyman P. Minsky's *Can "It" Happen Again? Essays on Instability and Finance* (Armonk: M.E. Sharpe, 1982). Also see Paul W. Drake's, ed., *Money Doctors, Foreign Debts, and Economic Reforms in Latin America from the 1890s to the Present* (Wilmington: SR Books, 1994) and Christian Suter's *Debt Cycles in the World Economy: Foreign Loans, Financial Crises, and Debt Settlements, 1820–1990* (Boulder: Westview Press, 1992). Michael Pettis's *The Volatility Machine: Emerging Economies and the Threat of Financial Collapse* (New York: Oxford University Press, 2001) identifies the specific events that set off liquidity booms in prior periods of global integration.

Frank Griffith Dawson's *The First Latin American Debt Crisis: The City of London and the 1822–25 Loan Bubble* (New Haven: Yale University Press, 1990) offers a wonderful account of an early era of globalization, although his description of events 180 years ago is too familiar for comfort. Matthew Josephson's famous *The Robber Barons: The Great American Capitalists, 1861–1901* (New York: Harcourt, Brace & World, 1962) is as good a place as any to start reading up on the railway booms of the late 19th century. Harold James's *The End of Globalization: Lessons from the Great Depression* (Cambridge: Harvard University Press, 2001) is one of the best recent books on economic history and discusses the globalizing period of the 1920s and the subsequent backlash.

Kevin H. O'Rourke and Jeffrey G. Williamson recently completed a major work on globalization at the end of the 19th century and the backlash against it in *Globalization and History: The Evolution of a Nineteenth-Century Atlantic Economy* (Cambridge: MIT Press, 1999). Finally, Niall Ferguson's *The House of Rothschild: Money's Prophets, 1798–1848* (New York: Viking Press, 1998) and *The World's Banker: The History of the House of Rothschild* (London: Weidenfeld & Nicolson, 1998) tell the story of a family intimately involved with every aspect of globalization.

• For links to relevant Web sites, as well as a comprehensive index of related FOREIGN POLICY articles, access **www.foreignpolicy.com**.

America's Two-Front Economic Conflict

C. Fred Bergsten

DOUBLE TROUBLE

SINCE THE END of the Cold War, the perceived threats to U.S. security have been mainly from "rogue states" such as Iraq and North Korea—none of which are superpowers or likely allies of each other in confronting the United States. But the United States now faces the real possibility of economic conflict with both Europe and East Asia—the commercial and financial equivalent of two-front combat. In this domain, both potential rivals are superpowers. Moreover, they have already demonstrated their ability to coalesce against the United States, as they did to help torpedo the Seattle ministerial meeting of the World Trade Organization (WTO) in December 1999.

Peaceful and effective resolution of these potential conflicts is one of the most important and difficult issues facing the new U.S. administration and the world. The American and global economies are slowing sharply, and their futures may be heavily affected by the outcomes. In a post–Cold War world in which economic issues are central to international relations, those outcomes will also be crucial for U.S. foreign policy and global stability. Compounding the complexity of the situation is the fact European and East Asian nations are not only the United States' economic competitors but also its economic partners—and many of them are close security allies as well.

CONTINENTAL DIVIDE

THE UNITED STATES and the European Union (EU) are on the brink of a major trade and economic conflict. Washington has already retaliated against European import restrictions on American beef and bananas—each retalia-

tion accounting for a hundred million dollars or so of annual trade—and has rejected all European efforts to resolve these disputes. Europe in turn threatens to retaliate against several billion dollars of U.S. export subsidies, as well as new U.S. trade laws that would channel the proceeds of antidumping penalties from the Treasury Department to the complaining industries and would force the president to continually change the products being retaliated against, thus intensifying the impact of U.S. punitive sanctions.

Still larger trade clashes loom. The troubled U.S. steel industry will likely file additional antidumping cases against European firms or even an industry-wide safeguard action that would restrict all European imports. In addition, a major dispute over commercial aircraft is brewing as the two sides quarrel over whether direct European governmental subsidies for Airbus or indirect Pentagon subsidies for Boeing are more egregious. Europe's outcry over U.S. sanctions against European firms that deal with American adversaries such as Cuba and Iran has only been swept under the rug. And just over the horizon lies the biggest battle of all: the debates over farm subsidies, genetically modified products, and overall agricultural trade that will explode in 2003, when the U.S.-EU "peace clause" (a moratorium on new complaints in the agricultural sector) expires.

The United States and Europe also differ on global trade issues for which they share leadership responsibility. They remain divided, for example, on whether to include competition policy and investment issues in new WTO negotiations. It was their opposing views on issues such as these that scuttled any prospect of launching a new round of trade talks at Seattle.

Furthermore, the United States and Europe are divided on energy and environmental issues. As energy prices soared and riots erupted on European roadways last fall, European resentment flared anew over Americans' penchant for cheap fuel and their profligate energy consumption. The recent Hague conference that sought to devise operational plans to check global warming broke up over fundamental disagreements about who bears responsibility for greenhouse gas emissions, how they should be cut back, and who should pay for doing so.

Financial relations are another potential land mine. When the European Central Bank intervened to halt the slide of the euro last September, the United States provided only grudging support. But now that the euro has rebounded, the shoe may soon be on the other foot as the dollar risks a sharp decline in the wake of a domestic economic slowdown and an annual trade deficit approaching $500 billion. Europe should be willing to help in such a circumstance, since it would not want to see the euro soar to levels that would jeopardize the price competitiveness of its exports. But it might be less enthusiastic to bolster the dollar if the net effect were to finance massive tax cuts à la President George W. Bush that would further reduce U.S. national savings and hence increase America's draw on foreign capital.

The accumulation of such potential conflicts poses high risks for both American and European economies. Moreover, the global impact of a commercial clash between these two titans could be severe including systemic damage to the WTO, especially its crucial but fragile dispute settlement mechanism. A transatlantic economic conflict may also exacerbate potential security tensions over issues such as a future policy toward the Balkans, American concern over European plans for an autonomous military force, and European anxieties that American proposals for a missile defense system will renew tensions with Russia and trigger another global arms race. All this calls for new basic strategies for managing globalization, especially in light of the developments simultaneously arising on the other side of the world.

ASIAN FUSION

THE POTENTIAL economic confrontation between the United States and East Asia is quite different from the transatlantic one. The sector-specific conflicts that have traditionally burdened U.S.-Asian trade relations (and that now burden U.S.-European ones) have diminished sharply. The problem now is that East Asia, for the first time in history, is creating its own economic bloc, which could include preferential trade arrangements and an Asian Monetary Fund (AMF).

Asian countries will shortly complete a Network of Bilateral Swap Arrangements, which will provide initially up to $50 billion and eventually as much as $100 billion in mutual currency supports among the "ASEAN + 3": the ten members of the Association of Southeast Asian Nations (ASEAN), plus Japan, China, and South Korea. In addition, they are contemplating cooperative exchange-rate systems to shield themselves from the huge fluctuations in the currencies of the major industrial countries—similar to Europe's moves toward monetary integration in the 1970s to defend itself against wide fluctuations of the dollar. These countries are also devising new "early warning systems" to help prevent future regional economic crises. Building on the 1998 Miyazawa Plan, under which Japan offered $30 billion to support the recovery of the nations hit hardest by the 1997–98 financial crisis, these countries are clearly headed toward creating their own monetary arrangements.

On the trade side, fundamental changes in the trade policies of the three main East Asian powers—Japan, South Korea, and China—have initiated a spate of subregional and bilateral free trade negotiations. Japan, which has traditionally relied on the multilateral frameworks of the General Agreement on Trade and Tariffs (GATT) and now the WTO, has begun to pursue bilateral trade agreements with Singapore, Mexico, and South Korea over the past two years. South Korea has made a similar policy shift and is now actively negotiating with Chile, as well. China, which had also previously eschewed regional approaches, stunned everyone at the fourth annual ASEAN + 3 summit in late 2000 by proposing a China-ASEAN free trade area—which the Southeast Asians, fearing Chinese domination, immediately broadened to include Japan and South Korea.

Thus a study of a possible East Asian free trade area, which would be a world-shaking development, was launched at the summit. The new study will build on the one already underway for the creation of a Northeast Asia free trade area comprising China, Japan, and South Korea, which itself is of major significance. In short, the East Asia Economic Group proposed a decade ago by Malaysian Prime Minister Mahathir bin Mohammed is beginning to take shape, albeit slowly and in subregional stages. ASEAN, for example, has already developed detailed plans to complete its own free trade area and currency network. No overarching political strategy drives Asian integration, as it did for the EU, and little coordination exists between the current financial and trade initiatives. But there can be little doubt that these new movements will result in the evolution of an East Asian economic bloc.

East Asian integration is not necessarily a bad thing and could in fact prompt new trade liberalization on the multilateral level. But an East Asian free trade area could also erect new discrimination against U.S. exports of at least $20 billion per year. And a unified East Asia could be an even more formidable competitor than Japan was in the past or China is today, though it should also be a more attractive market for both exports and investment from the United States.

On the financial side, the members of the AMF would hold monetary reserves of almost $1 trillion—the largest in the world and far larger than those of the United States or the countries of the eurozone. Japan and China would support each other's currencies with the two largest dollar hoards in the world, totaling more than $500 billion. The AMF could clearly rival the International Monetary Fund (IMF) and raise potential conflicts, including disputes over the conditions of country rescue packages.

As with Europe, the new economic developments in Asia carry foreign policy and security implications as well. A truly united East Asia could sharply reduce the risk of conflict in the region and hence be very much in the U.S. interest. On the other hand, a sense that America was being shunted aside by both Asia and Europe could reinforce isolationist tendencies within the United States.

A CHANGING GLOBAL SCENE

THE MAIN HISTORICAL underpinning of America's potential two-front economic conflict is the increasing multipolarization of the world economy. Despite America's prodigious economic performance in the 1990s, the EU is now the largest economic entity on the globe, and its lead will grow further as it expands its membership over the next few years. The euro, although still suffering numerous growing pains, has completed the region's economic integration.

East Asia has achieved an economic weight comparable to those of the United States and the EU, but it has learned that its disunity has precluded it from achieving equal status on the global scene. Its inferior position was made clear during the 1997–98 financial crisis, when the region became dependent on the international financial institutions directed by the Atlantic powers. The image of IMF Managing Director Michel Camdessus dictating terms to Indonesian President Suharto is bitterly seared into Asian memories, especially now that prominent Western economists argue that IMF programs actually made the crisis worse and point to how Malaysia has recovered effectively without IMF assistance. Under such circumstances, Asia's gross under-representation in the IMF and other key international institutions has suddenly attained great salience. The Asians have vowed to never again be in such thrall to the West.

In addition, the creation of the euro has prompted Asian countries to consider moving toward their own currency unit, albeit over a long period of time. More broadly, the traditional Asian repugnance toward "the huge bureaucracy in Brussels" has turned into widespread contemplation of emulating the basic European strategy of economic cooperation, despite recognition of the differences between the two regions and thus doubts about deep integration. Through biannual Asia-Europe Meetings, Asia is in fact seeking and receiving extensive European advice for its own coordination efforts.

The end of the Cold War has also contributed to the potential for a two-front economic conflict. The disappearance of the Soviet threat has reduced the importance of the American military umbrella over Europe and Asia. The security glue that traditionally encouraged the postwar allies to resolve their economic differences no longer exists. The semiannual U.S.-EU summits have been pitiful failures, and the Asia-Pacific Economic Cooperation forum (APEC), which seeks to prevent U.S.-Asia conflict by providing an institutional link across the Pacific, has only begun to address the issues posed by East Asian regionalism.

These changes in the global scene—Europe's and East Asia's achievement of rough economic parity with the United States, and the end of the Cold War—require a restructuring of global economic arrangements. Further delays in such reform will only heighten the risk of costly conflicts.

FOLLOW THE LEADER

A MORE SUBTLE CAUSE of the present crisis is the decline of effective U.S. leadership in the global economic system. This in turn stems from a domestic popular backlash against globalization and the resulting political stalemate in Washington.

America's international posture has been hurt by domestic backlash against globalization.

During the postwar period, the pervasive tension between regionalism and multilateralism (mainly as a result of increasing European integration) was generally resolved in favor of multilateralism due to steady American leadership in that direction. The United States insisted on a new round of global trade liberalization after each major step in the European integration process, which otherwise would have created additional trade discrimination and likely emulation around the world. Thus the primacy of GATT was maintained. Indeed, a positive dynamic between regional and global trade liberalization remained consistent for more than four decades. Even when the United States itself began to embrace regionalism—from bilateral free trade with Canada to the North American Free Trade Agreement to the proposed Free Trade Area of the Americas (FTAA)—it was careful to simultaneously pursue new multilateral initiatives to ensure an umbrella of global trade liberalization.

Washington's ability to maintain such leadership has been severely curtailed over the past five years, however.

Despite the strength of America's economy and the reduction of its unemployment rate to a 30-year low, the popular backlash against globalization has produced a political stalemate on most international economic issues. As a result, the president has had no effective authority to negotiate new trade agreements since 1994. Legislation to replenish the IMF languished for a year in the midst of the Asian crisis, until it was rescued fortuitously by the farm community's interest in restoring its exports to Asia. Even relatively straightforward issues—such as extending permanent normal trade relations to China or offering enhanced market access to Africa and the Caribbean—required lengthy, all-out presidential and business campaigns to persuade Congress.

Largely as a result of this domestic standstill, America's international economic posture has been compromised. The United States' initial refusal in 1997 to contribute to the IMF support package for Thailand for fear of further riling Congress, for example, earned lasting enmity throughout Asia. The main reason for the debacle at Seattle was the United States' inability to propose a new round of trade negotiations that would meet the legitimate interests of other major players. Lacking the domestic authority to lower its own trade barriers, Washington was forced to offer an agenda that sought to reduce protection only in other countries—a prospect that was understandably unappealing to the rest of the world. Similarly, in 1997–98 APEC negotiations, the United States unsuccessfully pushed a program of sector-specific liberalization that focused almost wholly on U.S. export interests. And six years after the idea of the FTAA was launched in Miami, little progress has been made toward hemispheric trade liberalization.

This international leadership vacuum has had two subtle but profound effects on the world economy. Like a bicycle on a hill, the global trading system tends to slip backwards in the absence of continual progress forward. Now, with no serious multilateral trade negotiations taking place anywhere in the world, the backsliding has come in the form of intensified regionalism (which is inherently discriminatory), as well as mercantilist and protectionist disputes across the Atlantic. An East Asian free trade area—and along with it, a three-bloc world—will likely emerge if the United States remains on the sidelines of international trade for another five years. Such U.S. impotence would also mean that the traditionally positive impact of regional liberalization on the multilateral process would give way to increasing antagonism and even hostility between the regional blocs.

The other chief effect of the leadership vacuum is increased international disregard of, or even hostility toward, the United States on the economic front. Because of its weight in the world economy, its dynamic growth, and its traditional leadership role, the United States remains the most important player in the global economic system. The other economic powers generally seek to avoid confronting it directly. The EU, for example, has tried to avoid overt battles, despite its escalating range of disputes with the United States. East Asian governments are careful to assure Washington that their new regional initiatives are fully consistent with existing global norms and institutions—a conciliatory stance that is in sharp contrast to Mahathir's shrill rhetoric of a decade ago and Japanese Vice Minister of Finance Eisuke Sakakibara's aggressive 1997 promotion of the AMF.

In reality, however, the United States is perceived as wanting to call the shots without putting up much of its own money or making changes in its own laws and practices. These specific economic complaints fuse with and feed on more general anti-American sentiments throughout the world. Hence, the two other economic superpowers are proceeding on their own. The EU has launched the euro, a new association agreement with Mexico, and negotiations with Mercosur (the trade bloc comprising Argentina, Brazil, Paraguay, and Uruguay); East Asia is pursuing the AMF and the East Asian free trade area. The result is a clear and steady erosion of both the United States' position on the global economic scene and the multilateral rules and institutions that it has traditionally championed. If not checked soon, this erosion could deteriorate into severe international conflicts and the disintegration of global economic links.

ALTERNATIVE MEDICINE

THE REMEDIES for this risky situation are intellectually straightforward but politically difficult. The cardinal requirement is to subsume the current bilateral disputes and evolving regional initiatives within a reinvigorated multilateral system that rests on an internationally shared vision of how to manage globalization. Such a system will have to restart the momentum of multilateral trade liberalization, provide a global umbrella that effectively reconciles the inevitable regional groupings, and negotiate rather than litigate the most politically sensitive disputes among the major powers. This remedy will require the restoration of a domestic consensus on globalization in the United States and considerable trade and financial reforms in Europe and East Asia.

The United States faces two tempting responses to the current tensions, each of which would be a mistake. One is to resurrect the mid-1990s proposal for a transatlantic free trade area (TAFTA) between the United States and the EU. TAFTA would erect new trade discrimination against East Asia and thus assure the acceleration of both its regional integration and its anti-Western orientation. In addition, TAFTA would discriminate against all developing countries—"the richest ganging up on the poorest"—and would end any prospect of their constructive participation in the WTO and other global institutions.

The second bad idea is for the United States to pursue the FTAA without simultaneously working toward a new round of multilateral negotiations at the WTO. The new

Bush administration has indicated interest in the FTAA and will have an early opportunity to pursue it at the third Summit of the Americas in Québec in late April. Absent a parallel multilateral effort, however, such an initiative would validate the regional emphases of both Europe and Asia and spark new trade discrimination.

Although President Bush will not have enough time to obtain fast-track negotiating authority before going to Québec, he must work out enough congressional support to provide credibility for any pledges he makes at the hemispheric summit. While doing so, he could also seek congressional blessing for a broad-based initiative in the WTO to get the multilateral process back on track. Such an initiative would ideally move toward global free trade in which all regional trade preferences would be eliminated. The United States could then begin working with Europe and Asia to launch a new round at the WTO ministerial conference later this year, while still proceeding with inter-American integration at Québec. To buy time for this strategy to be implemented, the United States and the EU should broaden their "peace clause" on agriculture by declaring a three-year freeze on all retaliatory actions and complaints in additional sectors.

America must adopt stronger safety nets to cushion the blows of globalization.

A useful adjunct to this strategy of renewed multilateralism would be cross-regional free trade agreements (CR-FTAs) that cut across East Asia, Europe, and the Americas. Such pacts are already being pursued at the bilateral level, such as the U.S.-Singapore and Japan-Mexico initiatives, and at the super-regional level, as with APEC and the EU-Mercosur talks. Although these arrangements still create new discrimination and potential trade conflict and are thus decidedly inferior to multilateral liberalization, they could nevertheless dilute the regional groupings that may otherwise solidify into rigid blocs. Thus CRFTAs represent a useful addition to renewed multilateral efforts, or at least a second-best fallback if that preferred course turns out to be unobtainable in the near term.

A renewal of multilateral efforts is also required on the financial side. The IMF has already made significant policy changes but must now take additional steps to buttress its ability to prevent and quickly respond to crises. The main institutional change needed at the IMF is to accord East Asia more voting shares and leadership assignments to account for its greatly increased economic weight—mostly at the expense of Europe, which is over-

represented. The IMF also needs to address the costly instability and prolonged misalignments among the dollar, the euro, and the yen, which contribute to the need felt by the Asians to create their own monetary zone.

The success of these remedies rests on the ability of the United States to overcome its crippling domestic resistance to globalization. This will be a difficult task for the new administration, but the potential threats to U.S. economic prosperity, its international leadership, and global stability should be enough to convince both the White House and Congress to agree on a new approach to the international economy.

This agreement should rest on several key elements. It must start from a clear consensus that globalization brings substantial net benefits to the American economy, including intensified competition that holds down inflation and thus permits the creation of millions of additional jobs. At the same time, Washington must acknowledge that globalization causes job and income losses in certain sectors, which exact significant psychological tolls. The government, therefore, has a responsibility to channel help from the winners to the losers, for humanitarian and equity reasons as well as to maintain political support for continued globalization efforts.

To fulfill that obligation, the country must adopt stronger safety nets, including more generous unemployment insurance eligibility criteria and compensation levels, portable health insurance and pensions, and perhaps a new program of wage insurance. Even more important, government and business leaders need to work together to provide better education and training programs to enable all Americans to benefit from globalization rather than feel victimized by it.

At the international level, the White House and Congress need to reassert American leadership in negotiating new agreements on both trade and finance, as described above, that will place the current conflicts into a broader global and strategic context. These agreements should, among other things, promote international labor and environmental standards that will avoid distorting either global competition or normal trade flows.

If Washington does not adopt such a strategy early on, the current situation could become much worse. The U.S. economy has slowed sharply, the unemployment rate will soon rise, and the annual trade deficit is approaching $500 billion (about five percent of GDP). Blame for such economic troubles will inevitably focus on foreign competition—especially if European and Asian countries raise new barriers against U.S. exports. Washington will feel intense pressure to retaliate against Europe and to thwart the rise of even tougher rivals in East Asia. It will also be tempted to adopt new unilateralist measures, such as withdrawal from international monetary cooperation and multilateral trade efforts.

Both the prospective global slowdown and any such U.S. reactions to it could accelerate the current trends in Europe and Asia. Tougher economic times will make it harder for Europe to resist its own protectionists, especially as French elections approach in 2002. Economic troubles will prod East Asians to speed their integration plans, as their financial crisis has already done. Any new protectionist or unilateralist steps taken by the United States would trigger parallel responses elsewhere. And any significant American slowdown would further embolden the Europeans and the Asians to overcome their humiliations over the initial fall of the euro and the shattering of the "economic miracle," respectively, and go their own ways.

This potential two-front economic conflict could severely threaten international prosperity and even global security. Restoration of both an effective global economic order and renewed U.S. leadership should be a top priority for the new administration and Congress.

C. FRED BERGSTEN is Director of the Institute for International Economics and former Assistant Secretary of the Treasury (1977–81) and Assistant for International Economic Affairs to the National Security Council (1969–71).

What's Wrong With This Picture?

THE RISE OF THE MEDIA CARTEL HAS BEEN A LONG TIME COMING. THE CULTURAL EFFECTS ARE NOT NEW IN KIND, BUT THE PROBLEM HAS BECOME CONSIDERABLY LARGER.

MARK CRISPIN MILLER

For all their economic clout and cultural sway, the ten great multinationals—AOL Time Warner, Disney, General Electric, News Corporation, Viacom, Vivendi, Sony, Bertelsmann, AT&T and Liberty Media—rule the cosmos only at the moment. The media cartel that keeps us fully entertained and permanently half-informed is always growing here and shriveling there, with certain of its members bulking up while others slowly fall apart or get digested whole. But while the players tend to come and go—always with a few exceptions—the overall Leviathan itself keeps getting bigger, louder, brighter, forever taking up more time and space, in every street, in countless homes, in every other head.

The rise of the cartel has been a long time coming (and it still has some way to go). It represents the grand convergence of the previously disparate US culture industries—many of them vertically monopolized already—into one global superindustry providing most of our imaginary "content." The movie business had been largely dominated by the major studios in Hollywood; TV, like radio before it, by the triune axis of the networks headquartered in New York; magazines, primarily by Henry Luce (with many independent others on the scene); and music, from the 1960s, mostly by the major record labels. Now all those separate fields are one, the whole terrain divided up among the giants—which, in league with Barnes & Noble, Borders and the big distributors, also control the book business. (Even with its leading houses, book publishing was once a cottage industry at both the editorial and retail levels.) For all the democratic promise of the Internet, moreover, much of cyberspace has now been occupied, its erstwhile wildernesses swiftly paved and lighted over by the same colossi. The only industry not yet absorbed into this new world order is the newsprint sector of the Fourth Estate—a business that was heavily shadowed to begin with by the likes of Hearst and other, regional grandees, flush with the ill-gotten gains of oil, mining and utilities—and such absorption is, as we shall see, about to happen.

Thus what we have today is not a problem wholly new in kind but rather the disastrous upshot of an evolutionary process whereby that old problem has become considerably larger—and that great quantitative change, with just a few huge players now co-directing all the nation's media, has brought about enormous qualitative changes. For one thing, the cartel's rise has made extremely rare the sort of marvelous exception that has always popped up, unexpectedly, to startle and revivify the culture—the genuine independents among record labels, radio stations, movie theaters, newspapers, book publishers and so on. Those that don't fail nowadays are so remarkable that they inspire not emulation but amazement. Otherwise, the monoculture, endlessly and noisily triumphant, offers, by and large, a lot of nothing, whether packaged as "the news" or "entertainment."

Of all the cartel's dangerous consequences for American society and culture, the worst is its corrosive influence on journalism. Under AOL Time Warner, GE, Viacom et al., the news is, with a few exceptions, yet another version of the entertainment that the cartel also vends nonstop. This is also nothing new—consider the newsreels of yesteryear—but the gigantic scale and thoroughness of the corporate concentration has made a world of difference, and so has made this world a very different place.

Let us start to grasp the situation by comparing this new centerfold with our first outline of the National Entertainment State, published in the spring of 1996. Back then, the national TV news appeared to be a tidy tetrarchy: two network news divisions owned by large appliance makers/weapons manufacturers (CBS by Westinghouse, NBC by General Electric), and the other two bought lately by the nation's top purveyors of Big Fun (ABC by Disney, CNN by Time Warner). Cable was still relatively immature, so that, of its many enterprises, only CNN competed with the broadcast networks' short-staffed newsrooms; and its buccaneering founder, Ted Turner, still seemed to call the shots from his new aerie at Time Warner headquarters.

Today the telejournalistic firmament includes the meteoric Fox News Channel, as well as twenty-six television stations owned outright by Rupert Murdoch's News Corporation (which holds majority ownership in a further seven). Although ultimately thwarted in his bid to buy DirecTV and thereby dominate the US satellite television market, Murdoch wields a pervasive influence on the news—and not just in New York, where he has two TV stations, a major daily (the faltering *New York Post*) *and* the Fox News Channel, whose inexhaustible platoons of shouting heads attracts a fierce plurality of cable-viewers. Meanwhile, Time Warner has now merged with AOL—so as to own the cyberworks through which to market its floodtide of movies, ball games, TV shows, rock videos, cartoons, standup routines and (not least) bits from CNN, CNN Headline News, CNNfn (devised to counter GE's CNBC) and CNN/Sports Illustrated (a would-be rival to Disney's ESPN franchise). While busily cloning CNN, the parent company has also taken quiet steps to make it more like Fox, with Walter Isaacson, the new head honcho, even visiting the Capitol to seek advice from certain rightist pols on how, presumably, to make the network even shallower and more obnoxious. (He also courted Rush Himself.) All this has occurred since the abrupt defenestration of Ted Turner, who now belatedly laments the overconcentration of the cable business: "It's sad we're losing so much diversity of thought," he confesses, sounding vaguely like a writer for this magazine.

Whereas five years ago the clueless Westinghouse owned CBS, today the network is a property of the voracious Viacom—matchless cable occupier (UPN, MTV, MTV2, VH1, Nickelodeon, the Movie Channel, TNN, CMT, BET, 50 percent of Comedy Central, etc.), radio colossus (its Infinity Broadcasting—home to Howard Stern and Don Imus—owns 184 stations), movie titan (Paramount Pictures), copious publisher (Simon & Schuster, Free Press, Scribner), a big deal on the web and one of the largest US outdoor advertising firms. Under Viacom, CBS News has been obliged to help sell Viacom's product—in 2000, for example, devoting epic stretches of *The Early Show* to what lately happened on *Survivor* (CBS). Of course, such synergistic bilge is commonplace, as is the tendency to dummy up on any topic that the parent company (or any of its advertisers) might want stifled. These journalistic sins have been as frequent under "longtime" owners Disney and GE as under Viacom and Fox [see Janine Jaquet, "The Wages of Synergy"]. They may also abound beneath Vivendi, whose recent purchase of the film and TV units of USA Networks and new stake in the satellite TV giant EchoStar—could soon mean lots of oblique self-promotion on *USAM News*, in *L'Express* and *L'Expansion*, and through whatever other news-machines the parent buys.

Such is the telejournalistic landscape at the moment—and soon it will mutate again, if Bush's FCC delivers for its giant clients. On September 13, when the minds of the American people were on something else, the commission's GOP majority voted to "review" the last few rules preventing perfect oligopoly. They thus prepared the ground for allowing a single outfit to own both a daily paper and a TV station in the same market—an advantage that was outlawed in 1975. (Even then, pre-existing cases of such ownership were grandfathered in, and any would-be owner could get that rule waived.) That furtive FCC "review" also portended the elimination of the cap on the percentage of US households that a single owner might reach through its TV stations. Since the passage of the Telecommunications Act of 1996, the limit had been 35 percent. Although that most indulgent bill was

dictated by the media giants themselves, its restrictions are too heavy for this FCC, whose chairman, Michael Powell, has called regulation per se "the oppressor."

And so, unless there's some effective opposition, the several-headed vendor that now sells us nearly all our movies, TV, radio, magazines, books, music and web services will soon be selling us our daily papers, too—for the major dailies have, collectively, been lobbying energetically for that big waiver, which stands to make their owners even richer (an expectation that has no doubt had a sweetening effect on coverage of the Bush Administration). Thus the largest US newspaper conglomerates—the New York Times, the Washington Post, Gannett, Knight-Ridder and the Tribune Co.—will soon be formal partners with, say, GE, Murdoch, Disney and/or AT&T; and then the lesser nationwide chains (and the last few independents) will be ingested, too, going the way of most US radio stations. America's cities could turn into informational "company towns," with one behemoth owning all the local print organs—daily paper(s), alternative weekly, city magazine—as well as the TV and radio stations, the multiplexes and the cable system. (Recently a federal appeals court told the FCC to drop its rule preventing any one company from serving more than 30 percent of US cable subscribers; and in December, the Supreme Court refused to hear the case.) While such a setup may make economic sense, as anticompetitive arrangements tend to do, it has no place in a democracy, where the people have to know more than their masters want to tell them.

That imperative demands reaffirmation at this risky moment, when much of what the media cartel purveys to us is propaganda, commercial or political, while no one in authority makes mention of "the public interest"—except to laugh it off. "I have no idea," Powell cheerily replied at his first press conference as chairman, when asked for his own definition of that crucial concept. "It's an empty vessel in which people pour in whatever their preconceived views or biases are." Such blithe obtuseness has marked all his public musings on the subject. In a speech before the American Bar Association in April 1998, Powell offered an ironic little riff about how thoroughly he doesn't get it: "The night after I was sworn in [as a commissioner], I waited for a visit from the angel of the public interest. I waited all night, but she did not come." On the other hand, Powell has never sounded glib about his sacred obligation to the corporate interest. Of his decision to move forward with the FCC vote just two days after 9/11, Powell spoke as if that sneaky move had been a gesture in the spirit of Patrick Henry: "The flame of the American ideal may flicker, but it will never be extinguished. We will do our small part and press on with our business, solemnly, but resolutely."

Certainly the FCC has never been a democratic force, whichever party has been dominant. Bill Clinton championed the disastrous Telecom Act of 1996 and otherwise did almost nothing to impede the drift toward oligopoly. (As *Newsweek* reported in 2000, Al Gore was Rupert Murdoch's personal choice for President. The mogul apparently sensed that Gore would happily play ball with him, and also thought—correctly—that the Democrat would win.)

What is unique to Michael Powell, however, is the showy superciliousness with which he treats his civic obligation to address the needs of people other than the very rich. That spirit has shone forth many times—as when the chairman genially compared the "digital divide" between the information haves and have-nots to a "Mercedes divide" between the lucky few who can afford great cars and those (like him) who can't. In the intensity of his pro-business bias, Powell recalls Mark Fowler, head of Reagan's FCC, who famously denied his social obligations by asserting that TV is merely "an appliance," "a toaster with pictures." And yet such Reaganite *bons mots*, fraught with the anti-Communist fanaticism of the late cold war, evinced a deadly earnestness that's less apparent in General Powell's son. He is a blithe, postmodern sort of ideologue, attuned to the complacent smirk of Bush the Younger—and, of course, just perfect for the cool and snickering culture of TV.

Although such flippancies are hard to take, they're also easy to refute, for there is no rationale for such an attitude. Take "the public interest"—an ideal that really isn't hard to understand. A media system that enlightens us, that tells us everything we need to know pertaining to our lives and liberty and happiness, would be a system dedicated to the public interest. Such a system would not be controlled by a cartel of giant corporations, because those entities are ultimately hostile to the welfare of the people. Whereas we need to know the truth about such corporations, they often have an interest in suppressing it (as do their advertisers). And while it takes much time and money to find out the truth, the parent companies prefer to cut the necessary costs of journalism, much preferring the sort of lurid fare that can drive endless hours of agitated jabbering. (Prior to 9/11, it was Monica, then *Survivor* and Chandra Levy, whereas, since the fatal day, we have had mostly anthrax, plus much heroic footage from the Pentagon.) The cartel's favored audience, moreover, is that stratum of the population most desirable to advertisers—which has meant the media's complete abandonment of working people and the poor. And while the press must help protect us against those who would abuse the powers of government, the oligopoly is far too cozy with the White House and the Pentagon, whose faults, and crimes, it is unwilling to expose. The media's big bosses want big favors from the state, while the reporters are afraid to risk annoying their best sources. Because of such politeness (and, of course, the current panic in the air), the US coverage of this government is just a bit more edifying than the local newscasts in Riyadh.

Against the daily combination of those corporate tendencies—conflict of interest, endless cutbacks, endless trivial pursuits, class bias, deference to the king and all his men—the

public interest doesn't stand a chance. Despite the stubborn fiction of their "liberal" prejudice, the corporate media have helped deliver a stupendous one-two punch to this democracy. (That double whammy followed their uncritical participation in the long, irrelevant *jihad* against those moderate Republicans, the Clintons.) Last year, they helped subvert the presidential race, first by prematurely calling it for Bush, regardless of the vote—a move begun by Fox, then seconded by NBC, at the personal insistence of Jack Welch, CEO of General Electric. Since the coup, the corporate media have hidden or misrepresented the true story of the theft of that election.

And having justified Bush/Cheney's coup, the media continue to betray American democracy. Media devoted to the public interest would investigate the poor performance by the CIA, the FBI, the FAA and the CDC, so that those agencies might be improved for our protection—but the news teams (just like Congress) haven't bothered to look into it. So, too, in the public interest, should the media report on all the current threats to our security—including those far-rightists targeting abortion clinics and, apparently, conducting bioterrorism; but the telejournalists are unconcerned (just like John Ashcroft). So should the media highlight, not play down, this government's attack on civil liberties—the mass detentions, secret evidence, increased surveillance, suspension of attorney-client privilege, the encouragements to spy, the warnings not to disagree, the censored images, sequestered public papers, unexpected visits from the Secret Service and so on. And so should the media not parrot what the Pentagon says about the current war, because such prettified accounts make us complacent and preserve us in our fatal ignorance of what people really think of us—and why—beyond our borders. And there's much more—about the stunning exploitation of the tragedy, especially by the Republicans; about the links between the Bush and the bin Laden families; about the ongoing shenanigans in Florida—that the media would let the people know, if they were not (like Michael Powell) indifferent to the public interest.

In short, the news divisions of the media cartel appear to work *against* the public interest—and *for* their parent companies, their advertisers and the Bush Administration. The situation is completely un-American. It is the purpose of the press to help us run the state, and not the other way around. As citizens of a democracy, we have the right and obligation to be well aware of what is happening, both in "the homeland" and the wider world. Without such knowledge we cannot be both secure and free. We therefore must take steps to liberate the media from oligopoly, so as to make the government our own.

Mark Crispin Miller is a professor of media studies at New York University, where he directs the Project on Media Ownership. He is the author of The Bush Dyslexicon: Observations on a National Disorder *(Norton).*

Overcoming Japan's China Syndrome

By Chi Hung Kwan

Introduction

More and more people in Japan have come to perceive the rise of China as a threat, prompted by the sharp contrast between the growth performances of the two countries in recent years. It should, however, be noted that the economic relations between Japan and China can be characterized as complementary rather than competitive, reflecting the prevailing gap in the level of development. Both sides can benefit by promoting a division of labor according to comparative advantage, with China specializing in labor-intensive products and Japan specializing in high-tech products.

Although Japan has lagged behind the United States and Europe in penetrating the Chinese market through foreign direct investment (FDI), it has benefited from importing cheap products from China. With China's import tariffs coming down after World Trade Organization (WTO) entry, Japanese companies now also have better access to the fast-growing Chinese market not only through FDI but also through exporting from their headquarters.

Do not Confuse "Made in China" with "Made by China"

Recently, Japanese imports of manufactured goods from China have surged and the reputation of Chinese products has improved substantially, giving rise to concern that China will soon replace Japan as the "factory of the world." An objective evaluation of China's industrial strength, however, suggests that there is still a long way to go before it will become a truly advanced industrial country on a par with Japan.

First of all, the high proportion of labor-intensive products in China's exports means that its trade structure is typical of a Newly Industrializing Economy (NIE). This is different from that of developed countries, where the major export items, such

as machinery, are technology-intensive. Although China is increasing its share of the global market for manufactured goods, including some information technology products that are classified as high-tech, Chinese exports are still highly concentrated in lower-end products. In the case of televisions, for instance, Japan specializes in high-definition and other higher-end models, while China produces standard models whose unit values are much lower.

> With China's import tariffs coming down after WTO entry, Japanese companies have better access to the fast-growing Chinese market through FDI and exporting from their headquarters

Reflecting China's emphasis on processing trade, goods "made in China" contain large numbers of foreign components, some of which are made in Japan. According to official Chinese statistics, increasing exports by $1 million requires importing intermediate goods and components worth over half a million dollars, which do not form part of China's gross domestic product (GDP). Moreover, the proportion of this imported content is higher for high-tech than for low-tech products. A computer labeled "made in China" is likely to contain a large portion of imported contents including an Intel CPU, Microsoft Windows operating system, and a liquid crystal display made in Japan or South Korea.

In addition, approximately half of China's exports are produced by subsidiaries of foreign companies, to which dividends, interest charges, royalties and other fees must be paid. Even among Chinese companies with no capital relations with foreign companies, the majority of their exports are processed

Figure 1 Competition between China and Japan

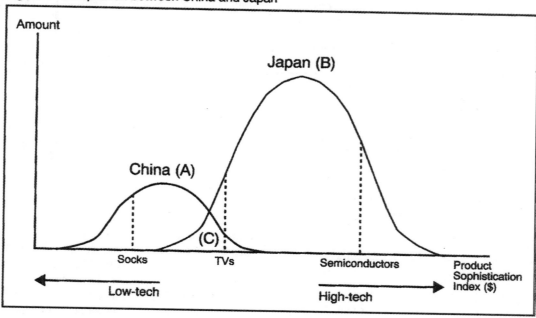

under original equipment manufacturing (OEM) contracts and sold with foreign brand names. Thus only a very small percentage of the added value of products labeled "made in China" is actually "made by China." The latter corresponds to the concept of China's gross national product (GNP), and excludes import charges on intermediate goods and investment income paid to foreign countries.

China is so heavily dependent on foreign partners that it has yet to develop its own cutting-edge technology and internationally recognized brand names. On the top of this, Chinese companies are inferior to their foreign counterparts in virtually every aspect, be it capital, human resources, or business management. As a result, China has no option but to look to cheap labor for its export competitiveness. Indeed, the majority of China's contribution to the added value of its exports lies with the cost of labor, and the very low wages in China averaging about $100 a month imply that this contribution must be very small.

As such, the common assumption that Chinese goods are competitive because the country's wage levels are low holds true only for labor-intensive products, and does not necessarily apply to industry as a whole. Instead, China's low wages should be interpreted as a reflection of the fact that its labor productivity is poor. It is when China's wage levels approach those of Japan, reflecting a rise in productivity, that China will really become a formidable competitor for Japan.

Complementarity between Japan and China

The recent economic relations between Japan and China can be explained in terms of Figure 1. The horizontal axis represents the level of sophistication of export items, and the vertical axis represents the amount of exports corresponding to export items at different levels of sophistication. A country's exports can then be represented by a distribution among products at different levels of sophistication ranging from low-tech products to high-tech products. Based on the assumption that high-value-added products are likely to be exported from high-income countries, while low-value-added products are likely to be exported from low-income countries, the product sophistication index for each product can be calculated as the weighted average of the per capita GDP of its exporters. The distribution for Japan's exports is expected to be larger than that of China, reflecting its larger volume. It should also be located more to the right, reflecting the fact that high-tech products make up a larger portion of Japan's total exports. The size of the part of the two distributions that overlap one another (C in Figure 1), as a proportion of each country's total exports (A for China and B for Japan), serves as an indicator of the degree of competition between the two countries. The greater the area of overlap between the two distributions as a percentage of Japanese exports (that is, C/B), the more China is a competitor of Japan. Conversely, the smaller the overlap, the more likely that China has an export structure complementary to that of Japan. For China, the degree of competition with Japan is given by C/A. (Figure 1)

There is no question that the size of exports from Japan is bigger than that from China, and that Japan's export structure is more advanced than that of China. However, there has been rising concern in Japan that the distribution representing China is expanding rapidly and moving fast to the right. In contrast, the Japanese distribution has been static and the prospect for restarting the engine of growth has remained dim. Against this background, many people in Japan have come to believe that China has already become a strong competitor for Japan, and

Table 1 Asian Countries' Competition with China in the U.S. Market			
	1990	1995	2000
Japan	3.0%	8.3%	16.3%
South Korea	24.0%	27.1%	37.5%
Taiwan	26.7%	38.7%	48.5%
Hong Kong	42.5%	50.5%	55.9%
Singapore	14.8%	19.2%	35.8%
Indonesia	85.3%	85.5%	82.8%
Malaysia	37.1%	38.9%	48.7%
Philppines	46.3%	47.8%	46.1%
Thailand	42.2%	56.3%	65.4%

Source: Calculated by the author based on U.S. Department of Commerce, *U.S. Import History*

that in the near future Japan will be eclipsed by China. Japan's China syndrome is merely an expression of this fear.

Although the total amount of exports from China has been increasing, labor-intensive products still feature largely in the export structure and the level of competition with Japan today is not necessarily high. It is clear that the export structures of China and Japan are complementary to, rather than competing with, each other—just as the big difference in their levels of economic development as well as their export volumes would lead one to expect. Based on the framework laid down in Figure 1, we use U.S. imports from individual countries with detailed breakdown by product (covering 10,000 manufactured goods according to the Harmonized System Commodity Classification) as proxies for their global exports to confirm this point. Our estimates show that China and Japan competed for only about 16.3% of their exports to the United States in value terms in 2000, although the percentage has been growing over time (from 3.0% in 1990 and 8.3% in 1995.)

These results show merely the extent to which products exported from Japan and China overlap, and two additional factors have to be considered to evaluate more accurately the degree of competition between the two countries. First of all, even though certain products are classified in the same category, in many cases Japan specializes in products for an upscale market and China specializes in low-priced products. TVs are a typical case in point, and the price tags for high-definition TVs exported by Japan are many times higher than those for the standard TVs made in China. Also, as noted above, Chinese exports include many more imported parts and components than Japanese exports.

Thus, the degree of actual competition between Japan and China is likely to be even lower than what the result of the calculations would indicate. In addition, the competition between Japan and China exists only in relatively low value-added products, in which Japan no longer enjoys any comparative advantage.

For the sake of comparison, we also calculate the level of competition with China for major Asian economies. Our estimates show that the Association of South-East Asian Nations (ASEAN) countries, whose income levels are still low, tend to compete more with China than Japan and the Asian NIEs, which are at a more advanced stage of economic development. (Table 1)

The Rise of China as a Business Opportunity for Japan

The potential complementarity between China and Japan, however, has not been fully exploited. Many Japanese companies view the expansion of China's production capacity as a threat, while at the same time finding little attractiveness in the Chinese market. As suggested by the GDP identity, income, and thus the size of the market, should grow at the same pace as output. Moreover, due to the high degree of complementarity between the two countries, Japan should enjoy advantages in penetrating the Chinese market. Market information may be biased as companies making money in China prefer to remain mute while those incurring losses are crying out in a loud voice, causing increased pessimism. If in fact market expansion is lagging behind production in China, it may reflect the following three factors.

First, foreign affiliates hold a large share of production as well as corporate earnings in China, so that the country's GNP is far below its GDP. The dividends paid by these foreign affiliates do not become income for the Chinese people, and may actually be transferred overseas. This makes China more attractive as a production base for exports than as a market. Among the foreign companies in China, if U.S. firms and European firms are making money while Japanese firms are not, then the latter should reexamine their corporate strategies.

Second, at the macro level aggregating the household, corporate and government sectors, China has a high rate of savings, so that expenditures are far lower than income. The difference does not translate into demand for goods, but rather is used by the monetary authorities to build up foreign exchange reserves. In this case, Japan should persuade China's monetary authorities to invest a larger proportion of its foreign exchange reserves in yen assets. Unfortunately, the lion's share of this "China money" is flowing to the United States instead of to Japan.

Third, China's terms of trade have been deteriorating as rising exports drive down export prices, so that the same amount of exports can be exchanged for less and less imports. This fall in purchasing power has been reflected in the yuan's sharp depreciation over time, and the slow growth in GDP in dollar terms. Under these circumstances, Japan should benefit by importing cheap products and components from China, which would allow lower prices for consumers while cutting costs at Japanese companies.

Thus even if the Chinese market is not growing as fast as production, there are various ways that Japanese companies can take advantage of China's growing economy. While importing goods produced in China through OEM and other schemes is going relatively well, Japanese investment in China has stayed at a very low level.

On the other hand, the U.S. and European countries view the emergence of China as a business opportunity, rather than as a threat, and success stories of their companies in China are on the rise. Indeed they now lead the list of the top ten foreign companies in China, which does not include a single Japanese company. Among automakers, for example, Germany's Volkswagen has a market share of 50%, while the mobile phone market has been dominated by Motorola of the United States, Nokia of Finland and Ericsson of Sweden. Even in China's electronics sector, Japanese companies are losing share as many Chinese firms emerge. Thus Japan should worry more about being left out of the fast-growing Chinese market, rather than about the hollowing out of its industry as more and more Japanese companies move to China.

Exports Versus Investment as a Means to Access the Chinese Market

One way for Japan to compromise between the concerns of hollowing out and losing the Chinese market is to put more emphasis on exporting to, than on producing in, China. Indeed, the sharp cut in tariffs following China's WTO accession should favor the former over the latter, particularly in areas where Japan enjoys comparative advantage while China does not.

One example is the automobile industry. Until now, the Chinese government has encouraged foreign automakers to produce in China by allowing auto sales in the domestic market on the one hand, while imposing high tariffs on auto imports on the other. Under this "swapping market for technology" strategy, foreign automakers would hit a wall of high tariffs when they tried to export to China, but the very same wall would protect them if their production were inside China. But things are changing. Following its entry to the WTO, China will abolish import quotas and lower import tariffs on finished cars from 80–100% to 25% by mid-2006. As a result, it may become cheaper to export finished cars to China rather than producing them there.

Should Japanese automakers opt for exports as a way to increase their presence in the Chinese market, they will be able to

shift their investment in China away from production facilities and into the reinforcement of sales networks, aftercare services and the establishment of research and development facilities for improving auto designs to better fit local tastes. Meanwhile, expanding production at home to satisfy the rising demand in China would reduce the pace of the hollowing out of Japanese industry.

Moreover, the Chinese auto industry faces so many handicaps that cheap labor in China does not necessarily translate into low production costs. Most automakers there are so small that they are unable to benefit from the economies of scale. Their productivity and research and development (R&D) capability also remain low and the quality of their products is far below global standards. As a consequence, the costs of local auto production are often higher than international prices for cars of equivalent quality. Utilizing their technology and financial resources, Japanese automakers may be able to overcome some of the drawbacks to producing in China. But it is doubtful whether they can compete with imported cars produced by their foreign rivals when China's import tariffs on auto imports come down to 25%.

The majority view among Japanese automakers is that they should produce close to the market so as to improve their brand image and meet the needs of consumers. But consumers rarely do their shopping by visiting car factories, and their needs can be served better by expanding local R&D facilities and aftercare services. Indeed, European automakers that are expanding their share of the Japanese market do not produce in Japan.

Also, for expensive goods like cars, other things being equal, Chinese consumers would certainly prefer a "made-in-Japan" to a "made-in-China" label. Above all, if Japanese automakers try to increase their production in China, they should also be prepared for the risk of overcapacity problems, which they may face in the near future as a result of too many foreign automakers rushing into the Chinese market. Instead of jumping on the bandwagon and building new plants in China, Japanese carmakers should consider the alternative option of exporting from their headquarters.

Chi Hung Kwan is a Senior Fellow at the Research Institute of Economy, Trade and Industry. He specializes in such fields as China's economic reform, regional integration in Asia and the yen bloc.

From the *Journal of Japanese Trade & Industry*, September/October 2002, pp. 26-29. © 2002 by Journal of Japanese Trade & Industry. Reprinted by permission.

LETTER FROM BOLIVIA

LEASING THE RAIN

The world is running out of fresh water, and the fight to control it has begun.

BY WILLIAM FINNEGAN

In April of 2000, in the central plaza of the beautiful old Andean city of Cochabamba, Bolivia, the body of Víctor Hugo Daza lay on a makeshift bier. Daza, a seventeen-year-old student, had been shot in the face by the Army during protests sparked by an increase in local water rates. These protests had been growing for months, and unrest had also erupted in other parts of the country. The national government had just declared martial law. In Cochabamba, a city of eight hundred thousand, the third largest in Bolivia, a good part of the population was now in the streets, battling police and soldiers in what people had started calling *la guerra del agua*—the Water War. Peasants from the nearby countryside manned barricades, sealing off all roads to the city. The protesters had captured the central plaza, where thousands milled around a tiled fountain and the catafalque of Víctor Daza. Some of their leaders had been arrested and taken to a remote prison in the Amazon; others were in hiding.

The chief demand of the water warriors, as they were called, was the removal of a private, foreign-led consortium that had taken over Cochabamba's water system. For the Bolivian government, breaking with the consortium—which was dominated by the United States-based Bechtel Corporation—was unthinkable, politically and financially.

Bolivia had signed a lucrative, long-term contract. Renouncing it would be a blow to the confidence of foreign investors in a region where national governments and economies depend on such confidence for their survival. (Argentina's recent bankruptcy was caused in large part by a loss of credibility with international bankers.) The rebellion in Cochabamba was setting off loud alarms, particularly among the major corporations in the global water business. This business has been booming in recent years—Enron was a big player, before its collapse—largely because of the worldwide drive to privatize public utilities.

For opponents of privatization, who believe that access to clean water is a human right, the Cochabamba Water War became an event of surpassing interest. There are many signs that other poor communities, especially in Third World cities, may start refusing to accept deals that put a foreign corporation's hand on the neighborhood pump or the household tap. Indeed, water auctions may turn out to test the limits of the global privatization gold rush. And while the number of populists opposing water privatization seems effectively inexhaustible—the leaders of the Cochabamba rebellion included peasant farmers and an unassuming former shoemaker named Óscar Olivera—the same cannot be said of the world's water supply. There was a great

deal more than local water rates riding on the outcome of this strange, passionate clash in Bolivia.

The world is running out of fresh water. There's water everywhere, of course, but less than three percent of it is fresh, and most of that is locked up in polar ice caps and glaciers, unrecoverable for practical purposes. Lakes, rivers, marshes, aquifers, and atmospheric vapor make up less than one percent of the earth's total water, and people are already using more than half of the accessible runoff. Water demand, on the other hand, has been growing rapidly—it tripled worldwide between 1950 and 1990—and water use in many areas already exceeds nature's ability to recharge supplies. By 2025, the demand for water around the world is expected to exceed supply by fifty-six percent.

Some of the resource depletion is visible from outer space. The Aral Sea, in central Asia, was until recently the world's fourth-largest lake. Then Soviet planners dammed and diverted its source waters for cotton irrigation. The Aral has since lost half its area and three-fourths of its volume. Its once great fisheries have vanished; all twenty-four species native to the lake are believed to be extinct. The local climate has changed, and dust storms now plague the region.

Aquifer depletion, though less visible, is an even more serious problem. There is sixty times as much fresh water stored underground as in lakes and rivers aboveground. And yet parts of northern China, to take one example, are approaching groundwater bankruptcy. Beijing's water table has dropped more than a hundred feet in the past forty years. In the United States, the Ogallala Aquifer, which reaches from Texas to South Dakota and is indispensable to farming on the Great Plains, is being drained eight times faster than it can naturally recharge. In vast areas of India, Mexico, the Middle East, and California's Central Valley the story is the same.

Meanwhile, more than a billion people have no access to clean drinking water, and nearly three billion live without basic sanitation. Five million people die each year from waterborne diseases such as cholera, typhoid, and dysentery. This enormous, slow-motion public-health emergency is, in large measure, a result of rapid, chaotic urbanization in the nations of the Global South. Traditional water sources have been polluted, destroyed, overtaxed, or abandoned.

Annual rainfall is not always a measure of water wealth. Poland, for instance, gets plenty of rain, but its lakes, rivers, and groundwater are so polluted that it has as little usable water as Bahrain. Arid regions with the means to pay (Southern California, the Persian Gulf States) already pipe water in from wetter areas. New technologies are being hurriedly developed: huge fabric bags holding millions of gallons of fresh water are being hauled by barges across the Mediterranean, and there are businessmen in Alaska who believe that the state's earnings from fresh water will eventually dwarf its earnings from oil.

For strategic planners at some of the world's largest corporations, the global freshwater shortage coincides opportunely with privatization. According to Johan Bastin, of the European Bank for Reconstruction and Development, "Water is the last infrastructure frontier for private investors." In the past fifteen years, municipal and regional water systems have been steadily coming onto the international market. Two French corporations, Vivendi Environment and Suez, lead the industry: Vivendi runs eight thousand systems in a hundred countries; Suez has operations in a hundred and thirty countries. The biggest American player, Bechtel, whose directors include former Secretary of State George P. Shultz, has always been notable for its political connections. The United States is itself a field for direct foreign investment in water. Suez is running Atlanta's water system, and Vivendi recently bought U.S. Filter, a national water-services group, for more than six billion dollars.

But the main push is in the Global South, where, over the past twenty years, the World Bank and the International Monetary Fund have effectively taken control of the economies of scores of nations that are heavily in debt. The Bank and the I.M.F. have been requiring these countries to accept "structural adjustment," which includes opening markets to foreign firms and privatizing state enterprises, including utilities. The Bank once had a quite different approach to public works: it was an enthusiastic financier of monumental projects, and would typically lend the money to build large dams. Many of the dams were spectacular failures, delivering few, if any, benefits (except to politicians and construction firms) while displacing millions of people and leaving behind environmental destruction and public debts. The Bank is now getting out of the dam business and into water privatization. It often works closely with the conglomerates, helping them to acquire the water assets of debtor nations.

The idea behind privatization is to bring market discipline and efficiency to bear on a crucial and frequently corrupt sector. Supporters argue that only private capital—which means, in practice, multinational corporations—can afford to expand water and sanitation networks sufficiently to reach the underserved poor. Since corporations are in business to make money, they often increase water rates. But, in theory, higher water rates can also help to promote conservation. Indeed, privatization advocates say, any valuable commodity—and this includes health care and education—that is provided free eventually gets taken for granted and wasted. According to this argument, turning water into a tradable commodity may even be the only practical way to avoid worldwide shortages and environmental disasters. Public subsidies for essential services such as water may sound like humane policy, but in the real world subsidies benefit the powerful, because they have the resources to manipulate them.

In Cochabamba, which has a chronic water shortage, this unintended consequence was grotesquely clear. Most of the poorest neighborhoods were not hooked up to the network, so state subsidies to the water utility went mainly to industries and middle-class neighborhoods; the poor paid far more for water of dubious purity from trucks and handcarts. In the World Bank's view, it was a city that was crying out for water privatization.

I went to Villa San Miguel, a ramshackle settlement on an arid hillside a couple of miles south of Cochabamba, to find out how the global water business looks from the ground. My guide was a student named Fredy Villagomez, who grew up in Villa San Miguel and helped organize an independent water coöperative for the barrio. The coöperative is one of dozens that have been formed in recent years, in part with international aid. We rode out of the old city in a clattering taxi, down a road jammed with trucks, buses, minivans, donkeys, and pedestrians. It was a hot, clear afternoon; Mt. Tunari, a seventeen-thousand-foot peak, glittered in the northwest. Cochabamba sits in a wide, fertile valley at eight thousand feet—only a middling elevation in Bolivia, where the capital, La Paz, is at twelve thousand feet. To the east, beyond the mountains, lies the Amazon rain forest. Cochabamba is more than four hundred years old, but a recent influx of migrants from the countryside has caused its population to quadruple since 1976. Today, Cochabamba is ringed by dozens of *barrios marginales*—dusty, impoverished settlements that have sprung up to house the newcomers. Basic services—electricity, transportation, sanitation, water—are catch-as-catch-can in the *barrios marginales*.

"We started digging our well in 1994," Villagomez told me. Stocky, soft-spoken, with Indian features and thick eyeglasses, he was leading me down a rocky path to a small cinder-block pump house. "The planning took years. It was an expensive project, and a lot of work. All the residents helped, and we finished in 1997." The coöperative received technical assistance from Danish aid workers; there is a dirt road in the barrio—the Avenida Dinamarca—named for them. Villagomez urged me to peer into the dim pump house, which contained a single electric pump. "The well is a hundred and twelve metres deep," he said.

I thought that sounded awfully deep. Villagomez agreed. "Before, the water under this valley was at only twenty metres."

The well made a major difference to Villa San Miguel. Clean water was suddenly plentiful and relatively cheap—households paid the water coöperative between two and five dollars a month to cover the costs of running the pump and maintaining the system. "It gives water to two hundred and ten families," Villagomez said. "We felt a lot of pride in this achievement."

Then, in 1999, the Bolivian government conducted an auction of the Cochabamba water system as part of its privatization program. The auction drew only one bidder: a consortium called Aguas del Tunari. The controlling partner in the consortium was International Water, a British engineering firm that was then wholly owned by Bechtel. (An Italian company later bought a half interest.) But the government, unfazed by its own weak bargaining position, decided to proceed.

The terms of the two-and-a-half-billion-dollar, forty-year deal reflected the lack of competition for the contract. Aguas del Tunari would take over the municipal water network and all the smaller systems—industrial, agricultural, and residential—in the metropolitan area, and would have exclusive rights to all the water in the district, even in the aquifer. The contract guaranteed the company a minimum fifteen-per-cent

annual return on its investment, which would be adjusted annually to the consumer price index in the United States. On coöperative wells such as Villa San Miguel's—which the government hadn't even helped build—the new water company could install meters and begin charging for water. Residents would also be charged for the installation of the meters. These expropriations were legal under a new water law that had been rushed through the Bolivian parliament.

The first Cochabambinos to question the terms of the water privatization were not the small water coöperatives that faced expropriation and crushing bills but local professionals—mainly engineers and environmentalists—and a federation of peasant farmers who rely on irrigation. They began calling public meetings to air their concerns. The government ignored them. At a contract-signing ceremony, the President of Bolivia and the mayor of Cochabamba drank champagne with consortium executives. The news of impending expropriations and rumors of big water-price hikes began to circulate. The list of alarmed groups—neighborhood associations, water coöperatives, the labor unions—grew. There were street protests, and a broad coalition emerged, called the Coördinator for the Defense of Water and Life, or simply La Coordinadora, led by Óscar Olivera.

Olivera, who is forty-six, at first seems an unlikely leader. He wears a black leather-billed cap, which makes him look like a gentle, would-be street hood from the Beat era. He is small and sad-faced, and he often looks downward, inward—as if he were thinking extremely hard about something unpleasant. It's an especially striking manner in Bolivia, where virtually everyone presents a placid, reserved face to the world. Olivera's grandfather worked in Bolivia's huge tin mines; his father went to work as a carpenter when he was a child. Olivera started out as a machinist in a shoe factory, then went on to work for the factory's union. He is now the head of a confederation of factory workers' unions. Oddly, his air of distraction doesn't seem to unsettle anyone around him. Working people volubly engage

him wherever he goes, calling him Oscarito.

"*Compañeros*," Olivera would tell crowds. "It's become a fight between David and Goliath, between poor people and a multinational corporation. They have a lot of money, and they want to take away our water."

The World Bank warmly calls Bolivia an "early adjuster." Other poor, indebted countries have had to be forced to accept structural adjustment, but in Bolivia the World Bank and the I.M.F. have enjoyed a deep understanding with successive governments since 1985. Public enterprises—the railways, the telephone system, the national airlines, the great tin mines of Oruro and Potosí—have been sold, mainly to foreign investors. (This fire sale goes on: a Bolivian government Web site lists dozens of factories, refineries, cement plants, paper mills, and municipal utilities that are still available.)

The tin mines, as it happens, had been nationalized after a popular revolution in 1952, which also destroyed the semi-feudal hacienda system that had been in place in Bolivia for centuries. The United States played an unlikely role in that revolution. The Eisenhower Administration, already busy undermining left-wing governments in Guatemala and Guyana, accepted the new government's plea that it was not Communist (even if some of its allies were) while demanding, and getting, a new investment code that permitted American companies to start operating in Bolivia's eastern oil fields. The United States increased food aid to help the new government survive, and saw to it that "state capitalism" became the official economic model.

An Army coup overthrew the elected government in 1964, leading to the first in a long string of military regimes. (Many of the officers involved had received "counter-insurgency" training in the United States.) They were not, except during a period in the late nineteen-seventies, as violent as those in Chile, Argentina, and Brazil. The unions and the left were repressed, but not so severely as to engender a guerrilla movement. By 1982, however, when civilian rule was finally restored, the Bolivian economy, plundered by the generals, was in ruins, and hyperinflation soon took hold, hitting an annual rate of twenty-five thousand percent in 1985.

Enter "the Boys," also known as "the Chicago Boys," after a group of economists, educated at the University of Chicago, who implemented free-market policies (known in Latin America as neoliberalism) in Pinochet's Chile. In Bolivia they were led by Gonzalo Sánchez de Lozada, a wealthy mine owner who was then Minister of Planning (he was later President). Sánchez worked closely with Jeffrey Sachs, the Harvard economist who became famous for the "shock therapy" he designed for post-Communist Poland.

Bolivia's shock treatment was ferocious. The currency was devalued, all price and wage controls were abolished, government spending was cut, and the state-owned tin mines were effectively closed. The economy went into instant recession; unemployment soared. The inflationary spiral, however, was broken, and good relations between the government and the I.M.F. were restored, initiating a new flow of foreign investment and loans.

And for the past sixteen years Bolivia has dutifully followed the dictates of the World Bank and the I.M.F. Most of its people, however, have nothing to show for it. Poverty was never significantly reduced. This is not unusual in Latin America, where the poverty rate is higher today than it was in 1980—after a full generation of nominal democracy and ever-increasing free trade. But Bolivia, like Argentina, really put on what Thomas L. Friedman, the *Times* columnist, calls "the golden straitjacket" of liberalized economic policies. The predicted foreign investment materialized, but the prosperity did not. Landlocked, with a population of eight million and a wretched infrastructure, Bolivia remains the poorest country in South America.

Hugo Banzer, who was Bolivia's military dictator in the nineteen-seventies, became President again after the last election, before stepping down last August in favor of his Vice-President, Jorge Quiroga, because of illness. Quiroga now sits at the head of a "megacoalition." The

political class in Bolivia has always been small, rich, and overwhelmingly white, but rarely have the major parties, all business-aligned, shared a ruling philosophy so peaceably as in recent years. The consensus was that nothing and no one should be exempt from the discipline of the market. Then Bechtel came to Cochabamba and, as the local peasants put it, tried to "lease the rain."

When the first monthly bills from Aguas del Tunari arrived, in January, 2000, stunned business owners and middle-class householders began to join the Coordinadora's protests. Some bills had doubled, and ordinary workers now had water bills that amounted to a quarter of their monthly income.

Aguas del Tunari seemed to have given little thought to how its plans would be received in Cochabamba. The International Water executives who were actually doing the work in the city were engineers, not marketers, and, being newly arrived from abroad, they were not attuned to the problems or passions of the Bolivian public. Geoffrey Thorpe, the company's manager, simply said that if people didn't pay their water bills their water would be turned off.

In truth, the price hikes were not as arbitrary as they seemed. The consortium had agreed, in its contract, to expand the city's water system. This was going to be expensive, as was the large-scale repair job required by the deterioration of the existing system. "We were confident that we could implement this program in a shorter period of time than the one required by the contract," Didier Quint, the managing director of International Water, said. He added, however, "We had to reflect in the tariff increase all the increases that had never been implemented before." The consortium had also agreed to finish a stalled dam project known as Misicuni, which would pipe water through the mountains. This aspect of the deal seemed to make little sense—the World Bank had commissioned studies that pronounced Misicuni uneconomic. But the dam project had less to do with how privatization works in theory than with the reality of how multinational cor-

porations must come to terms with local politics.

Plans for the Misicuni Dam have been around for decades. The mayor of Cochabamba, Manfred Reyes Villa, had campaigned hard to complete it and was undeterred by questions about whether it was worth building. Reyes Villa was a popular mayor. People liked to say that his good looks got him the women's vote—he is better known by the nickname Bombón (Sweetie)—but his political instincts are sharp. Despite the widespread poverty in his city, he had pulled off major vanity projects, including a gargantuan white statue of Christ on a hilltop. (This Cristo de la Concordia is supposedly six feet taller than its rival in Rio de Janeiro.) Reyes Villa had been a real-estate developer before he became mayor, and everyone in Cochabamba was quick to note the proximity of most new roads and parks to Bombón's properties. His political party, the New Republican Force, was a personal vehicle as well as a formidable municipal patronage machine, and people said that N.F.R., its Spanish abbreviation, stood for "Nueva Forma de Robar" ("New Form of Robbery"). Reyes Villa lived in conspicuous splendor, and when I visited, soldiers were guarding his estate. There was a wall-size oil painting of his family. "Poverty is growing here, with wealth being concentrated in very few hands," he told me earnestly while we sat in his mansion. Still, Bombón won elections, his party was in President Banzer's megacoalition, and he had national ambitions. More to the point, some of his main financial backers stood to profit fabulously from the Misicuni Dam's construction. When the central government first tried to lease Cochabamba's water system to foreign bidders, in 1997, and did not include Misicuni in the tender, Bombón stopped it cold. It was only the inclusion of the project in the Aguas del Tunari contract that got the Mayor on board.

The Water War began in earnest in February, 2000. "The people went to the plaza to demonstrate against the contract," Fredy Villagomez told me. "The young men were in the city center, trying to hold the plaza. Others were here, maintaining a barricade across the highway. The women were cooking for those on the barricade. There were many campesinos passing by, walking to Cochabamba to join the rebellion." Crowds of *regantes*—peasant irrigators—arrived in the city, marching under village banners, or *wiphala*. The women, *cholitas*, wore fine straw hats and shimmering pleated velvet dresses, and their black hair was braided in long pigtails woven together in back with bright ribbons.

Most of the troops that Óscar Olivera could personally call out for demonstrations were *jubilados*—retired factory workers, old union men shuffling under battered fedoras, like faded figures on a postcard from some antipodal workers' state of another era. For a great many Bolivians, the labor unions retain an association with a time when workers were organized and proud, and when Bolivia's railroads and airlines and mines belonged to Bolivians. It made symbolic sense, then, that Olivera, the leader of the Cochabamba rebellion, was a union man, even though only a minority of Cochabamba's factory workers are still union members. (In the union confederation's dingy offices on the corner of Cochabamba's central plaza, soccer trophies fill the shelves, but when I asked about them the receptionist explained that the unions could no longer field a team.)

The growing crowds of Coordinadora supporters were drawn instead from the informal sector of pieceworkers, sweatshop employees, and street venders that has expanded enormously since the advent of structural adjustment and the closing of the tin mines. These men and women, most of them young, had more flexible schedules than workers in regular factories. Students from the University of Cochabamba, some of them middle-class anarchists, also joined in, carrying banners denouncing the World Bank, the I.M.F., and neoliberalism. But the front-line troops, particularly after the conflict sharpened and the authorities began to fire live bullets, proved to be the city's street children—an adolescent army of the homeless which has been growing in recent years.

The government's response escalated steadily. After the first protests, a ministerial delegation was sent to Cochabamba. By the end of the month, the water-price hikes had been rolled back, but the protests continued. Then the government sent in troops from Oruro and La Paz. Nearly two hundred protesters were arrested, and seventy civilians and fifty-one policemen were wounded. The Catholic archbishop of Cochabamba tried to mediate. In March, the Coordinadora held an unofficial referendum, counted nearly fifty thousand votes, and announced that ninety-six percent favored the cancellation of the contract with Aguas del Tunari. "There is nothing to negotiate," the government replied.

In April, protesters again occupied Cochabamba's central plaza, and when the Coordinadora's leaders, including Óscar Olivera, arrived at the governor's office for a meeting they were arrested. After they were released the following day, some went into hiding. Then more leaders were arrested, and some were taken to a jungle prison in the Amazon. The house of Óscar Olivera's parents was searched four times.

Protests had begun to break out in other parts of the country—in La Paz, Oruro, and Potosí, and in many rural communities—and national peasant organizations held demonstrations. By this point, most of the country's major highways were blocked.

Bolivia's rulers have always harbored a deep fear that the country's Indian majority might one day rise up and kill them in their beds—or, more realistically, trap them in their cities. In 1781, an Indian rebel army, having killed all the Spaniards in a regional capital, laid siege to La Paz for several months. The rebellion was ultimately defeated by troops brought from Buenos Aires, but white Bolivia's fear of a horizon suddenly filling with angry Indians has never fully dissipated, and on April 8, 2000, the Banzer government declared a national state of siege. This meant martial law, and it allowed for mass arrests. The minister who announced the decree also said—in a remark that Bechtel's spokesman in London quickly picked up—that the uprising in Cochabamba was being financed by *narcotraficantes*.

The state of siege, along with the comments about drug traffickers, backfired in Cochabamba. The small coca-leaf farmers, known as *cocaleros*, from the lowlands east of Cochabamba, had indeed joined the protesters, but ordinary Bolivians draw a sharp distinction between *cocaleros* and the wealthy, Army-bribing, customs-bribing *narcotraficantes*. Many of the *cocaleros* are ex-miners. Water is not their issue—they are more concerned about a coca-eradication program sponsored by the United States—but their natural sympathy was with the protesters. And the *cholitas* in their velvet dresses, and the *jubilados* marching in their rumpled fedoras, and the water warriors in their bandannas did not appreciate the suggestion that they were insincere.

The day the state of siege was declared, the main plaza in Cochabamba was filled with people. The Army fired tear gas into the narrow streets of the old city, where protesters had built barricades. Demonstrators blocked all roads into the city; the government cut off power to local radio and television stations. Middle-class matrons took wounded protesters into their homes and beauty salons to nurse them, and bowls of vinegar mixed with water and baking powder—useful for soaking bandannas for protection from tear gas—appeared outside a thousand respectable doorways.

Then, from behind a line of military police, a sharpshooter in civilian clothes fired a rifle into a crowd of unarmed civilians. He was caught on video by a Bolivian television crew, and was later identified as Robinson Iriarte de la Fuente, a Bolivian Army captain who had been trained in the United States. Víctor Hugo Daza, the seventeen-year-old student, who was on his way home from a part-time job, was, according to eyewitnesses, among the crowd that Iriarte fired into. He was hit in the face and died instantly. Dozens of other people were treated for bullet wounds. By the time Daza had been raised onto his bier and the police and the Army had been repeatedly prevented from seizing his body, there was clearly no future for Aguas del Tunari in Cochabamba.

The company's executives were told that the police could no longer guarantee their safety, and fled Cochabamba for the lowland city of Santa Cruz. They may have noted that, several weeks earlier, Mayor Reyes Villa had left their side. When the people took to the streets en masse, Bombón had assessed their mood and stepped away from Aguas del Tunari so fast that it was as if he had never seen these foreigners before. He was not the only one trying to distance himself; when water privatization collapsed in Cochabamba, the World Bank's representatives insisted that the fiasco had nothing to do with them. The government informed Aguas del Tunari that, because the company had "abandoned" its concession, its contract was revoked. (The company argued that it had not left voluntarily but had been pushed out.) The day after Víctor Hugo Daza's funeral, Óscar Olivera announced the consortium's departure to thousands of exhausted, disbelieving demonstrators from the balcony of his union's offices above the plaza.

The Coordinadora had swept the field so completely that a new national water law was immediately passed—"written from below," as the water-rights campaigners say. Banged together by parliamentarians and water specialists from the Coordinadora who gathered in La Paz, the new law gave legal recognition to *usos y costumbres*—traditional communal practices—by protecting small independent water systems, guaranteeing public consultation on rates, and giving social needs priority over financial goals. This triumph seemed to the water warriors too good to be true, and it was. Laws in Bolivia are implemented—if, indeed, they are ever implemented—only after bylaws have been attached and approved, and the government soon made it clear that, in the case of the new water law, this process could take years.

After the Bechtel consortium's exit, the management of Cochabamba's water was returned to the old public utility known as SEMAPA, which was thoroughly overhauled. The new board of directors included Coordinadora representatives, who vowed to treat water as a "social good" and not as an ordinary commodity. But so far the fervently envisioned transformation of Cochabamba's water system has been fitful at best. Service is still poor. Even within the existing network, many neighborhoods have service only occasionally, and the valley's aquifer continues to sink. Corruption has reportedly been reduced, but an intolerable situation persists: the poor in Cochabamba, those who are not on the network and who have no well, pay ten times as much for their water as the relatively wealthy residents who are hooked up. The new SEMAPA, having driven away international capitalists, desperately needs new capital. Since simply raising water rates across the board is politically impossible, that means new partners or new loans.

The Bolivian government has little interest in seeing the new SEMAPA succeed. Neither is it likely to get much help from the World Bank. In recent years, the Bank has been widely accused of not fulfilling (or even seriously pursuing) its self-proclaimed mission of fighting poverty, and, in response, has changed its description of its mandate, emphasizing "empowerment" and "pro-poor coalitions" over fiscal discipline. What its officials do when confronted by an actual pro-poor coalition is, of course, another matter. In Bolivia, its representatives have never met with the leaders of the Coordinadora.

The Coordinadora's leadership knows that an old-fashioned state-centered solution to Cochabamba's water crisis would be unsatisfactory. The new SEMAPA has not won the hearts of many in Cochabamba—it is, after all, still the water company. But, its supporters point out, at least it is Bolivian.

"The tragedy is that the solution to Cochabamba's water problem has been pushed off for at least another five years," Michael Curtin, a Washington-based executive who became president of Aguas del Tunari after its removal from Cochabamba, told me. Curtin had been involved in the deal as a consultant for International Water, and represents the interests of Bechtel and its partners in talks with the Bolivian government. In November, 2001, after negotiations had deadlocked, the consortium filed a complaint against the Bolivian government in a World Bank trade court in Washington. The complainant was Aguas del Tunari, but the political weight was supplied by Bechtel. The company's claim is being made under a bilateral investment treaty between Bolivia and, of all places, the Netherlands. It seems that International Water, which was originally registered in the Cayman Islands (which has no comparable investment treaty with Bolivia), moved its registration to Amsterdam soon after the Cochabamba contract was signed. Bechtel and its partners are demanding at least twenty-five million dollars in compensation for the broken contract. While trade-court proceedings are notoriously slow, nobody seems to believe that the Bolivians can afford that kind of money.

Still, the prospect of one more financial burden is the least of the government's worries in connection with what people in Bolivia refer to as the Bechtel case. The truly frightening part is the impact that the Water War has had on the foreign-investment climate. "Right now the situation is 'You can't trust Bolivia,'" Curtin said. The United States Embassy agrees. "It was a pretty significant blow," an Embassy official told me. A subsequent auction to sell the La Paz telephone company drew no bidders at all. The United States has not yet decided, the official added, whether to formally designate the breaking of the contract in Cochabamba an expropriation. That will depend on the settlement, if there is one, that is reached in the case. Having the United States label the episode an expropriation would be a blow to Bolivia's hope of seeing any new foreign investment in the near future.

And yet Jorge Quiroga, Bolivia's new President, is bullish on his country's prospects. Quiroga, who is forty-one, is tall and fair. He graduated in 1981 from the University of Texas, where he studied industrial engineering, and worked for I.B.M. in Austin as a self-described "corporate yuppie" before moving back to Bolivia with his American wife, Virginia. "We will be the vital heart of South America," Quiroga predicted repeatedly when I visited him in his ornate, republican-era office in La Paz. Gas exports will lead the way. A long-awaited transcontinental highway connecting Brazil and Chile will pass through Cochabamba. Fibre-optic cables will be laid. One of the biggest things delaying Bolivia's economic progress? The hypocrisy of the United States and Europe on free trade. "Bolivia is the most open economy in Latin America," Quiroga said. Meanwhile, American and European farm subsidies, along with tariffs on textiles and agricultural products, make it impossible for Bolivia to sell its exports in the Global North. "They tell us to be competitive while tying our arms behind our backs."

When I asked him about the Water War, he looked uncomfortable. "A lot of things certainly could have been different along the way, from a lot of different actors," he said. But, like Michael Curtin, he was certain of one point: "The net effect is that we have a city today with no resolution to the water problem." In the end, he said, it will be "necessary to bring in private investment to develop the water."

Quiroga insists that the World Bank and the I.M.F. are not running Bolivia. If it sometimes looks as if they are, that is because technocrats in the government (he has been one) use what he calls "the I.M.F. blackmail" on politicians—warning them that loans will stop if hard fiscal choices are not made, and thus giving them some cover, when in truth there have been no direct threats from the Bank or the I.M.F.

The World Bank now requires all its client governments to submit a "poverty-reduction strategy," and Quiroga gave me a sketch of Bolivia's. He had recently been to a meeting of the World Economic Forum—he goes almost every year—and his remarks struck me as having an up-to-the-minute cosmopolitan gloss. The Bolivian state has "socialized" its spending away from production toward education and health, he said, because that's the one sure route to real development. Quiroga sighed. "But we'll still be talking about poverty reduction twenty-five years from now, when one of

my daughters is sitting here being interviewed."

Some things, certainly, never change. Last month, Captain Robinson Iriarte was acquitted of all responsibility for the death of young Víctor Hugo Daza. The case had been transferred from the civilian criminal-justice system—no judge was willing to hear it—to a military tribunal, which has final jurisdiction over the cases it hears. Upon his acquittal, Iriarte was promoted to the rank of major.

Claudia Vargas, a lawyer in La Paz who worked for Bolivia's fledgling utility-regulation body, travelled often to Cochabamba. The leaders of the new SEMAPA, she told me, were unnecessarily preoccupied with the idea that the central government was going to make another attempt to privatize the city's water. "I tell them 'Don't *worry*. Nobody is going to try that again for a long time.' Really, they have a lot of problems in Cochabamba now, and that is not one of them."

Vargas is young and chic, and she had recently finished a postgraduate course in development administration in thoroughly deregulated Britain. She nonetheless has an unfashionable enthusiasm for regulation. It's still poorly understood in Bolivia, she thinks, and it got a very bad name in Cochabamba when her boss (now retired) appeared to shill both for the government and for Bechtel's consortium, which he was supposed to regulate. Vargas was appalled both by the Aguas del Tunari contract and by the consortium's performance. She thinks that the members of the Coordinadora "are still the good guys in the film," but that some of them are, at best, misguided. "In the name of *usos y costumbres*, a lot of terrible things are done." Water-truck operators, for instance, "drill polluted water and sell it. They waste a lot of water. But the world is changing. Within five years, they will be charged for water—and they will be regulated."

Trying to find a path outside the binary of state or market control, the Coordinadora has no model. Its leaders vow to treat water as a "human right," and yet they know it cannot be provided free. They also know that their water com-

pany—and Bolivia generally—cannot survive without foreign investment. Meanwhile, Manfred Reyes Villa has started making his move to regain control of SEMAPA. In elections to be held soon, his party will run candidates for all the seats on the water company's board that are now occupied by the Coordinadora. Their campaigns will be far more extravagant and crowd-pleasing than anything Óscar Olivera and his colleagues are likely to stage. Bombón knows his constituency. He has also announced his candidacy for President.

The juggernaut of water privatization has hardly slowed. Bechtel, through International Water, has closed two major water deals in the past year, winning a thirty-year concession for the port city of Guayaquil, in Ecuador, and a controlling stake in the water company of Tallinn, the capital of Estonia. Many water privatizations seem likely to deliver eventually on their backers' promises of improved service. In Chile, fierce opposition to concessions has been overcome in several cities by innovative price structures, including water vouchers, that assure poor residents of an adequate supply of clean water.

At the same time, water privatizations have been backfiring all over Latin America. In Panama, popular anger about an attempted privatization helped cost the President his bid for reëlection. Vivendi, the French multinational, had its thirty-year water contract with the Argentine province of Tucumán terminated after two years because of alleged poor performance. Major water privatizations in Lima and Rio de Janeiro have had to be cancelled because of popular opposition. Trinidad recently allowed a management contract with a British water giant to expire. Protests against water privatization have also erupted in Indonesia, Pakistan, India, South Africa, Poland, and Hungary.

One large-scale Bolivian water privatization that the World Bank still points to with pride took place a few years ago in La Paz. The concession was awarded to Suez, which honored its commitment to expand the La Paz water network to several poor neighborhoods just outside the city. This area, known as El Alto, is home to nearly three quarters of a million people, virtually all of them Indians recently arrived from the countryside. But a problem emerged. It seemed that the people in El Alto weren't using enough water. Accustomed to Andean peasant life, they were extremely careful with water, never wasting a drop, and they continued to be so even after they had taps installed in their homes. This was good conservation, but it was bad for Suez's bottom line, and the corporation was disappointed in the return on its investment. After it appeared to raise its rates, which were pegged to the dollar, when the local currency was devalued the general happiness with the contract evaporated and residents began to complain about the service. When I was in La Paz, the people of El Alto were marching against Suez. When I asked a World Bank official about the situation, she agreed that there was a basic problem: those Indians needed to learn to use more water.

From *The New Yorker,* April 8, 2002, pp. 43-53. © 2002 by The Condé Nast Publications, Inc.

Going Cheap

Slave labour plugs neatly into the global economy. **Kevin Bales** argues that it's time to look beyond the cost-benefit analysis.

This past April the world's media zoomed in on the 'slave ship' of Benin. The ship, reported to be carrying 200 enslaved children, was refused entry to Gabon and Cameroon. For two days it disappeared while a search was mounted and fears grew over the fate of the children. When the ship finally reappeared and docked in Benin, it had on board only 43 children and 100 or so adults. After questioning, it was found that most of the children were being trafficked to work in Gabon. In spite of this the ship's captain denied any involvement. The Benin Government then suggested that there was another ship with child slaves, but none was located. Were there other child slaves? Was there another ship? At this point no-one knows.

What we do know is that this confusing incident is just a small part of the regular human traffic between Benin and Gabon. What was news to the world's media is well known in West Africa—on what was once called the Slave Coast, the trade continues. Increasingly, children are bought and sold within and across national borders, forced into domestic work, work in markets or as cheap farm labour. UNICEF estimates there are more than 200,000 children trafficked in West and Central Africa each year.

Child slavery is a significant money-maker in countries like Benin and Togo. Destitute parents are tricked into giving their children to slave-traders. A local UNICEF worker explains: 'People come and offer the families money and say that their children will work on plantations and send money home. They give the family a little money, from $15 to $30— and then they never see their children again.'

While a slave ship off the African Coast is shocking at the turn of the 21st century, it represents only a tiny part of the world's slavery which has seen a rapid escalation since 1945 and a dramatic change in character.

If she is ill or injured, she is disposable

Three things have sparked this rapid change. Firstly, the world's population has tripled since 1945 with the bulk of the growth in the Majority World. Secondly, economic change and globalization have driven rural people in poor countries to the cities and into debt. These impoverished and vulnerable people are a bumper crop of potential slaves. Finally, government corruption is essential. When those responsible for law and order can be made to turn a blind eye through bribes, the slave-takers can operate unchecked.

This new slavery is marked by a dramatic shift in the basic economic equation of exploitation—slaves are cheaper today than at any other time in human history. The agricultural slave that cost $1,000 in Alabama in 1850 ($50,000 at today's prices) can be purchased for around $100 today. This fall in price has altered not only the profits to be made from slavery, but the relationship between slave and master as well. The expensive slave of the past was a protected investment; today's slave is cheap and disposable.

A good example is a 14-year-old girl sold into a working-class brothel in Thailand. Her initial purchase price might be less than $1,000. In the brothel she will be told she must repay four times that to gain her freedom—plus rent, food and medicine costs. Even if she has sex with 10-15 men a night, her debt will keep expanding through false accounting and she will never be allowed to leave.

The profit that her 'owners' make from her is very large, as high as 800 per cent. Her annual turnover, the amount men pay for her, is more than $75,000— though she won't see a penny. These profits buy protection from the police, influence with local government, as well as social prestige. Her owners will be lucky to get five years' use from her since HIV is common in the brothels. But because she was so cheap, she is easily replaced. If she is ill or injured or just troublesome, she's disposable.

The brothels of Thailand are just one of the places where new slavery can be found. Slaves tend to be used in simple, non-technological and traditional work. Most work in agriculture. But they are also found in brick-making, mining and quarrying, textiles, leather-working, prostitution, gem-working and jewellery-making, cloth and carpet-making. Or they may work as domestic servants, clear forests, make charcoal or work in shops. Most of this work is aimed at local sale and consumption but slave-made goods filter through the entire global economy and may even end up in western homes.

Studies have documented the slave origins of several international products such as carpets, sugar and jewellery. We may be using slave-made goods or investing in slavery without knowing it. Slave-produced cocoa, for example, goes into the chocolate we buy. Rugs made by slave children in India, Pakistan and Nepal are mainly exported to Europe and the US. The value of global slavery is estimated at $12.3 billion per year, including a significant amount of international trade in slave-produced goods. Despite this outrage few Northern businesses or organizations are taking action. Most trade associations argue that it is impossible to trace the twisted path to a product's origin or, more bluntly, that it's simply not their responsibility. The World Trade Organization has the power to introduce a 'social clause' to block products of forced labour, but it has not done so. And while 'fair trade' programmes are important alternatives to exploitation, they do not directly address the needs of enslaved workers. Obviously, there are many questions yet to be answered both about the economics of slavery and about the most effective strategies for abolition.

Recent studies show that human trafficking is increasing. The US Central Intelligence Agency estimates more than 50,000 persons a year are trafficked into the US. The UN Centre for International Crime Prevention says trafficking is now the third largest money earner for organized crime after drugs and guns. But a lack of reliable information means that governments are scrambling to build databases, develop effective interdiction, work out ways to free and rehabilitate trafficking victims, develop laws and conduct the research needed to address the issue.

Business is also pressed to deal with recent revelations of slavery amongst their suppliers. The filming of slaves on cocoa plantations in Cote d'Ivoire last year led to calls for a boycott of chocolate. Cote d'Ivoire produces about half of the world's cocoa. Some local activists claim that up to 90 per cent of the country's plantations use slave labour. Chocolate-producing companies have promised their own investigation.

The situation in Cote d'Ivoire encapsulates much of contemporary slavery. Slaves on the cocoa plantations are mostly from Mali. Desperate for work and tricked by promises of a good job, they can be purchased in village markets for $40 per person. The plantation owners who enslave them are facing a dramatic fall in the world price of cocoa as a result of the World Bank forcing an end to the state marketing monopoly. Meanwhile, Cote d'Ivoire carries $13.5 billion in debt to the Bank and other lenders. With debt payments five times greater than the nation's healthcare budget, there are few resources to protect the enslaved migrants producing its key cash crop.

As well as importing, traffickers in West Africa export slaves to richer countries. Educated young women from Ghana and Cameroon, lured with a chance of further study in the US, have been enslaved as domestics in Washington DC. Large numbers of Nigerian women have been forced into prostitution in Italy. This human traffic into and out of the African coast is mirrored in many countries of the developing and developed world.

In Pakistan and India, across North Africa, in Southeast Asia and in Central and South America, more traditional forms of debt bondage enslave up to 20 million people. These slaves who may be in their third or fourth generation of bondage contribute little to export markets. Laws on bonded labour are either not strict enough, or not enforced. Police are often ignorant of those laws or, as in Brazil and Thailand, they may be profiting from bonded labour themselves.

The result is that underfunded non-governmental organizations bear the brunt of liberating slaves, sometimes in the face of government resistance. And liberation is just the first step in returning slaves to a life of freedom.

Think for a moment about the 43 children rescued from the Benin slave ship. Questions about their future are every bit as perplexing as questions about their recent past. Many child slaves have suffered physical and psychological abuse and require help. Nearly all have to adjust both to freedom and the challenge of earning a living. With luck, rehabilitation programmes will help them. But few governments are involved in this work.

If there is any good news about modern slavery, it is the dramatic growth in media interest and public awareness. The global coverage of the slave ship was just one example. The UN has several new initiatives on slavery and trafficking, as does the European Union. At the same time, anti-slavery organizations are experiencing an upturn in interest. As one representative of Anti-Slavery International explained recently: 'It is heartening, after years of neglect, to be part of a global movement against slavery. It is still in its infancy, but it is growing everyday.

Kevin Bales is Director of Free the Slaves, the North American sister organization of Anti-Slavery International and the author of *Disposable People: New Slavery in the Global Economy* (University of California Press, 1999), which won the 2000 Viareggio Prize.

From *New Internationalist*, August 2001, pp. 14-15. © 2001 by New internationalist Publications Ltd.

UNIT 5
Conflict

Unit Selections

Key Points to Consider

- Are violent conflicts and warfare increasing or decreasing today? Explain your response.

- What changes have taken place in recent years in the types of conflicts and who participates?

- How is military doctrine changing to reflect new political realities?

- How is the nature of terrorism different than conventional warfare? What new threats do terrorists pose?

- How is the national security policy of the United States likely to change? What about Russia, India, and China?

 **Links: www.dushkin.com/online/**
These sites are annotated in the World Wide Web pages.

DefenseLINK
   http://www.defenselink.mil
Federation of American Scientists (FAS)
   http://www.fas.org
ISN International Relations and Security Network
   http://www.isn.ethz.ch
The NATO Integrated Data Service (NIDS)
   http://www.nato.int/structur/nids/nids.htm

Do you lock your doors at night? Do you secure your personal property to avoid theft? These are basic questions that have to do with your sense of personal security. Most individuals take steps to protect what they have, including their lives. The same is true for groups of people, including countries.

In the international arena, governments frequently pursue their national interest by entering into mutually agreeable "deals" with other governments. Social scientists call these types of arrangements "exchanges" (i.e., each side gives up something it values in order to gain something in return that it values even more). On an economic level, it functions like this: "I have the oil that you need. I will sell it to you. In turn I will buy from you the agricultural products that I lack." Whether on the governmental level or the personal level ("If you help me with my homework, then I will drive you home this weekend"), this is the process used by most individuals and groups to "secure" and protect what is of value. The exchange process, however, can break down. When threats and punishments replace mutual exchanges, conflict ensues. Neither side benefits and there are costs to both. Further, each may use threats and hope that the other will capitulate, but if efforts at intimidation and coercion fail, the conflict may escalate into violent confrontation.

With the end of the cold war, issues of national security and the nature of international conflict have changed. In the late 1980s agreements between the former Soviet Union and the United States led to the elimination of superpower support for participants in low-intensity conflicts in Central America, Africa, and Southeast Asia. Fighting the cold war by proxy is now a thing of the past. In addition, cold war military alliances have either collapsed or have been significantly redefined. Despite these historic changes, there is no shortage of conflicts in the world today. The dramatic events of September 11, 2001, have made people around the world fully aware that new threats exist and that the strategy and tactics of warfare have changed.

Many experts initially predicted that the collapse of the Soviet Union would decrease the arms race and diminish the threat of nuclear war. However, some analysts now believe that the threat of nuclear war has in fact increased as control of nuclear weapons has become less centralized and the command structure less reliable. In addition, the proliferation of nuclear weapons into South Asia (India and Pakistan) is a new security issue. Further, there are growing concerns about both dictatorial governments and terrorists obtaining a variety of different types of weapons of mass destruction. What these changing circumstances mean for U.S. policy is a topic of considerable debate.

The unit begins with a timely discussion of two interrelated topics: the war on terrorism (highlighted by the attacks on the World Trade Center and the Pentagon on September 11, 2001) and the conflict in the Middle East. What should be the level of involvement of the United States in addressing these issues and

can it lead unilaterally? In addition what are some of the future dangers that terrorism poses and what can be done to counteract them?

In addition to the pressing issues of terrorism and the conflicts in the Middle East, there are strategic issues to be considered, including the growing military role of China. The unit concludes with a broad overview of the rapidly changing technology of warfare with a specific discussion of space-based weapons.

Like all the other global issues described in this anthology, international conflict is a dynamic problem. It is important to understand that conflicts are not random events, but that they follow patterns and trends. Forty-five years of cold war established discernable patterns of international conflict as the superpowers contained each other with vast expenditures of money and technological know-how. The consequence of this stalemate was often a shift to the developing world for conflict by superpower proxy.

The changing circumstances of the post–cold war era generate a series of important new policy questions: Will there be more nuclear proliferation? Is there an increased danger of so-called "rogue" states destabilizing the international arena? Is the threat of terror a temporary or permanent feature of world affairs? Will there be a growing emphasis on low-intensity conflicts related to the interdiction of drugs, or will some other unforeseen issue determine the world's hot spots? Will the United States and its European allies lose interest in security issues that do not directly involve their economic interests and simply look the other way as age-old ethnic conflicts become brutally violent? Can the international community develop viable institutions to mediate and resolve disputes before they become violent? The answers to these and related questions will determine the patterns of conflict in the twenty-first century.

The Reluctant Imperialist

Terrorism, Failed States, and the Case for American Empire

Sebastian Mallaby

Lawrence Summers, the dominant professor-politician of the Clinton years, used to say that the United States is history's only nonimperialist superpower. But is this claim anything to boast about today? The war on terrorism has focused attention on the chaotic states that provide profit and sanctuary to nihilist outlaws, from Sudan and Afghanistan to Sierra Leone and Somalia. When such power vacuums threatened great powers in the past, they had a ready solution: imperialism. But since World War II, that option has been ruled out. After more than two millennia of empire, orderly societies now refuse to impose their own institutions on disorderly ones.

This anti-imperialist restraint is becoming harder to sustain, however, as the disorder in poor countries grows more threatening. Civil wars have grown nastier and longer. In a study of 52 conflicts since 1960, a recent World Bank study found that wars started after 1980 lasted three times longer than those beginning in the preceding two decades. Because wars last longer, the number of countries embroiled in them is growing. And the trend toward violent disorder may prove self-sustaining, for war breeds the conditions that make fresh conflict likely. Once a nation descends into violence, its people focus on immediate survival rather than on the longer term. Saving, investment, and wealth creation taper off; government officials seek spoils for their cronies rather than designing policies that might build long-term prosperity. A cycle of poverty, instability, and violence emerges.

There is another reason why state failures may multiply. Violence and social disorder are linked to rapid population growth, and this demographic pressure shows no sign of abating. In the next 20 years, the world's population is projected to grow from around six billion to eight billion, with nearly all of the increase concentrated in poor countries. Some of the sharpest demographic stresses will be concentrated in Afghanistan, Pakistan, Saudi Arabia, Yemen, and the Palestinian territories—all

Islamic societies with powerful currents of anti-Western extremism. Only sub-Saharan Africa faces a demographic challenge even sharper than that of the Muslim world. There, an excruciating combination of high birth rates and widespread AIDs infection threatens social disintegration and governmental collapse—which in turn offer opportunities for terrorists to find sanctuary.

Terrorism is only one of the threats that dysfunctional states pose. Much of the world's illegal drug supply comes from such countries, whether opium from Afghanistan or cocaine from Colombia. Other kinds of criminal business flourish under the cover of conflict as well. Sierra Leone's black-market diamonds have benefited a rogues' gallery of thugs, including President Charles Taylor of Liberia and Lebanon's Hezbollah. Failed states also challenge orderly ones by boosting immigration pressures. And those pressures create a lucrative traffic in illegal workers, filling the war chests of criminals.

None of these threats would conjure up an imperialist revival if the West had other ways of responding. But experience has shown that nonimperialist options—notably, foreign aid and various nation-building efforts—are not altogether reliable.

RICH MAN'S BURDEN

Take the chief alternative to imperialism, foreign aid. It is no coincidence that the main multilateral organizations for dispensing it—the United Nations and the World Bank—were set up at the end of World War II as the European empires started to unravel. For decades, the aid intelligentsia was certain that it had the solution to chaos. In the 1950s and 1960s, it thought that simply providing capital would ensure self-sustaining growth in poor countries. In the 1970s, the focus shifted to relieving poverty directly by building health clinics and schools. In the 1980s, donors sought to tie their aid to economic reforms. In the 1990s, they added on demands for anticorruption measures and other improvements in governance. Along the way, development theorists flirted with the idea that population control might hold the key. But no magic key has yet been found. An obstinate group of dysfunctional countries has refused to respond to these approaches.

This is not to say that aid has failed. Since 1960, life expectancy in poor countries has risen from 45 to 64 years. The global illiteracy rate has fallen from 47 to 25 percent over the last three decades. And the number of poor people has fallen by about 200 million in the last two decades—at a time when the world population has increased by 1.6 billion. Development institutions deserve more credit than they get, whether from antiglobalization protesters or from aid critics within Congress and the Bush administration. But aid donors must face up to their inability to shake the most dysfunctional countries out of poverty, especially in regions such as sub-Saharan Africa.

The World Bank has convened an internal task force to confront its record on failed states, and the group will undoubtedly come up with some suggestions. Routing aid around dysfunctional governments is likely to be one of them. But the bank has tried this kind of thing before. In Chad, for example, it spent years devising a plan to develop the country's oil fields while preventing the profits from being wasted by corrupt rulers. An elaborate accounting system was designed in which oil revenues would go into a special fund to pay for health, education, and other worthwhile causes. When the system was unveiled in 2000, the bank even suggested that the "Chadian model" might point the way forward for other resource-rich developing nations. Within six months, however, Chad's government found a way of diverting $4.5 million of oil money to finance unauthorized arms purchases.

In countries such as China and India, which have functioning governments broadly committed to development, aid and technical advice have greatly accelerated the escape from poverty. In Uganda, which has weaker institutions but a fierce dedication to development, aid has helped cut poverty by 40 percent in one decade. But in countries such as Chad, Haiti, or Angola, aid cannot accomplish much. Such places are beyond the reach of economists who prescribe policies from afar. If outsiders want to make a difference in this kind of environment, they must begin by building the institutions that make development possible. They must engage, in other words, in the maligned business of nation building.

NO QUICK FIX

Modern nation building is partly an offshoot of the development business. In the late 1980s, development theorists began to acknowledge that the main alternative to imperialism—economic aid—could not stabilize the weakest states. A political supplement was needed, starting with transparency and other principles of decent governance. This recognition coincided with the collapse of authoritarian regimes—first in Latin America and East Asia, then, more spectacularly, in communist countries. Suddenly dozens of nations found themselves in a state of uncertain transition. The need to focus on the political components of international stability was clear, and a new era of nation building began.

The post–Cold War history of this experiment resembles the history of development aid since World War II. The simple-sounding goal of building stable democracies has proved maddeningly elusive. In turn, nation builders have pushed their strategy through successive stages of elaboration. The sudden collapse of authoritarian regimes first encouraged hopes that democratization might be quick, requiring little more than a simple effort to organize and monitor elections. Later, as transition turned out to be hard, donors sought to build political parties, police forces, law courts, tax offices, central banks, and customs systems—not to mention newspapers, community organizations, and the entire penumbra of independent groups that make up civil society. Within each of these categories, donor efforts have grown more elaborate as well. Rather than simply monitoring elections, for example, nation builders now seek to assess the preceding campaigns to ensure that the playing field is level.

As with development aid, democratization efforts have succeeded in some promising settings. Except for the Balkans, eastern Europe has done well thanks to peace, educated populations, and proximity to the rich European Union. But in the toughest countries, where state failure threatens the export of chaos, nation building has been hard. Perhaps the closest to a success story in a war-torn country is Mozambique, which has remained reasonably stable since foreign peacekeepers pulled out in 1995 after organizing multiparty elections (although that achievement now looks shaky). More typical is Angola, where a 1992 election under U.N. auspices proved worthless because defeated rebels refused to respect the verdict of the polls. In Cambodia, the loser in a 1993 U.N.-supervised election, Hun Sen's Cambodian People's Party, ignored the results and stayed in power through force. In Bosnia, Kosovo, and East Timor, meanwhile, nation builders are making some headway but are not yet successful enough to withdraw.

To their credit, nation builders have tried to confront this discouraging record. Lakhdar Brahimi, the Algerian diplomat now overseeing the U.N.'s efforts toward Afghanistan, recently produced a report on beefing up the peacekeeping department at the U.N.'s New York headquarters. Out in the field, some peacekeeping operations have been reinforced. The U.N. went into Sierra Leone with an inadequate contingent of 6,000 in 1999; it now has more than 17,000 troops there. The world has also tried to make up for the U.N.'s peacekeeping inadequacies by sending in other types of forces, with or without a U.N. umbrella. The West winked at Nigeria's failed effort to impose order on Sierra Leone in 1997, even though Nigeria intervened without a U.N. mandate. Some observers even argue that mercenaries might carry out more nation-building tasks, as seen in brief deployments in Sierra Leone and Angola.

As with the World Bank's task force on failed states, however, these efforts to beef up nation building are more interesting as implied admissions of failure than as signs of decisive progress. In the absence of greatly increased commitment from the U.N.'s leading member states, a wide gap will remain between nation builders' aspiration to create stable democratic states and what the world's institutions can deliver. Yet the Brahimi report and the occasional calls for a standing U.N. army recognize the dilemma posed by the end of empire. The rich world increasingly realizes that its interests are threatened by chaos, and that it lacks the tools to fix the problem.

SOMETIMES A GREAT NATION

Might an imperial America arise to fill the gap? Most people would dismiss this as utterly implausible. The United States, it is assumed, has a strong inhibition against external adventurism. Look at the no-passport crowd in Congress, Washington's occasional isolationist fits, and the Bush administration's repeated denunciation of nation building. From the failure to occupy Iraq at the end of the Gulf War to the refusal to commit peacekeeping troops in Afghanistan, the United States has not exactly displayed latent imperialist tendencies.

Yet these inhibitions are less than they appear. U.S. history includes an isolationist tradition, but it is by no means dominant. Other traditions, such as the urge to go forth and improve the world or open up foreign markets, have been present throughout American history as well, and the tradition that prevails at any time is the one that best matches the circumstances. Until the attack on Pearl Harbor in 1941, security seemed assured by brute geography; potential enemies were far across the seas, so foreign policy was often regarded as a luxury. During World War II and the Cold War, that presumption changed. Fascist expansionism and nuclear weapons threatened U.S. interests in obvious ways, and the United States responded with unusual vigor. It fought wars both hot and cold, deploying troops all over the globe. It still spends much more on defense than do the European governments that routinely protest American isolationism.

Now U.S. foreign policy must again respond to circumstance—this time to the growing danger of failed states. The Bush administration's denigration of nation building and its refusal to participate in a peacekeeping force for Afghanistan are not the final words on this subject. By launching his war on terrorism, the president has at least acknowledged the urgency of the threat. For all the grumbling over Balkan commitments, the administration has pulled out of neither Bosnia nor Kosovo. The logic of neoimperialism is too compelling for the Bush administration to resist. The chaos in the world is too threatening to ignore, and existing methods for dealing with that chaos have been tried and found wanting.

MANIFEST DESTINY?

Empires are not always planned. The original American colonies began as the unintended byproduct of British religious strife. The British political class was not so sure it wanted to rule India, but commercial interests dragged it in there anyway. The United States today will be an even more reluctant imperialist. But a new imperial moment has arrived, and by virtue of its power America is bound to play the leading role. The question is not whether the United States will seek to fill the void created by the demise of European empires but whether it will acknowledge that this is what it is doing. Only if Washington acknowledges this task will its response be coherent.

The first obstacle to acknowledgment is the fear that empire is infeasible. True, imposing order on failed states is expensive, difficult, and potentially dangerous. Between 1991 and 2000 the United States spent $15 billion on military intervention in the Balkans. A comparable effort in Afghanistan, a much bigger area with deeper traditions of violence, would cost far more. But these expenses need to be set against the cost of fighting wars against terrorists, drug smugglers, and other international criminals. Right after September 11, Congress authorized $40 billion in emergency spending—and that was just a down payment in the struggle against terrorism. The estimated cost to the U.S. economy ranges from $100 billion to $300 billion.

The second obstacle to facing the imperial challenge is the stale choice between unilateralism and multilateralism. Neither option, as currently understood, provides a robust basis for responding to failed states. Unilateralists rightly argue that weak allies and cumbersome multilateral arrangements undercut international engagement. Yet a purely unilateral imperialism is no more likely to work than the sometimes muddled multilateral efforts assembled in the past. Unilateralists need to accept that chaotic countries are more inclined to accept foreign nation builders if they have international legitimacy. And U.S. opinion surveys suggest that international legitimacy matters domestically as well. The American public's support for the Persian Gulf War and the Afghan conflict reflected the perception that each operation was led by the United States but backed by the court of world opinion.

The best hope of grappling with failed states lies in institutionalizing this mix of U.S. leadership and international legitimacy. Fortunately, one does not have to look far to see how this could be accomplished. The World Bank and the International Monetary Fund (IMF) already embody the same hybrid formula: both institutions reflect American thinking and priorities yet are simultaneously multinational. The mixed record of both institutions—notably the World Bank's failure on failed states—should not obscure their organizational strengths: they are more professional and less driven by national patronage than are U.N. agencies.

A new international body with the same governing structure could be set up to deal with nation building. It would be subject neither to the frustrations of the U.N. Security Council, with its Chinese and Russian vetoes, nor to those of the U.N. General Assembly, with its gridlocked one-country-one-vote system. A new international reconstruction fund might be financed by the rich countries belonging to the Organization for Economic Cooperation and Development and the other countries that currently contribute to the World Bank's subsidized lending program to the poorest nations. It would assemble nation-building muscle and expertise and could be deployed wherever its American-led board decided, thus replacing the ad hoc begging and arm-twisting characteristic of current peacekeeping efforts. Its creation would not amount to an imperial revival. But it would fill the security void that empires left—much as the system of mandates did after World War I ended the Ottoman Empire.

The new fund would need money, troops, and a new kind of commitment from the rich powers—and it could be established only with strong U.S. leadership. Summoning such leadership is immensely difficult, but America and its allies have no easy options in confronting failed states. They cannot wish away the problem that chaotic power vacuums can pose. They cannot fix it with international institutions as they currently exist. And they cannot sensibly wish for a unilateral American imperium. They must either mold the international machinery to address the problems of their times, as their predecessors did in creating the U.N., the World Bank, and the IMF after World War II. Or they can muddle along until some future collection of leaders rises to the challenge.

SEBASTIAN MALLABY is an editorial writer and columnist for *The Washington Post*.

Nasty, Brutish and Long: America's War on Terrorism

"A critical question as the United States enters this new 'cold war' is whether it has learned the lessons of the last—or whether it is destined to repeat its mistakes."

Ivo H. Daalder and James M. Lindsay

The post-Cold War era ended abruptly on the morning of September 11, 2001. From the moment terrorists turned jetliners into weapons of mass destruction, the United States was inescapably engaged in a new "war" against global terrorism. The Bush administration now intends to make that war the central organizing principle of America's foreign and defense policies.

This war is not like the one against Iraq a decade ago, when the United States and its allies had a clear territorial objective that could be swiftly achieved. It is also not like the war over Kosovo in 1999, in which the Serbs relented after 78 days of bombing Yugoslavia and NATO suffered no combat deaths. And while the attacks on New York and Washington immediately brought to mind memories of Pearl Harbor, the United States campaign against terrorism will not be like America's effort to force Japan's unconditional surrender.

The campaign against terrorism is instead much more like the cold war of the past century. Like the fight against Soviet communism, today's campaign against terrorism is likely to be nasty, brutish, and long. Because of the diverse nature of the threat, the United States has no clear vision of when or how the war will end. Complete success in the military operations in Afghanistan will not necessarily mean victory. Osama bin Laden's Al Qaeda network of terrorists extends well beyond Afghanistan. It could easily reconstitute itself even if the United States captures or kills bin Laden and his lieutenants. Future attacks might even involve the use of sophisticated germ warfare or radiological weapons, if not nuclear weapons.

As at the start of the cold war, the United States response has begun with the arduous task of assembling a global coalition. President Harry Truman's rousing call in 1947 "to support free peoples who are resisting attempted subjugation by armed minorities or by outside pressures" set the course of United States history for the next four decades. President Bush's invitation to every nation to join the United States in "civilization's fight" was phrased as expansively—and intended to be as enduring. In the new war against terrorism the United States also faces ideologically motivated foes who do not shrink from death. America's fight will end only when, as Defense Secretary Donald Rumsfeld said, Americans can once again get on with their daily lives without fear or thought of a possible terrorist attack. That is a tall order.

A critical question as the United States enters this new "cold war" is whether it has learned the lessons of the last—or whether it is destined to repeat its mistakes. Will Washington again overemphasize military force to achieve its goals and give short shrift to the non-military instruments of statecraft? Will it again focus so narrowly on battle that it forgets other important foreign-policy goals? Will it cut deals today to gain support from other nations that will return to haunt it down the road—in much the same way that supporting the shah led to a deeply hostile Iran and arming Afghan rebels to fight the Soviet Union contributed to the terrorist threat the United States faces today? Will it repudiate its own values at home as it tries to fend off an enemy abroad?

It is crucial that the United States fight its new war against terrorism with the dedication and vigor that President Bush has promised. It is also crucial that it fight that war wisely. Washington must recognize the complexities of its new fight—and the pitfalls that lie before it.

THE FIRST PHASE

The Bush administration's campaign against terrorism will occur in stages and on multiple fronts. Much of the fight will be conducted through diplomatic pressure; economic, financial, and political sanctions; and intelligence and law enforcement cooperation. But the first phase—capturing or killing bin Laden, destroying his Al Qaeda network in Afghanistan, and deposing the Taliban regime—will be predominantly military.

The administration launched the first phase of its military campaign on October 7, when United States and British forces struck from the air targets in Afghanistan. Administration officials understandably declined to spell out their military strategy in any detail, but early on it appeared to have three components: a Kosovo-style strategic bombing of military targets to weaken the Taliban's hold on power, Somalia-like commando raids to wipe out the terrorists holed up in the unforgiving countryside, and Nicaragua-like support for the Taliban's adversaries (especially the Northern Alliance).

The first weeks of the campaign showed just how difficult things could be. Although United States and British forces quickly destroyed obvious Taliban and Al Qaeda targets, they also hit several civilian sites. These accidents raised tensions within the international coalition the White House had painstakingly assembled in support of the operation, and especially with neighboring countries, such as Pakistan, that fear pro-Taliban sentiments within their own societies. Both bin Laden and Taliban leader Mullah Mohammed Omar escaped capture, perhaps by fleeing to remote caves and valleys. But they may also have taken refuge somewhere in the city of Khandahar, calculating that they would be safe from attack because of the American desire to avoid hitting civilians.

At the same time, United States and British forces initially refrained from attacking front-line Taliban troops around Kabul for fear that doing so would enable Northern Alliance forces to capture Afghanistan's capital before Washington could assemble a new pan-Afghan government. But given the fractious nature of Afghani politics, both within and across ethnic groups, it was far from clear that the political campaign could ever catch up with the military one. When attacks on front-line Taliban forces finally began in the third week of the campaign, they did not produce immediate gains by Northern Alliance forces.

With luck, the military campaign against Afghanistan will end in a matter of weeks or months—with bin Laden and his network inside the country eliminated and the Taliban regime toppled from power. But achieving this set of objectives will require a major and prolonged undertaking with significant costs. And when it succeeds, the campaign against terrorism that Bush promised will only have begun. Just as the Korean War blunted communist expansion but did not end it, the administration will need to turn to a long, grinding, difficult and expensive campaign to disrupt, deter, and defeat terrorist operations elsewhere in the world. And while military force will continue to play some role in this effort, it will be a distinctly secondary role.

MAINTAINING THE INTERNATIONAL COALITION

Ultimately, for the Bush administration to succeed in its campaign against terrorism it must push ahead on three other fronts. First, it must maintain the anti-terrorism coalition it has assembled in support of military operations in Afghanistan. The coalition is critical because the United States cannot defeat terrorism on its own: it needs other countries to share information about terrorist activities; impose tighter controls over illicit money, weapons and technology flows; isolate and pressure states that sponsor and support terrorists; and strike militarily if targets for action present themselves.

Unfortunately for the Bush administration, the anti-terrorism coalition is not robust. In the Persian Gulf war, more than two dozen countries, including several Arab nations, contributed troops to the fighting. In contrast, American and British forces carried out the initial military operations in Afghanistan alone. Four other countries—Australia, Canada, France, and Germany—have offered to contribute forces at some future point. But significantly, no troops from the Arab or Islamic world participated in the fighting. Only Oman and Pakistan allowed their territory to be used as staging areas for thrusts into Afghan territory. Saudi Arabia, America's main ally in the gulf and host to a large United States Air Force strike force, refused to allow the United States to use its territory as a base for attack (although the air war is coordinated from the United States air control facility at the Prince Sultan airbase located in the middle of the Arabian desert).

The coalition's lack of robustness reflects concerns among the coalition partners over what the campaign against terrorism means for them. Middle Eastern and Islamic governments are crucial to the coalition's success, if only because so many terror groups originate on their soil. These governments fear, however, that joining with Washington will inflame anti-American sentiment in their own societies. Nor are Islamic countries the only ones unsure of how far to follow Washington. Although NATO invoked the organization's Article V provision on mutual defense for the first time 24 hours after the September 11 attacks—ironically, turning an alliance designed to ensure a United States defense of Europe into one in which Europe would help defend the United States—some European countries worry that the United States will go too far in its fight against terrorism.

The issue most likely to fracture the coalition is Iraq. Before September 11, critics speculated that the Bush administration was spoiling for an opportunity to clear up unfinished business from the elder Bush's presidency and would seek to remove Saddam Hussein from power. In the days immediately following the attacks on the World Trade Center and the Pentagon, key administration officials argued for a broad military response that would include Saddam's removal as one of its objectives. President Bush quickly ruled out that option. But in doing so, he embraced an "Afghanistan-first" approach—not the "Afghanistan-only" policy that many in Europe and elsewhere support.

Making Iraq the subject of military attack in a second phase of the campaign against terrorism poses problems for the Bush administration. The United States would almost certainly have to carry out the attack on its own and perhaps even without access to bases in the Persian Gulf area, making it far more difficult to win. Middle Eastern countries argue that attacking Iraq will inflame Arab public opinion and make bin Laden's case that the United States is waging war against Islam. Russia, which has provided Washington with considerable intelligence cooperation since the crisis began, has good relations with Baghdad. Most European governments have long opposed Washington's Iraqi policy. To make matters worse for the administration, these same constraints hold with respect to using military force against Iran or Syria, two countries that have actively sponsored and harbored terrorist groups like Lebanon-based Hezbollah.

Ultimately, the fight against global terrorism is one that the United States cannot win on its own.

Iraq, Iran, and Syria are not the only issues that could fracture the antiterrorism coalition. Should military operations in Afghanistan drag on, or result in large numbers of civilian deaths, the Bush administration could find itself under increasing pressure abroad to end the mission prematurely.

The challenge facing the Bush administration in the near term is to strike the proper balance between its short-term military objectives in Afghanistan and elsewhere and its longer term objective to sustain the international cooperation necessary to conduct a successful fight against global terrorism. In doing so, it will attempt to make the mission define the coalition, rather than letting the coalition define the mission. But it may then find itself confronting what every administration fears: what it wants to do, and perhaps should do, does not mesh with what it can do on its own. There may be times in the campaign against terrorism—as during the cold war itself—when going it alone is both necessary and desirable, but this should be the exception rather than the rule. Ultimately, the fight against global terrorism is one that the United States cannot win on its own.

SECURING THE HOMELAND

The second step Washington must take is to improve homeland security. Much of the focus will understandably be on spending more money on the problem, but the immediate challenge will be to ensure that money is spent wisely. And here the core challenge is to organize the government so that it is more effective in providing homeland security. As Dwight D. Eisenhower aptly noted, "although organization cannot make a genius out of an incompetent. . .disorganization can scarcely fail to result in inefficiency and can easily lead to disaster."

President Bush moved swiftly to address the organizational issue. In a September 20 address to Congress, he announced that he would appoint Pennsylvania Governor Tom Ridge head of a new Office of Homeland Security in the White House. The executive order detailing Ridge's duties also created a Homeland Security Council (HSC), modeled after the National Security Council. The HSC's members will consist of the president, vice president, and key cabinet members and agency heads who will advise and assist the president on all aspects of homeland security.

Critics countered that a White House coordinator, even one who was a friend of the president, could not begin to meet the challenge facing the country. They argued that Ridge would have clout only if he were given control of agency budgets or was put in charge of a newly created, cabinet-level department for homeland security that consolidated existing government operations. Proposals along these lines work better in theory than in practice, however. Contrary to the critics, control of budgets or command of an agency is not necessary, or even sufficient, to exercise power in the federal government. National security advisers possess neither, yet no one doubts their authority.

Nor is centralization necessarily the proper prescription. Homeland security to a considerable extent requires decentralization—where the decisions made by people on the "front lines" matter as much if not more than what is decided in Washington. Customs agents need to know what to look for at the border, Coast Guard commanders need to know which ships to interdict, and Immigration and Naturalization Service officers need to know who is to be barred entry. Intelligence officers need to know which pieces of information culled from an overload of data fit together to enable pre-emptive actions. Hospital emergency-room doctors need to know what symptoms indicate possible exposure to a biological attack. Trying to cram these various agencies, and their diverse missions, into a single organization could make the government less effective in battling terrorism, not more.

Another problem with centralization proposals is the sheer number of federal agencies with a stake in counterterrorism—a number that ranges from 46 to 151, depending on who counting. Short of making the entire federal government a counterterrorism agency, that means that any consolidation must be selective. Some agencies and functions critical to the counterterrorism task cannot, by their very nature, be consolidated. The Federal Bureau of Investigation must remain in the Justice Department (where, incidentally, it often resists the attorney general's direction). Further complicating matters is that formal consolidation does not guarantee effective integration. The Department of Energy is the classic cautionary lesson: it was created in 1977 to bring a variety of units under one umbrella, but a quarter of a century later its integration remains far from complete and its effectiveness often questioned. Finally, giving

Ridge command of his own agency would likely destroy his ability to be an "honest broker" who could coordinate conflicting agency demands. Instead, he would, in the eyes of other agencies, simply become another bureaucratic competitor for money and influence.

Thus, the Bush approach of having Ridge coordinate domestic agencies much as the national security adviser coordinates foreign policy agencies makes more sense. His job certainly is more difficult in one key respect: the national security adviser must worry about getting coordinated information to the president in a timely fashion, whereas Ridge must concern himself mainly with how the agencies operate in the field. One factor working in his favor is that September 11 made counterterrorism a priority across all agencies. They not only know that it is a critical mission but also that it is the key to bigger budgets and more authority. The challenge facing Ridge is to forge the channels of formal and informal agency cooperation where they do not exist today—both among domestic agencies and between them and the national security apparatus. Within this coordination framework some agency consolidation may make sense (for example, combining agencies with closely related functions such as the Customs Service, Border Patrol and Coast Guard).

Even if Washington gets organizational matters right, it will not be enough. It must also take major steps to reduce the country's vulnerability to attack. One obvious need is better control over America's borders. Several of the September 11 hijackers were in the United States on expired student visas. Others entered even though they were suspected of ties to Al Qaeda. But to speak of better border enforcement is to acknowledge the difficulty of the task. Millions of people enter the United States each year, legally and illegally, and only a few have any interest in committing terrorist attacks. The United States—Canada border is 4,000 miles long and in most places is uncontrolled-the Border Patrol had only one officer for every 12 miles of border before September 11. America's 2,000-mile-long border with Mexico is also notorious for its porosity.

Another obvious need is to make the country's transportation networks more secure. Congress and the White House took initial steps in this direction immediately after September 11 by tightening aviation security. Much more remains to be done to make rail and vehicular traffic less vulnerable. Reports that Al Qaeda operatives had obtained licenses to drive trucks carrying hazardous materials points only to the beginning of the havoc that terrorists could wreak using ordinary ground transport. And the United States needs to improve the transportation system's equivalent in cyberspace. The Bush administration took an initial organizational step in this direction by appointing a special adviser to the president for cybersecurity, but more must be done to persuade private actors to make their computer networks more secure. That raises the difficult question of who should pay for this "security tax" and whether protection will be best generated by government mandates or incentives.

Washington also needs to improve the ability of federal, state, and local governments to respond once a terrorist attack occurs—especially chemical, biological, and radiological attacks. The anthrax incidents that followed the September 11 attacks broke the taboo against using such weapons and possibly foreshadowed much more devastating future attacks. Congress and the Bush administration have already responded to this need to some extent, especially by deciding to stockpile additional vaccines and antibiotics for biological attacks using communicable diseases such as smallpox. But perhaps just as important, the initial anthrax attacks have made clear the importance of creating a more effective organizational structure for responding. The government's initial response to the release of anthrax in Senate Majority Leader Tom Daschle's office was marked by inconsistent and conflicting statements about the extent of the danger. And federal credibility suffered when two postal workers died of pulmonary anthrax after government officials failed, regrettably though perhaps understandably, to anticipate that Daschle's anthrax-laden letter might have contaminated mailrooms along the way.

ADDRESSING ANTI-AMERICANISM ABROAD

The campaign against terrorism must also address the sources of the intense anti-Americanism that now roils the Arab and Islamic world and forms the backdrop for Al Qaeda attacks. Hatred of the United States is not peculiar to the Middle East, nor does it translate directly into a desire to launch terrorist attacks. The relationship between the two is more complicated and indirect, akin in many ways to that between oxygen and fire. Oxygen does not cause fires—the spark must come from somewhere else—but fire requires oxygen to rage. In the same fashion, terrorists need anti-American sentiment. It provides them with recruits, and more important, it provides them with people willing to give aid and comfort.

But how can the United States cut off oxygen to the fires of anti-Americanism, especially when the justifiable military operation in Afghanistan and the support it has received from ruling elites in the Arab and Islamic world is likely to feed them? One strategy is to redouble United States efforts to limit and resolve conflicts around the world, especially the one between Israel and the Palestinians. Again, these conflicts did not cause the attacks on the World Trade Center and the Pentagon. They do, however, contribute to the anger that terrorists manipulate to their own, despicable ends.

The payoff from this strategy is questionable; it is easy to call for conflict resolution and hard to deliver it. A Middle East peace deal has been the holy grail of the last six presidential administrations. The escalating violence of the past twelve months has only made it more difficult to persuade Israelis and Palestinians to speak of concession and compromise. The conflict between India and Pakistan over Kashmir—an issue of great concern to Pakistani militants—has dragged on for more than half a century. New Delhi and Islamabad are not about to toss aside their longstanding differences simply because Washington thinks they should. Nor will being seen as actively pursuing peace necessarily do much to deflect Arab and Islamic anger. Former President Bill Clinton's feverish and ultimately unsuccessfully effort in 2000 to broker a Middle East peace deal passed largely unnoticed in the Arab and Islamic world. The frequent complaints about United States policy seldom mention that Washington liberated Kuwait, saved hundreds of thousands of starving Somalis in the early 1990s, fought not one but two wars to protect Muslims in the Balkans, and provided more humanitarian aid than any other country to the people of Afghanistan.

Washington will also need to intensify its support for democracy and economic development—especially in areas like Central Asia and the Arab world, where repression and poverty provide breeding grounds for anti-American sentiments. Prosperous democratic countries are America's best allies against terrorism. But again, this strategy is easier to urge than to carry out. One problem is that while Washington generally knows how to promote economic development, its success in promoting political development is spotty—witness the record in Vietnam, Somalia, and Haiti. The other problem is whether the United States can gracefully extract itself from its current political commitments in the Arab and Islamic world. Calls are likely to mount in the coming months for the United States to distance itself from Cairo and Riyadh unless they enact democratic reforms. Yet that policy could well endanger other important United States foreign policy objectives—not the least securing Egyptian and Saudi cooperation in the fight against terrorism. And it may produce regime change, though not necessarily one that comports with American values. To judge by the slogans of dissidents in Saudi Arabia, greater mass political participation will not deliver a Westminster-style democracy but rather an Iranian-style theocracy steeped in anti-Americanism.

> *Americans must [recognize] that military force alone is not enough; pretending that it is takes us down a dangerous road.*

The focus on big-picture policies like conflict resolution and political development should not obscure small-picture policies that take aim at anti-Americanism. A key here is a concerted public diplomacy campaign, much like the one the United States waged vigorously in the early years of the cold war. Then United States used the media and exchange programs to refute the lies of communist rulers. Today, as President Bush noted in his October 11 press conference, it needs to make a better case for America and to argue that bin Laden represents a perversion of Islam and threatens the stability of all civilizations.

This public diplomacy should use the public relations tools of the cold war, including radio broadcasts, magazines, and cultural centers. But the strategy should also adapt to the times. Bin Laden shrewdly used the Arab-language satellite television channel Al Jazeera to broadcast his hatred to Arabs around the world. Before September 11, however, American officials seldom sought to appear on Al Jazeera and other media outlets

in the Arab world to present their case. And the United States needs to press its Middle Eastern allies to do their share in discrediting anti-Americanism. In recent years countries such as Egypt and Saudi Arabia have bought social peace for themselves by ignoring and even encouraging growing hatred toward the United States. Criticizing President Hosni Mubarak on the streets of Cairo will get you arrested; criticizing America will get you applause. Washington must press these governments to confront dangerous distortions of the truth rather than to stoke anti-Americanism.

Hard Lessons

Because the fight against terrorism is a new cold war, some key lessons of the old one are worth remembering. Although the United States ultimately triumphed in that conflict, it made critical mistakes along the way that it must now seek to avoid.

Americans must begin by recognizing that military force alone is not enough; pretending that it is takes us down a dangerous road. The militarization of containment—which elevated military responses over all other tools of policy and culminated in the disastrous United States engagement in Vietnam—undermined the American public's trust in its government. It also weakened the very alliances on which the United States depended to win its confrontation with the Soviet Union.

In calling on the nation to conduct a "war" against terrorism, the Bush administration has helped create the impression that America's victory will be a military one. Again, there is a role for military force in today's fight—to destroy the terrorist infrastructure in Afghanistan, compel an end to continued state-sponsorship of terrorism, and pre-empt any planned terrorist attacks. But the force of arms alone cannot defeat terrorism. The United States also needs better law enforcement, enhanced intelligence, focused diplomacy, and targeted sanctions to succeed.

Second, the United States must avoid creating new threats even as it seeks to defeat the current ones. The most immediate need is to end the cycle of violence in Afghanistan. In the 1980s, the United States armed the mujahideen to help them to defeat the Soviet invaders. When Moscow finally withdrew its troops in 1989, Washington walked away from the scene. We are now living with the consequences.

Washington must not repeat this mistake. The threat of further destabilization in Afghanistan is real. Nearly constant war for more than two decades—first against the Soviet Union and then among Afghans themselves—has created more than 1.5 million refugees and left many hundreds of thousands of others lacking sufficient food or adequate shelter. United States military operations will make matters worse, notwithstanding the efforts to drop relief supplies to those in need. Should Afghanistan's unrest spread, the consequences for neighboring Pakistan—an internally divided and failing state in possession of nuclear weapons—could be severe. The last thing the United States wants or needs is for Islamic fundamentalist sympathizers of bin Laden to take over Pakistan.

United States military operations need to be followed by a concerted effort to stabilize Afghanistan if the United States hopes to discourage its future use as a terrorist haven. Fortunately, the Bush administration, despite a deep-seated hostility to nation-building, has signaled that it understands it must be engaged in Afghanistan. Even if the White House insists that it is not engaged in "nation building" but rather the "stabilization of a future government," these efforts are necessary to increase the prospects for regional stability. The Bush administration wants the United Nations to play a major role in reconstructing Afghanistan, thereby spreading responsibility and perhaps making success more likely, but extensive American involvement is inescapable.

But avoiding policies that inadvertently create new threats also means not carelessly sacrificing other important foreign policy interests and values to serve the cause of defeating global terrorism. During the cold war, Washington made battling the spread of communism an all-consuming fight. Other priorities and interests were jettisoned when they conflicted with the objective of holding Soviet expansion at bay. As a result, the United States embraced unsavory characters (from Spain's Franco to Zaire's Mobutu and Chile's Pinochet), engaged in highly questionable conduct (from assassinations to secret coups), wasted billions of dollars on dead-end interventions and superfluous weapon systems, and ignored a long list of other foreign policy challenges (from human rights to weapons proliferation to the environment).

The same risks exist today. To solicit support for its anti-terror coalition, the Bush administration has lifted sanctions imposed on Pakistan for testing nuclear weapons, begun to side rhetorically with Russia in its brutal fight in Chechnya, and sought assistance from key state sponsors of terrorism such as Iran and Sudan. These and other steps may be needed to address a short-term emergency, but they may come at a hefty price in the long term.

Third, the United States must not needlessly sacrifice its civil liberties as it combats the terrorist threat. The willingness with which Washington and the country as a whole trampled on cherished civil liberties during the McCarthy years of the early cold war is too well known to merit repeating. Perhaps most remarkable about how Americans reacted in the first month after the September 11 attacks was how quickly they acknowledged the importance of not forfeiting America's basic principles as the country met its new challenge. Politicians, the media, and the public emphasized the importance of tolerance. Civil libertarians challenged the merits of some of the administration's proposed changes to law enforcement authority. The question that remains is whether this commitment to fundamental principles of liberty will withstand future terrorist attacks.

In the end, America's campaign to restore the margin of security it enjoyed before September 11 will be neither easy nor quick. The defeat of terrorism will not be achieved or celebrated in one grand moment. There will be no V-E or V-J day, no ticker-tape parade along Fifth Avenue. America's victory will be piecemeal. Every day the United States goes without a terrorist attack will be a triumph. But even that limited achievement requires waging the fight against terrorism with a clear memory that the last war demanded much more than just battlefield bravery. Otherwise, any victory will be tarnished by the new problems the United States will reap.

Reprinted from *Current History*, December 2001, pp. 403-409. © 2001 by Current History, Inc. Reprinted by permission.

Nuclear Nightmares

**Experts on terrorism and proliferation agree on one thing:
Sooner or later, an attack will happen here. When
and how is what robs them of sleep.**

By Bill Keller

The panic that would result from contaminating the Magic Kingdom with a modest amount of cesium would probably shut the place down for good and constitute a staggering strike at Americans' sense of innocence.

Not If But When Everybody who spends much time thinking about nuclear terrorism can give you a scenario, something diabolical and, theoretically, doable. Michael A. Levi, a researcher at the Federation of American Scientists, imagines a homemade nuclear explosive device detonated inside a truck passing through one of the tunnels into Manhattan. The blast would crater portions of the New York skyline, barbecue thousands of people instantly, condemn thousands more to a horrible death from radiation sickness and—by virtue of being underground—would vaporize many tons of concrete and dirt and river water into an enduring cloud of lethal fallout. Vladimir Shikalov, a Russian

nuclear physicist who helped clean up after the 1986 Chernobyl accident, envisioned for me an attack involving highly radioactive cesium-137 loaded into some kind of homemade spraying device, and a target that sounded particularly unsettling when proposed across a Moscow kitchen table—Disneyland. In this case, the human toll would be much less ghastly, but the panic that would result from contaminating the Magic Kingdom with a modest amount of cesium—Shikalov held up his teacup to illustrate how much—would probably shut the place down for good and constitute a staggering strike at Americans' sense of innocence. Shikalov, a nuclear enthusiast who thinks most people are ridiculously squeamish about radiation, added that personally he would still be happy to visit Disneyland after the terrorists struck, although he would pack his own food and drink and destroy his clothing afterward.

Another Russian, Dmitry Borisov, a former official of his country's atomic energy ministry, conjured a suicidal pilot. (Suicidal pilots, for obvious reasons, figure frequently in these fantasies.) In Borisov's scenario, the hijacker dive-bombs an Aeroflot jetliner into the Kurchatov Institute, an atomic research center

in a gentrifying neighborhood of Moscow, which I had just visited the day before our conversation. The facility contains 26 nuclear reactors of various sizes and a huge accumulation of radioactive material. The effect would probably be measured more in property values than in body bags, but some people say the same about Chernobyl. Maybe it is a way to tame a fearsome subject by Hollywoodizing it, or maybe it is a way to drive home the dreadful stakes in the arid-sounding business of nonproliferation, but in several weeks of talking to specialists here and in Russia about the threats an amateur evildoer might pose to the homeland, I found an unnerving abundance of such morbid creativity. I heard a physicist wonder whether a suicide bomber with a pacemaker would constitute an effective radiation weapon. (I'm a little ashamed to say I checked that one, and the answer is no, since pacemakers powered by plutonium have not been implanted for the past 20 years.) I have had people theorize about whether hijackers who took over a nuclear research laboratory could improvise an actual nuclear explosion on the spot. (Expert opinions differ, but it's very unlikely.) I've been instructed how to disperse plutonium into the ventilation system of an office building.

The realistic threats settle into two broad categories. The less likely but far more devastating is an actual nuclear explosion, a great hole blown in the heart of New York or Washington, followed by a toxic fog of radiation. This could be produced by a black-market nuclear warhead procured from an existing arsenal. Russia is the favorite hypothetical source, although Pakistan, which has a program built on shady middlemen and covert operations, should not be overlooked. Or the explosive could be a homemade device, lower in yield than a factory nuke but still creating great carnage.

The second category is a radiological attack, contaminating a public place with radioactive material by packing it with conventional explosives in a "dirty bomb" by dispersing it into the air or water or by sabotaging a nuclear facility. By comparison with the task of creating nuclear fission, some of these schemes would be almost childishly simple, although the consequences would be less horrifying: a panicky evacuation, a gradual increase in cancer rates, a staggeringly expensive cleanup, possibly the need to demolish whole neighborhoods. Al Qaeda has claimed to have access to dirty bombs, which is unverified but entirely plausible, given that the makings are easily gettable.

Nothing is really new about these perils. The means to inflict nuclear harm on America have been available to rogues for a long time. Serious studies of the threat of nuclear terror date back to the 1970's. American programs to keep Russian nuclear ingredients from falling into murderous hands—one of the subjects high on the agenda in President Bush's meetings in Moscow this weekend—were hatched soon after the Soviet Union disintegrated a decade ago. When terrorists get around to trying their first nuclear assault, as you can be sure they will, there will be plenty of people entitled to say I told you so.

All Sept. 11 did was turn a theoretical possibility into a felt danger. All it did was supply a credible cast of characters who hate us so much they would thrill to the prospect of actually doing it—and, most important in rethinking the probabilities, would be happy to die in the effort. All it did was give our nightmares legs.

Tom Ridge cupped his hands prayerfully and pressed his fingertips to his lips. "Nuclear," he said simply.

And of the many nightmares animated by the attacks, this is the one with pride of place in our experience and literature—and, we know from his own lips, in Osama bin Laden's aspirations. In February, Tom Ridge, the Bush administration's homeland security chief, visited The Times for a conversation, and at the end someone asked, given all the things he had to worry about—hijacked airliners, anthrax in the mail, smallpox, germs in crop-dusters—what did he worry about most? He cupped his hands prayerfully and pressed his fingertips to his lips. "Nuclear," he said simply.

My assignment here was to stare at that fear and inventory the possibilities. How afraid should we be, and what of, exactly? I'll tell you at the outset, this was not one of those exercises in which weighing the fears and assigning them probabilities laid them to rest. I'm not evacuating Manhattan, but neither am I sleeping quite as soundly. As I was writing this early one Saturday in April, the floor began to rumble and my desk lamp wobbled precariously. Although I grew up on the San Andreas Fault, the fact that New York was experiencing an earthquake was only my second thought. The best reason for thinking it won't happen is that it hasn't hap-

pened yet, and that is terrible logic. The problem is not so much that we are not doing enough to prevent a terrorist from turning our atomic knowledge against us (although we are not). The problem is that there may be no such thing as "enough."

25,000 Warheads, and It Only Takes One My few actual encounters with the Russian nuclear arsenal are all associated with Thomas Cochran. Cochran, a physicist with a Tennessee lilt and a sense of showmanship, is the director of nuclear issues for the Natural Resources Defense Council, which promotes environmental protection and arms control. In 1989, when glasnost was in flower, Cochran persuaded the Soviet Union to open some of its most secret nuclear venues to a roadshow of American scientists and congressmen and invited along a couple of reporters. We visited a Soviet missile cruiser bobbing in the Black Sea and drank vodka with physicists and engineers in the secret city where the Soviets first produced plutonium for weapons.

Not long ago Cochran took me cruising through the Russian nuclear stockpile again, this time digitally. The days of glasnost theatrics are past, and this is now the only way an outsider can get close to the places where Russians store and deploy their nuclear weapons. On his office computer in Washington, Cochran has installed a detailed United States military map of Russia and superimposed upon it high-resolution satellite photographs. We spent part of a morning mouse-clicking from missile-launch site to submarine base, zooming in like voyeurs and contemplating the possibility that a terrorist could figure out how to steal a nuclear warhead from one of these places.

"Here are the bunkers," Cochran said, enlarging an area the size of a football stadium holding a half-dozen elongated igloos. We were hovering over a site called Zhukovka, in western Russia. We were pleased to see it did not look ripe for a hijacking.

"You see the bunkers are fenced, and then the whole thing is fenced again," Cochran said. "Just outside you can see barracks and a rifle range for the guards. These would be troops of the 12th Main Directorate. Somebody's not going to walk off the street and get a Russian weapon out of this particular storage area."

In the popular culture, nuclear terror begins with the theft of a nuclear weapon. Why build one when so many are lying around for the taking? And stealing tends to make better drama than engineering. Thus the stolen nuke has been a staple in the literature at least since 1961, when Ian Fleming published "Thunderball," in which the malevolent Spectre (the Special Executive for Counterintelligence, Terrorism, Revenge and Extortion, a strictly mercenary and more technologically sophisticated precursor to al Qaeda) pilfers a pair of atom bombs from a crashed NATO aircraft. In the movie version of Tom Clancy's thriller "The Sum of All Fears," due in theaters this week, neo-Nazis get their hands on a mislaid Israeli nuke, and viewers will get to see Baltimore blasted to oblivion.

Eight countries are known to have nuclear weapons—the United States, Russia, China, Great Britain, France, India, Pakistan and Israel. David Albright, a nuclear-weapons expert and president of the Institute for Science and International Security, points out that Pakistan's program in particular was built almost entirely through black markets and industrial espionage, aimed at circumventing Western export controls. Defeating the discipline of nuclear nonproliferation is ingrained in the culture. Disaffected individuals in Pakistan (which, remember, was intimate with the Taliban) would have no trouble finding the illicit channels or the rationalization for diverting materials, expertise—even, conceivably, a warhead.

But the mall of horrors is Russia, because it currently maintains something like 15,000 of the world's (very roughly) 25,000 nuclear warheads, ranging in destructive power from about 500 kilotons, which could kill a million people, down to the one-kiloton land mines that would be enough to make much of Manhattan uninhabitable. Russia is a country with sloppy accounting, a disgruntled military, an audacious black market and indigenous terrorists.

It's easier to take the fuel and build an entire weapon from scratch than it is to make one of these things go off.

There is anecdotal reason to worry. Gen. Igor Valynkin, commander of the 12th Main Directorate of the Russian Ministry of Defense, the Russian military sector in charge of all nuclear weapons outside the Navy, said recently that twice in the past year terrorist groups were caught casing Russian weapons-storage facilities. But it's hard to know how seriously to take this. When I made the rounds of nuclear experts in Russia earlier this year, many were skeptical of these near-miss anecdotes, saying the security forces tend to exaggerate such incidents to dramatize their own prowess (the culprits are always caught) and enhance their budgets. On the whole, Russian and American military experts sound not very alarmed about the vulnerability of Russia's nuclear warheads. They say Russia takes these weapons quite seriously, accounts for them rigorously and guards them carefully. There is no confirmed case of a warhead being lost. Strategic warheads, including the 4,000 or so that President Bush and President Vladimir Putin have agreed to retire from service, tend to be stored in hard-to-reach places, fenced and heavily guarded, and their whereabouts are not advertised. The people who guard them are better paid and more closely vetted than most Russian soldiers.

Eugene E. Habiger, the four-star general who was in charge of American strategic weapons until 1998 and then ran nuclear antiterror programs for the Energy Department, visited several Russian weapons facilities in 1996 and 1997. He may be the only American who has actually entered a Russian bunker and inspected a warhead *in situ*. Habiger said he found the overall level of security comparable to American sites, although the Russians depend more on people than on technology to protect their nukes.

The image of armed terrorist commandos storming a nuclear bunker is cinematic, but it's far more plausible to think of an inside job. No observer of the unraveling Russian military has much trouble imagining that a group of military officers, disenchanted by the humiliation of serving a spent superpower, embittered by the wretched conditions in which they spend much of their military lives or merely greedy, might find a way to divert a warhead to a terrorist for the right price. (The Chechen warlord Shamil Basayev, infamous for such ruthless exploits as taking an entire hospital hostage, once hinted that he had an opportunity to buy a nuclear warhead from the stockpile.) The anecdotal evidence of desperation in the military is plentiful and disquieting. Every year the Russian press provides stories like that of the 19-year-old sailor who went on a rampage aboard an Akula-class nuclear submarine, killing eight people and threatening to blow up the boat and its nuclear reactor; or the five soldiers at Russia's nuclear-weapons test site who killed a guard, took a hostage and tried to hijack an aircraft, or the officers who reportedly stole five assault helicopters, with their weapons pods, and tried to sell them to North Korea.

The Clinton administration found the danger of disgruntled nuclear caretakers worrisome enough that it considered building better housing for some officers in the nuclear rocket corps. Congress, noting that the United States does not build housing for its own officers, rejected the idea out of hand.

If a terrorist did get his hands on a nuclear warhead, he would still face the problem of setting it off. American warheads are rigged with multiple PAL's ("permissive action links")—codes and self-disabling devices designed to frustrate an unauthorized person from triggering the explosion. General Habiger says that when he examined Russian strategic weapons he found the level of protection comparable to our own. "You'd have to literally break the weapon apart to get into the gut," he told me. "I would submit that a more likely scenario is that there'd be an attempt to get hold of a warhead and not explode the warhead but extract the plutonium or highly enriched uranium." In other words, it's easier to take the fuel and build an entire weapon from scratch than it is to make one of these things go off.

Then again, Habiger is not an expert in physics or weapons design. Then again, the Russians would seem to have no obvious reason for misleading him about something that important. Then again, how many times have computer hackers hacked their way into encrypted computers we were assured were impregnable? Then again, how many computer hackers does al Qaeda have? This subject drives you in circles.

The most troublesome gap in the generally reassuring assessment of Russian weapons security is those tactical nuclear warheads—smaller, short-range weapons like torpedoes, depth charges, artillery shells, mines. Although their smaller size and greater number makes them ideal candidates for theft, they have gotten far less attention simply because, unlike all of our long-range weapons, they happen not to be the subject of any formal treaty. The first President Bush reached an informal understanding with President Gorbachev and then with President Yeltsin that both sides would gather and destroy thousands of tactical nukes. But the agreement included no inventories of the stockpiles, no outside monitoring, no verification of any kind. It was one of those trust-me deals that, in the hindsight of Sept. 11, amount to an enormous black hole in our security.

Did I say earlier there are about 15,000 Russian warheads? That number includes, alongside the scrupulously counted strategic warheads in bombers, missiles and submarines, the commonly used estimate of 8,000 tactical warheads. But that figure is at best an educated guess. Other educated guesses of the tactical nukes in Russia go as low as 4,000 and as high as 30,000. We just don't know. We don't even know if the Russians know, since they are famous for doing things off the books. "They'll tell you they've never lost a weapon," said Kenneth Luongo, director of a private antiproliferation group called the Russian-American Nuclear Security Advisory Council. "The fact is, they don't know. And when you're talking about warhead counting, you don't want to miss even one."

And where are they? Some are stored in reinforced concrete bunkers like the one at Zhukovka. Others are deployed. (When the submarine Kursk sank with its 118 crewmen in August 2000, the Americans' immediate fear was for its nuclear armaments. The standard load out for a submarine of that class includes a couple of nuclear torpedoes and possibly some nuclear depth charges.) Still others are supposed to be in the process of being dismantled under terms of various formal and informal arms-control agreements. Some are in transit. In short, we don't really know.

The other worrying thing about tactical nukes is that their anti-use devices are believed to be less sophisticated, because the weapons were designed to be employed in the battlefield. Some of the older systems are thought to have no permissive action links at all, so that setting one off would be about as complicated as hot-wiring a car.

Efforts to learn more about the state of tactical stockpiles have been frustrated by reluctance on both sides to let visitors in. Viktor Mikhailov, who ran the Russian Ministry of Atomic Energy until 1998 with a famous scorn for America's nonproliferation concerns, still insists that the United States programs to protect Russian nuclear weapons and material mask a secret agenda of intelligence-gathering. Americans, in turn, sometimes balk at reciprocal access, on the grounds that we are the ones paying the bills for all these safety upgrades, said the former Senator Sam Nunn, co-author of the main American program for securing Russian nukes, called Nunn-Lugar.

People in the field talk of a nuclear 'conex' bomb, using the name of those shack-size steel containers—2,000 of which enter America every hour on trains, trucks and ships. Fewer than 2 percent are cracked open for inspection.

"We have to decide if we want the Russians to be transparent—I'd call it cradle-to-grave transparency with nuclear material and inventories and so forth," Nunn told me. "Then we have to open up more ourselves. This is a big psychological breakthrough we're talking about here, both for them and for us."

The Garage Bomb One of the more interesting facts about the atom bomb dropped on Hiroshima is that it had never been tested. All of those spectral images of nuclear coronas brightening the desert of New Mexico— those were to perfect the more complicated plutonium device that was dropped on Nagasaki. "Little Boy," the Hiroshima bomb, was a rudimentary gunlike device that shot one projectile of highly enriched uranium into another, creating a critical mass that exploded. The mechanics were so simple that few doubted it would work, so the first experiment was in the sky over Japan.

The closest thing to a consensus I heard among those who study nuclear terror was this: building a nuclear bomb is easier than you think, probably easier than stealing one. In the rejuvenated effort to prevent a terrorist from striking a nuclear blow, this is where most of the attention and money are focused.

A nuclear explosion of any kind "is not a sort of high-probability thing," said a White House official who follows the subject closely. "But getting your hands on enough fissile material to build an improvised nuclear device, to my mind, is the least improbable of them all, and particularly if that material is highly enriched uranium in metallic form. Then I'm really worried. That's the one."

To build a nuclear explosive you need material capable of explosive nuclear fission, you need expertise, you need some equipment, and you need a way to deliver it.

Delivering it to the target is, by most reckoning, the simplest part. People in the field generally scoff at the mythologized suitcase bomb; instead they talk of a "conex bomb," using the name of those shack-size steel containers that bring most cargo into the United States. Two thousand containers enter America every hour, on trucks and trains and especially on ships sailing into more than 300 American ports. Fewer than 2 percent are cracked open for inspection, and the great majority never pass through an X-ray machine. Containers delivered to upriver ports like St. Louis or Chicago pass many miles of potential targets before they even reach customs.

"How do you protect against that?" mused Habiger, the former chief of our nuclear arsenal. "You can't. That's scary. That's very, very scary. You set one of those off in Philadelphia, in New York City, San Francisco, Los Angeles, and you're going to kill tens of thousands of

people, if not more." Habiger's view is "It's not a matter of *if*; it's a matter of *when*"—which may explain why he now lives in San Antonio.

The Homeland Security office has installed a plan to refocus inspections, making sure the 2 percent of containers that get inspected are those without a clear, verified itinerary. Detectors will be put into place at ports and other checkpoints. This is good, but it hardly represents an ironclad defense. The detection devices are a long way from being reliable. (Inconveniently, the most feared bomb component, uranium, is one of the hardest radioactive substances to detect because it does not emit a lot of radiation prior to fission.) The best way to stop nuclear terror, therefore, is to keep the weapons out of terrorist hands in the first place.

Fabricating a nuclear weapon is not something a lone madman—even a lone genius—is likely to pull off in his hobby room.

The basic know-how of atom-bomb-building is half a century old, and adequate recipes have cropped up in physics term papers and high school science projects. The simplest design entails taking a lump of highly enriched uranium, about the size of a cantaloupe, and firing it down a big gun barrel into a second lump. Theodore Taylor, the nuclear physicist who designed both the smallest and the largest American nuclear-fission warheads before becoming a remorseful opponent of all things nuclear, told me he recently looked up "atomic bomb" in the World Book Encyclopedia in the upstate New York nursing home where he now lives, and he found enough basic information to get a careful reader started. "It's accessible all over the place," he said. "I don't

mean just the basic principles. The sizes, specifications, things that work."

Most of the people who talk about the ease of assembling a nuclear weapon, of course, have never actually built one. The most authoritative assessment I found was a paper, "Can Terrorists Build Nuclear Weapons?" written in 1986 by five experienced nuke-makers from the Los Alamos weapons laboratory. I was relieved to learn that fabricating a nuclear weapon is not something a lone madman—even a lone genius—is likely to pull off in his hobby room. The paper explained that it would require a team with knowledge of "the physical, chemical and metallurgical properties of the various materials to be used, as well as characteristics affecting their fabrication; neutronic properties; radiation effects, both nuclear and biological; technology concerning high explosives and/or chemical propellants; some hydrodynamics; electrical circuitry; and others." Many of these skills are more difficult to acquire than, say, the ability to aim a jumbo jet.

The schemers would also need specialized equipment to form the uranium, which is usually in powdered form, into metal, to cast it and machine it to fit the device. That effort would entail months of preparation, increasing the risk of detection, and it would require elaborate safeguards to prevent a mishap that, as the paper dryly put it, would "bring the operation to a close."

Still, the experts concluded, the answer to the question posed in the title, while qualified, was "Yes, they can."

David Albright, who worked as a United Nations weapons inspector in Iraq, says Saddam Hussein's unsuccessful crash program to build a nuclear weapon in 1990 illustrates how a single bad decision can mean a huge setback. Iraq had extracted highly enriched uranium from research-reactor fuel and had, maybe, barely enough for a bomb. But the manager in charge of casting the metal was so afraid the stuff would spill or get contaminated that he decided to melt it in tiny batches. As a result, so much of the uranium was wasted that he ended up with too little for a bomb.

"You need good managers and organizational people to put the elements together," Albright said. "If you do a straight-line extrapolation, terrorists will all get nuclear weapons. But they make mistakes."

On the other hand, many experts underestimate the prospect of a do-it-yourself bomb because they are thinking too professionally. All of our experience with these weapons is that the people who make them (states, in other words) want them to be safe, reliable, predictable and efficient. Weapons for the American arsenal are designed to survive a trip around the globe in a missile, to be accident-proof, to produce a precisely specified blast.

But there are many corners you can cut if you are content with a big, ugly, inefficient device that would make a spectacular impression. If your bomb doesn't need to fit in a suitcase (and why should it?) or to endure the stress of a missile launch; if you don't care whether the explosive power realizes its full potential; if you're willing to accept some risk that the thing might go off at the wrong time or might not go off at all, then the job of building it is immeasurably simplified.

"As you get smarter, you realize you can get by with less," Albright said. "You can do it in facilities that look like barns, garages, with simple machine tools. You can do it with 10 to 15 people, not all Ph.D.'s, but some engineers, technicians. Our judgment is that a gun-type device is well within the capability of a terrorist organization."

All the technological challenges are greatly simplified if terrorists are in league with a country—a place with an infrastructure. A state is much better suited to hire expertise (like dispirited scientists from decommissioned nuclear installations

in the old Soviet Union) or to send its own scientists for M.I.T. degrees.

Thus Tom Cochran said his greatest fear is what you might call a bespoke nuke—terrorists stealing a quantity of weapons-grade uranium and taking it to Iraq or Iran or Libya, letting the scientists and engineers there fashion it into an elementary weapon and then taking it away for a delivery that would have no return address.

That leaves one big obstacle to the terrorist nuke-maker: the fissile material itself.

To be reasonably sure of a nuclear explosion, allowing for some material being lost in the manufacturing process, you need roughly 50 kilograms—110 pounds—of highly enriched uranium. (For a weapon, more than 90 percent of the material should consist of the very unstable uranium-235 isotope.) Tom Cochran, the master of visual aids, has 15 pounds of depleted uranium that he keeps in a Coke can; an eight-pack would be plenty to build a bomb.

> **O**nly 41 percent of Russia's weapon-usable material has been secured... So the barn door is still pretty seriously ajar. We don't know whether any horses have gotten out.

The world is awash in the stuff. Frank von Hippel, a Princeton physicist and arms-control advocate, has calculated that between 1,300 and 2,100 metric tons of weapons-grade uranium exists—at the low end, enough for 26,000 rough-hewed bombs. The largest stockpile is in Russia, which Senator Joseph Biden calls "the candy store of candy stores."

Until a decade ago, Russian officials say, no one worried much about the safety of this material. Vik-

tor Mikhailov, who ran the atomic energy ministry and now presides over an affiliated research institute, concedes there were glaring lapses.

"The safety of nuclear materials was always on our minds, but the focus was on intruders," he said. "The system had never taken account of the possibility that these carefully screened people in the nuclear sphere could themselves represent a danger. The system was not designed to prevent a danger from within."

Then came the collapse of the Soviet Union and, in the early 90's, a few frightening cases of nuclear materials popping up on the black market.

If you add up all the reported attempts to sell highly enriched uranium or plutonium, even including those that have the scent of security-agency hype and those where the material was of uncertain quality, the total amount of material still falls short of what a bomb-maker would need to construct a single explosive.

But Yuri G. Volodin, the chief of safeguards at Gosatomnadzor, the Russian nuclear regulatory agency, told me his inspectors still discover one or two instances of attempted theft a year, along with dozens of violations of the regulations for storing and securing nuclear material. And as he readily concedes: "These are the detected cases. We can't talk about the cases we don't know." Alexander Pikayev, a former aide to the Defense Committee of the Russian Duma, said: "The vast majority of installations now have fences. But you know Russians. If you walk along the perimeter, you can see a hole in the fence, because the employees want to come and go freely."

The bulk of American investment in nuclear safety goes to lock the stuff up at the source. That is clearly the right priority. Other programs are devoted to blending down the highly enriched uranium to a diluted product unsuitable for weapons but good as reactor fuel. The Nuclear Threat Initiative, financed by Ted Turner and led by Nunn, is studying ways to double the rate of this diluting process.

Still, after 10 years of American subsidies, only 41 percent of Russia's weapon-usable material has been secured, according to the United States Department of Energy. Russian officials said they can't even be sure how much exists, in part because the managers of nuclear facilities, like everyone else in the Soviet industrial complex, learned to cook their books. So the barn door is still pretty seriously ajar. We don't know whether any horses have gotten out.

And it is not the only barn. William C. Potter, director of the Center for Nonproliferation Studies at the Monterey Institute of International Studies and an expert in nuclear security in the former Soviet states, said the American focus on Russia has neglected other locations that could be tempting targets for a terrorist seeking bomb-making material. There is, for example, a bomb's worth of weapons-grade uranium at a site in Belarus, a country with an erratic president and an anti-American orientation. There is enough weapons-grade uranium for a bomb or two in Kharkiv, in Ukraine. Outside of Belgrade, in a research reactor at Vinca, sits sufficient material for a bomb—and there it sat while NATO was bombarding the area.

"We need to avoid the notion that because the most material is in Russia, that's where we should direct all of our effort," Potter said. "It's like assuming the bank robber will target Fort Knox because that's where the most gold is. The bank robber goes where the gold is most accessible."

Weapons of Mass Disruption The first and, so far, only consummated act of nuclear terrorism took place in Moscow in 1995, and it was scarcely memorable. Chechen rebels obtained a canister of cesium, possibly from a hospital they had commandeered a few months before. They hid it in a Moscow park famed for its weekend flea market and called the press. No one was hurt. Authorities treated the incident discreetly, and a surge of panic quickly passed.

The story came up in virtually every conversation I had in Russia about nuclear terror, usually to illustrate that even without splitting atoms and making mushroom clouds a terrorist could use radioactivity—and the fear of it—as a potent weapon.

The idea that you could make a fantastic weapon out of radioactive material without actually producing a nuclear bang has been around since the infancy of nuclear weaponry. During World War II, American scientists in the Manhattan Project worried that the Germans would rain radioactive material on our troops storming the beaches on D-Day. Robert S. Norris, the biographer of the Manhattan Project director, Gen. Leslie R. Groves, told me that the United States took this threat seriously enough to outfit some of the D-Day soldiers with Geiger counters.

No country today includes radiological weapons in its armories. But radiation's limitations as a military tool—its tendency to drift afield with unplanned consequences, its long-term rather than short-term lethality—would not necessarily count against it in the mind of a terrorist. If your aim is to instill fear, radiation is anthrax-plus. And unlike the fabrication of a nuclear explosive, this is terror within the means of a soloist.

If your aim is to instill fear, radiation is anthrax-plus. And unlike the fabrication of a nuclear explosive, this is terror within the means of a soloist.

That is why, if you polled the universe of people paid to worry about weapons of mass destruction (W.M.D., in the jargon), you would find a general agreement that this is probably the first thing we'll see. "If there is a W.M.D. attack in the next year, it's likely to be a radiological attack," said Rose Gottemoeller, who handled Russian nuclear safety in the Clinton administration and now follows the subject for the Carnegie Endowment. The radioactive heart of a dirty bomb could be spent fuel from a nuclear reactor or isotopes separated out in the process of refining nuclear fuel. These materials are many times more abundant and much, much less protected than the high-grade stuff suitable for bombs. Since Sept.11, Russian officials have begun lobbying hard to expand the program of American aid to include protection of these lower-grade materials, and the Bush administration has earmarked a few million dollars to study the problem. But the fact is that radioactive material suitable for terrorist attacks is so widely available that there is little hope of controlling it all.

The guts of a dirty bomb could be cobalt-60, which is readily available in hospitals for use in radiation therapy and in food processing to kill the bacteria in fruits and vegetables. It could be cesium-137, commonly used in medical gauges and radiotherapy machines. It could be americium, an isotope that behaves a lot like plutonium and is used in smoke detectors and in oil prospecting. It could be plutonium, which exists in many research laboratories in America. If you trust the security of those American labs, pause and reflect that the investigation into the great anthrax scare seems to be focused on disaffected American scientists.

Back in 1974, Theodore Taylor and Mason Willrich, in a book on the dangers of nuclear theft, examined things a terrorist might do if he got his hands on 100 grams of plutonium—a thimble-size amount. They calculated that a killer who dissolved it, made an aerosol and introduced it into the ventilation system of an office building could deliver a lethal dose to the entire floor area of a large skyscraper. But plutonium dispersed outdoors in the open air, they estimated, would be far less effective. It would blow away in a gentle wind.

The Federation of American Scientists recently mapped out for a Congressional hearing the consequences of various homemade dirty bombs detonated in New York or Washington. For example, a bomb made with a single footlong pencil of cobalt from a food irradiation plant and just 10 pounds of TNT and detonated at Union Square in a light wind would send a plume of radiation drifting across three states. Much of Manhattan would be as contaminated as the permanently closed area around the Chernobyl nuclear plant. Anyone living in Manhattan would have at least a 1-in-100 chance of dying from cancer caused by the radiation. An area reaching deep into the Hudson Valley would, under current Environmental Protection Agency standards, have to be decontaminated or destroyed.

Frank von Hippel, the Princeton physicist, has reviewed the data, and he pointed out that this is a bit less alarming than it sounds. "Your probability of dying of cancer in your lifetime is already about 20 percent," he said. "This would increase it to 20.1 percent. Would you abandon a city for that? I doubt it."

Indeed, some large portion of our fear of radiation is irrational. And yet the fact that it's all in your mind is little consolation if it's also in the minds of a large, panicky population. If the actual effect of a radiation bomb is that people clog the bridges out of town, swarm the hospitals and refuse to return to live and work in a contaminated place, then the impact is a good deal more than psychological. To this day, there is bitter debate about the actual health toll from the Chernobyl nuclear accident. There are researchers who claim that the people who evacuated are actually in worse health over all from the trauma of relocation, than those who stayed put and marinated in the residual radiation. But the fact is, large swaths of developed land around the Chernobyl site still lie abandoned,

much of it bulldozed down to the subsoil. The Hart Senate Office Building was closed for three months by what was, in hindsight, our society's inclination to err on the side of alarm.

There are measures the government can take to diminish the dangers of a radiological weapon, and many of them are getting more serious consideration. The Bush administration has taken a lively new interest in radiation-detection devices that might catch dirty-bomb materials in transit. A White House official told me the administration's judgment is that protecting the raw materials of radiological terror is worth doing, but not at the expense of more catastrophic threats.

"It's all over," he said. "It's not a winning proposition to say you can just lock all that up. And then, a bomb is pretty darn easy to make. You don't have to be a rocket scientist to figure about fertilizer and diesel fuel." A big fertilizer bomb of the type Timothy McVeigh used to kill 168 people in Oklahoma City, spiced with a dose of cobalt or cesium, would not tax the skills of a determined terrorist.

"It's likely to happen, I think, in our lifetime," the official said. "And it'll be like Oklahoma City plus the Hart Office Building. Which is real bad, but it ain't the World Trade Center."

The Peril of Power Plants Every eight years or so the security guards at each of the country's 103 nuclear power stations and at national weapons labs can expect to be attacked by federal agents armed with laser-tag rifles. These mock terror exercises are played according to elaborate rules, called the "design basis threat," that in the view of skeptics favor the defense. The attack teams can include no more than three commandos. The largest vehicle they are permitted is an S.U.V. They are allowed to have an accomplice inside the plant, but only one. They are not allowed to improvise. (The mock assailants at one Department of Energy

lab were ruled out of order because they commandeered a wheelbarrow to cart off a load of dummy plutonium.) The mock attacks are actually announced in advance. Even playing by these rules, the attackers manage with some regularity to penetrate to the heart of a nuclear plant and damage the core. Representative Edward J. Markey, a Massachusetts Democrat and something of a scourge of the nuclear power industry, has recently identified a number of shortcomings in the safeguards, including, apparently, lax standards for clearing workers hired at power plants.

One of the most glaring lapses, which nuclear regulators concede and have promised to fix, is that the design basis threat does not contemplate the possibility of a hijacker commandeering an airplane and diving it into a reactor. In fact, the protections currently in place don't consider the possibility that the terrorist might be willing, even eager, to die in the act. The government assumes the culprits would be caught while trying to get away.

A nuclear power plant is essentially a great inferno of decaying radioactive material, kept under control by coolant. Turning this device into a terrorist weapon would require cutting off the coolant so the atomic furnace rages out of control and, equally important, getting the radioactive matter to disperse by an explosion or fire. (At Three Mile Island, the coolant was cut off and the reactor core melted down, generating vast quantities of radiation. But the thick walls of the containment building kept the contaminant from being released, so no one died.)

One way to accomplish both goals might be to fly a large jetliner into the fortified building that holds the reactor. Some experts say a jet engine would stand a good chance of bursting the containment vessel, and the sheer force of the crash might disable the cooling system—rupturing the pipes and cutting off electricity that pumps the water through the core. Before nearby residents had begun to evacuate, you could have a meltdown that would spew a volcano of radioactive isotopes into the air, causing fatal radiation sickness for those exposed to high doses and raising lifetime cancer rates for miles around.

This sort of attack is not as easy, by a long shot, as hitting the World Trade Center. The reactor is a small, low-lying target, often nestled near the conspicuous cooling towers, which could be destroyed without great harm. The reactor is encased in reinforced concrete several feet thick, probably enough, the industry contends, to withstand a crash. The pilot would have to be quite a marksman, and somewhat lucky. A high wind would disperse the fumes before they did great damage.

Invading a plant to produce a meltdown, even given the record of those mock attacks, would be more complicated, because law enforcement from many miles around would be on the place quickly, and because breaching the containment vessel is harder from within. Either invaders or a kamikaze attacker could instead target the more poorly protected cooling ponds, where used plutonium sits, encased in great rods of zirconium alloy. This kind of sabotage would take longer to generate radiation and would be far less lethal.

Discussion of this kind of potential radiological terrorism is colored by passionate disagreements over nuclear power itself. Thus the nuclear industry and its rather tame regulators sometimes sound dismissive about the vulnerability of the plants (although less so since Sept.11), while those who regard nuclear power as inherently evil tend to overstate the risks. It is hard to sort fact from fear-mongering.

Nuclear regulators and the industry grumpily concede that Sept. 11 requires a new estimate of their defenses, and under prodding from Congress they are redrafting the so-called design basis threat, the one plants are required to defend against. A few members of Congress have proposed installing ground-to-air missiles at nuclear plants, which most experts think is a recipe for a disastrous mishap.

"Probably the only way to protect against someone flying an aircraft into a nuclear power plant," said Steve Fetter of the University of Maryland, "is to keep hijackers out of cockpits."

Being Afraid For those who were absorbed by the subject of nuclear terror before it became fashionable, the months since the terror attacks have been, paradoxically, a time of vindication. President Bush, whose first budget cut $100 million from the programs to protect Russian weapons and material (never a popular program among conservative Republicans), has become a convert. The administration has made nuclear terror a priority, and it is getting plenty of goading to keep it one. You can argue with their priorities and their budgets, but it's hard to accuse anyone of indifference. And resistance—from scientists who don't want security measures to impede their access to nuclear research materials, from generals and counterintelligence officials uneasy about having their bunkers inspected, from nuclear regulators who worry about the cost of nuclear power, from conservatives who don't want to subsidize the Russians to do much of anything—has become harder to sustain. Intelligence gathering on nuclear material has been abysmal, but it is now being upgraded; it is a hot topic at meetings between American and foreign intelligence services, and we can expect more numerous and more sophisticated sting operations aimed at disrupting the black market for nuclear materials. Putin, too, has taken notice. Just before leaving to meet Bush in Crawford, Tex., in November, he summoned the head of the atomic energy ministry to the Kremlin on a Saturday to discuss nuclear security. The subject is now on the regular agenda when Bush and Putin talk.

These efforts can reduce the danger but they cannot neutralize the fear, particularly after we have been so vividly reminded of the hostility some of the world feels for us, and of our vulnerability.

Fear is personal. My own—in part, because it's the one I grew up with, the one that made me shiver through the Cuban missile crisis and "On the Beach"—is the horrible magic of nuclear fission. A dirty bomb or an assault on a nuclear power station, ghastly as that would be, feels to me within the range of what we have survived. As the White House official I spoke with said, it's basically Oklahoma City plus the Hart Office Building. A nuclear explosion is in a different realm of fears and would test the country in ways we can scarcely imagine.

A mushroom cloud of irradiated debris would blossom more than two miles into the air. Then highly lethal fallout would begin drifting back to earth, riding the winds into the Bronx or Queens or New Jersey.

As I neared the end of this assignment, I asked Matthew McKinzie, a staff scientist at the Natural Resources Defense Council, to run a computer model of a one-kiloton nuclear explosion in Times Square, half a block from my office, on a nice spring workday. By the standards of serious nuclear weaponry, one kiloton is a junk bomb, hardly worthy of respect, a fifteenth the power of the bomb over Hiroshima.

A couple of days later he e-mailed me the results, which I combined with estimates of office workers and tourist traffic in the area. The blast and searing heat would gut buildings for a block in every direction, incinerating pedestrians and crushing people at their desks. Let's say 20,000 dead in a matter of seconds. Beyond this, to a distance of more than a quarter mile, anyone directly exposed to the fireball would die a gruesome death from radiation sickness within a day—anyone, that is, who survived the third-degree burns. This larger circle would be populated by about a quarter million people on a workday. Half a mile from the explosion, up at Rockefeller Center and down at Macy's, unshielded onlookers would expect a slower death from radiation. A mushroom cloud of irradiated debris would blossom more than two miles into the air, and then, 40 minutes later, highly lethal fallout would begin drifting back to earth, showering injured survivors and dooming rescue workers. The poison would ride for 5 or 10 miles on the prevailing winds, deep into the Bronx or Queens or New Jersey.

A terrorist who pulls off even such a small-bore nuclear explosion will take us to a whole different territory of dread from Sept. 11. It is the event that preoccupies those who think about this for a living, a category I seem to have joined.

"I think they're going to try," said the physicist David Albright. "I'm an optimist at heart. I think we can catch them in time. If one goes off, I think we will survive. But we won't be the same. It will affect us in a fundamental way. And not for the better."

Bill Keller is a Times columnist and a senior writer for the magazine.

ROBERT HARRISON—STAFF

PROUD FATHER: In a Pakistan pharmacy, pictured above, Amirul Haq (r.), says he is 'satisfied' that his Muslim son was killed in the Kashmir. He's 'against America, because it doesn't care about those who die in Pakistan.'

'Why do they hate us?'

By Peter Ford

Staff writer

asked President Bush in his speech to Congress last Thursday night. It is a question that has ached in America's heart for the past two weeks. Why did those 19 men choose to wreck the icons of US military and economic power?

Most Arabs and Muslims knew the answer, even before they considered who was responsible. Retired Pakistani Air Commodore Sajad Haider—a friend of the US—understood why. Radical Egyptian-born cleric and US enemy Abu Hamza al-Masri understood. And Jimmy Nur

Zamzamy, a devout Muslim and advertising executive in Indonesia, understood.

They all understood that this assault was more precisely targeted than an attack on "civilization." First and foremost, it was an attack on America.

In the United States, military planners are deciding how to exact retribution. To many people in the Middle East and beyond, where US policy has bred widespread anti-Americanism, the carnage of Sept. 11 *was* retribution.

And voices across the Muslim world are warning that if America doesn't wage its war on terrorism in a way that the Muslim world considers just, America risks creating even greater animosity.

Mr. Haider is a hero of Pakistan's 1965 war against India, and a sworn friend of America. But he and his neighbors in one of Islamabad's toniest districts are clear about why their warm feelings toward the US are not widely shared in Pakistan.

In his dim office in a north London mosque, Abu Hamza al-Masri sympathizes with the goals of Osama bin Laden, fingered by US officials as the prime suspect behind the Sept. 11 attacks. Abu Hamza has himself directed terrorist operations abroad, according to the British police, although for lack of evidence, they have never brought him to trial.

Mr. Zamzamy, a 30-something advertising executive in Jakarta, knew what was behind the attack, too. Trying to give his ads some zip and still stay within the bounds of his Muslim faith, he is keenly aware of the tensions between Islam and American-style global capitalism.

The 19 men—who US officials say hijacked four American passenger jets and flew them on suicide missions that left more than 7,000 people dead or missing—were all from the Middle East. Most of the hijackers have been identified as Muslims.

The vast majority of Muslims in the Middle East were as shocked and horrified as any American by what they saw happening on their TV screens. And they are frightened of being lumped together in the popular American imagination with the perpetrators of the attack.

But from Jakarta to Cairo, Muslims and Arabs say that on reflection, they are not surprised by it. And they do not share Mr. Bush's view that the perpetrators did what they did because "they hate our freedoms."

Rather, they say, a mood of resentment toward America and its behavior around the world has become so commonplace in their countries that it was bound to breed hostility, and even hatred.

And the buttons that Mr. bin Laden pushes in his statements and interviews—the injustice done to the Palestinians, the cruelty of continued sanctions against Iraq, the presence of US troops in Saudi Arabia, the repressive and corrupt nature of US-backed Gulf governments—win a good deal of popular sympathy.

The resentment of the US has spread through societies demoralized by their recent history. In few of the world's 50 or so Muslim countries have governments offered their citizens either prosperity or democracy. Arab nations have lost three wars against their arch-foe—and America's closest ally—Israel. A sense of failure and injustice is rising in the throats of millions.

Three weeks ago, a leading Arabic newspaper, Al-Hayat, published a poem on its front page. A long lament about the plight of the Arabs, addressed to a dead Syrian poet, it ended:

"Children are dying, but no one makes a move.
Houses are demolished, but no one makes a move.
Holy places are desecrated, but no one makes a move....
I am fed up with life in the world of mortals.
Find me a hole near you. For a life of dignity is in those holes."

It sounds as if it could have been written by a desperate and hopeless man, driven by frustration to seek death, perhaps martyrdom. A young Palestinian refugee planning a suicide bomb attack, maybe. In fact, it was written by the Saudi Arabian ambassador to London, a member of one of the wealthiest and most influential families in the kingdom that is Washington's closest Arab ally.

Against the background of that humiliated mood, America's unchallenged military, economic, and cultural might be seen as an affront even if its policies in the Middle East were neutral. And nobody voices that view.

From one end of the region to the other, the perception is that Israel can get away with murder—literally—and that Washington will turn a blind eye. Clearly, the US and Israel have compelling reasons for their actions. But little that US diplomats have done in recent years to broker a peace deal between Israel and the Palestinians has persuaded Arabs that the US is a fair-minded and equitable judge of Middle Eastern affairs.

Over the past year, Arab TV stations have broadcast countless pictures of Israeli soldiers shooting at Palestinian youths, Israeli tanks plowing into Palestinian homes, Israeli helicopters rocketing Palestinian streets. And they know that the US sends more than $3 billion a year in military and economic aid to Israel.

"You see this every day, and what do you feel?" asks Rafiq Hariri, the portly prime minister of Lebanon, who is not an excitable man. "It hurts me a lot. But for hundreds of thousands of Arabs and Muslims, it drives them crazy. They feel humiliated."

RESENTMENT RISES, AND A RADICAL IS BORN

Ask Sheikh Abdul Majeed Atta why Palestinians may not like the United States, and he does not immediately answer. Instead, he pads barefoot across the red swirls of his living room carpet and reaches for three framed photographs on the floor beside a couch.

The black-and-white prints show dusty, rock-strewn hills dotted with tiny tents and cinderblock houses: the early days of Duheisheh refugee camp, south of Bethlehem in the West Bank. It was where Mr. Atta was born, and where his family has lived for more than half a century. Atta's family village was destroyed in the struggle between Palestinian Arabs and Jews after Britain divided Palestine between them in 1948. For 10 years his family of

13 lived in a tent. The year Atta was born, the United Nations gave them a one-room house.

It doesn't matter to Atta that the United States was not directly involved in "the catastrophe," as Palestinians refer to the events of 1948. Washington averted its eyes when it could have helped, he says, and since then has been firmly on Israel's side.

Heavyset, solid, with a neatly trimmed full beard, Atta is the preacher at a nearby mosque. He looks the part of the community leader, always meticulously turned out in crisp shirts and pressed trousers, gold-rimmed reading glasses tucked into a pocket.

In the past year of the Palestinian-Israeli conflict, Atta has joined Hamas, the radical group responsible for recently sending most of the suicide bombers into Israeli towns. Frustration at watching the rising Palestinian death toll at the hands of the Israeli army played a large part in his decision, he says.

His resentment at Israel, though, dates back to his infancy, and the stories he heard of his village, Ras Abu Amar, which he never knew. That village is still alive for him, just as millions of Palestinians in the West Bank and Gaza Strip, and throughout the Middle East cherish photos, house keys, and deeds to homes that no longer exist or which have housed Israelis for generations.

Today he lives in his own house in Duheisheh, a sprawling tangle of densely packed concrete buildings that crowd snaking, narrow alleys. But he still dreams of the home he never knew, and recalls who took it from him, and remembers who they rely on for their strength.

What happened on Sept. 11 "was an awful thing, a tragedy, and since we live a continuous tragedy, we felt like this touched us," he adds. "But when we see something like this in Israel or the US, we feel a contradiction. We see it's a tragedy, but we remember that these are the people behind our tragedy."

"Even small children know that Israel is nothing without America," says Atta. "And here America means F-16, M-16, Apache helicopters, the tools Israelis use to kill us and destroy our homes."

SUPERPOWER SWAGGER

Such weapons are very much the visible face of American policy in the Middle East, where military might has held the balance of power for 50 years. Thousands of US soldiers stationed in the Gulf, and billions of US dollars each year in military aid to Israel, Egypt, and other allies, have shored up Washington's interests in the strategically crucial, oil-rich region.

That military presence and power looks like swagger to some in the Muslim world, even far from the flashpoints. "Now America is ready with its airplanes to bomb this poor nation [Afghanistan], and most people in Indonesia don't like arrogance," says Imam Budi Prasodjo, an Indonesian sociologist and talk-show host.

"You are a superpower, you are a military superpower, and you can do whatever you want. People don't like that, and this is dangerous," he adds.

"America should spread its culture, rather than weapons or tanks," adds Mohammed el-Sayed Said, deputy director of Cairo's influential Al Ahram think tank. "They need to act like any respectable commander or leader of an army. They can't just project an image of contempt for those they wish to lead."

Ten years ago, at the head of a broad coalition of Western and Arab countries, the United States used its superpower status to kick the Iraqi army out of Kuwait. Since then, however, Washington has found itself alone—save for loyal ally Britain—in its determination to keep bombing Iraq, and to keep imposing strict economic sanctions that the United Nations says are partly responsible for the deaths of half a million Iraqi children.

'We wish the American people could see what their governments are doing in the rest of the world.'

—Saniya Ghussein, whose daughter, Raafat, was killed in the 1986 US bombing raid against Libya

Those deaths, and those bombs (which US and British planes drop regularly, but without fanfare), are felt keenly among fellow Arabs. And Saniya Ghussein knows all about bombs.

A DAUGHTER DIES, AND PARENTS WAIT FOR US APOLOGY

In the middle of the night of April 16, 1986, the deafening sound of anti-aircraft guns woke Saniya Ghussein with a sudden start. "My God," she thought, "there's a war being fought above my house."

She slipped out of bed and ran into the bedroom where her husband Bassem and their 7-year-old daughter Kinda had fallen asleep earlier in the evening. "Bassem, the Americans are here," she said urgently. "It looks like they're going to hit us."

She checked on her other daughter, Raafat. She had been suffering from her annual bout of hay fever, and the 18-year-old art student was in the television room next to the humidifier so she could breathe easier.

Raafat was still sleeping, completely oblivious of all the commotion going on around her, due to the medication she had taken earlier. There was little Saniya felt she could do. She climbed back into bed and pulled the sheets tight around her.

Bassem lay awake on the bed, listening to the appalling noise in the night sky above.

A Palestinian-born Lebanese national, Bassem had worked in Libya as an engineer for Occidental, the American oil giant, for 20 years, helping exploit the country's

ROBERT HARRISON—STAFF

'When Bush talked of a crusade…it was not a slip of the tongue.'
—Sajad Haider, retired Pakistani air force officer

massive oil reserves. He and his family lived in the up-market Ben Ashour neighborhood of Tripoli, the Libyan capital, on the ground floor of a two-story apartment block.

Bassem never heard the explosion. Instead, he watched in astonishment as the window frame suddenly flew into the room, and the roof collapsed on top of him and his daughter.

Kinda was screaming in the darkness near him. Bassem tried to move, but was pinned by the rubble. He groped in the blackness for Kinda. "Don't worry," he said, squeezing his daughter's hand. "Daddy's here, don't cry, it will be okay."

The blast had knocked Saniya unconscious. She woke to hear Bassem calling from the next room and Kinda screaming. She stumbled in the darkness, barefoot across the rubble and glass shards, choking on the fumes from the missile blast, as she called her daughter's name "Raafat! Raafat!" for several minutes. But there was no response, and Saniya knew with a terrible certainty that her daughter was dead.

"Bassem," she cried. "Raafat has gone."

Pinned beneath the rubble, Bassem heard his wife's words, and he felt a deep sense of anger and resentment well up inside him. His life and that of his family had been shattered, and nothing would ever be the same again.

It took them eight hours to dig Raafat out from under the ruins of the house. "Our pain and agony, which I cannot describe, started at that moment," Saniya says.

Raafat was one of an estimated 55 victims of an air raid mounted by US warplanes against a series of targets in Tripoli and another Libyan city, Benghazi.

The attacks were in retaliation for the bombing of a disco in Berlin, Germany, 10 days earlier in which 200 people were injured, 63 of them US soldiers; one soldier

and one civilian were killed. The Reagan administration blamed Libyan leader Muammar Qaddafi.

Bassem and Saniya Ghussein are not natural anti-Americans. Bassem studied in the US before going to work for Esso and then Occidental. He sent Raafat to an American Catholic school, and on family trips to the US, Saniya would take Raafat to Disney World in Florida. "We did all the typical American things," she says.

But since that terrible night 16 years ago, neither Bassem nor Saniya have stepped foot in America. They returned to Beirut in 1994 when Bassem retired.

In 1989, the Libyan government enlisted the help of Ramsey Clark, an attorney general during the Carter administration, to file a lawsuit against President Ronald Reagan and British Prime Minister Margaret Thatcher for the civilian deaths during the air raids. "When Clark came to collect our documents and evidence, I asked him if he thought we had a case," Bassem recalls. "He said 'Oh, definitely. This was murder.' "

But US district court judge Thomas Penfield Jackson disagreed. He dismissed the suit, and fined Clark for presenting a "frivolous" case that "offered no hope whatsoever of success."

Twelve years later, the court's decision still rankles with Bassem. "I will only return to America when I know someone will listen to me and say: 'yes, it was our fault your daughter died, and I am sorry.' So long as they think my daughter's death is 'frivolous,' I won't go back," Bassem says.

The Ghusseins have no sympathy for religious extremism and thoroughly condemn the Sept. 11 suicide bombings in New York and Washington. Yet they both maintain that the devastating attack was a result of America's "arrogant" policies in the Middle East and elsewhere. "We wish the American people could see what their governments are doing in the rest of the world," Saniya says.

A FEELING OF BETRAYAL AMONG FRIENDS

On the other side of Asia, in Pakistan, Air Commodore Haider would sympathize with the Ghusseins' wish. He has always been a friend of the United States, and not just because he enjoyed the 10 years he spent in Washington as his country's military attaché. Like most other members of the ruling elite in Pakistan, in the armed forces, in business, and in the political parties, he sees America as a natural ally.

But not a reliable one. The prevailing mood in Pakistan of anger and suspicion toward the United States springs from a deeply rooted perception that the US has been a fickle friend, Haider says, and not just to Pakistan, but to other nations in the Muslim world.

If there was a moment of betrayal for Haider, it was the 1965 war between India and Pakistan, largely over the future of Kashmir. As Indian tanks advanced on the Pakistani metropolis of Lahore, Haider was head of a squadron of F-86 Sabre jets sent to destroy them. India's Soviet allies helped with money, arms, and diplomatic support. But at a crucial moment, Pakistan's ally, the US, refused to send more weapons. As it turned out, Pakistan was able to defeat the Indian attack on Lahore and elsewhere without US help. Haider's squadron decimated the column of Indian tanks that had reached to within six miles of Lahore. But the lesson lingered: America cannot be trusted.

"There is a feeling of being betrayed, it's a feeling of being let down, and you can only be let down by somebody you care for," says Haider, out for an evening stroll in a tony Islamabad neighborhood.

"They said you will be the bulwark of America and of the free world against Communism. But then they dropped a friend for no good reason."

Today, Haider sees a "convergence of interests" between the United States and Pakistan in the fight against terrorism. But he says that President Bush will need to watch his language when he talks about the Muslim world. "When Bush talked of a Crusade... it was not a slip of the tongue. It was a mindset. When they talk of terrorism, the only thing they have in mind is Islam."

Ultimately, Haider does see a way for America and Muslim nations to become lasting friends, but only if the US begins to give as much weight to the interests of Muslim nations as it does to Israel.

"When you deny justice to people, which you have been doing for several decades in Palestine, and they are intelligent, sensitive people, they are going to find something to do," warns Haider. "They might take shelter in Islam, in fatalism, and some will come to despise you."

AN EGYPTIAN 'INSPIRED' TO JOIN AFGHAN FIGHTERS

Sheikh Abu Hamza al-Masri, the radical Muslim cleric who runs a mosque in a shabby district of north London, has certainly come to despise America.

Abu Hamza says he used to admire the West when he was a young man—so much so that he dropped out of university in his native Alexandria, Egypt, to study in Britain. And he clearly had nothing against the British government when he took a job as a civil engineer at Sandhurst, the British equivalent of West Point, after he graduated.

But as he immersed himself more and more in religious studies, and came into contact with more and more Arab mujahideen, who had travelled from the mountains of Afghanistan to England for medical treatment, he began to change his outlook.

"When you see how happy they are, how anxious to just have a new limb so they can run again and fight again, not thinking of retiring, their main ambition is to get killed in the cause of God... you see another dimension in the verses of the Koran," says Abu Hamza.

How the world views a US military response

In your opinion, once the identity of the terrorists is known, should the American government launch a military attack on the country or countries where the terrorists are based, or should the American government seek to extradite the terrorists to stand trial?

	Launch attack	Try the terrorists	Don't know
Israel	77%	19%	4%
India	72	28	0
United States	54	30	16
Korea	38	54	9
France	29	67	4
Czech Republic	22	64	14
Italy	21	71	8
South Africa	18	75	7
United Kingdom (excluding N. Ireland)	18	75	7
Germany	17	77	6
Bosnia	14	80	6
Colombia	11	85	4
Pakistan	9	69	22
Greece	6	88	6
Mexico	2	94	3

Source: Gallup International surveys Sept. 14 to 17.

Inspired by their example, he took his family to Afghanistan in 1990, to work there as a civil engineer, building roads, tunnels, and "anything I could do." And he also fought with the mujahideen against Afghan President Mohammad Najibullah (seen as a Russian stand-in supported by the Soviets), until he blew both his hands off and lost the sight in his left eye, in a mine explosion.

What transformed him and his comrades-in-arms from anti-Soviet to anti-American militants, he says, was the way Washington abandoned them at the end of the war in Afghanistan, and sought to disarm and disperse them.

"It was when the Americans took the knife out of the Russians and stabbed it in our back, it's as simple as that," says Abu Hamza. "It was a natural turn, not a theoretical one.

"In the meantime, they were bombarding Iraq and occupying the [Arabian] peninsula," he says, referring to the US troops stationed in Saudi Arabia after the Gulf War, "and then with the witch-hunt against the mujahideen, all of it came together, that was a full-scale war, it was very clear."

Abu Hamza would rather see Islamic militants fight corrupt or secular Arab governments before they take on America (indeed, the Yemeni government has sought his extradition from Britain for plotting to overthrow the government in Sana). But he is in no doubt that the American government brought the events of Sept. 11 on its own head.

"The Americans wanted to fight the Russians with Muslim blood, and they could only justify that by triggering the word 'jihad,' " he argues. "Unfortunately for everybody except the Muslims, when that button is pushed, it does not come back that easy. It only keeps going on and on until the Muslim empire swallows every empire existing."

Can he understand the motivation behind the assault on New York and Washington? "The motivation is everywhere," he says, with the current US administration. "When a president stands up before the planet and says an American comes first, he is only preaching hatred. When a president stands up and says we don't honor our missile treaty with the Russians, he is only preaching arrogance. When he refuses to condemn what's happening in Palestine, he is only preaching tyranny.

"American foreign policy has invited everybody, actually, to try to humiliate America, and to give it a bloody nose," he adds.

IN JAKARTA, COUNTERING AMERICAN CULTURE WITHOUT VIOLENCE

You wouldn't catch Rizky "Jimmy" Nur Zamzamy justifying violence that way, though he professes just as deep an attachment to Islam as Abu Hamza.

Mr. Zamzamy, a rangy young Indonesian advertising executive in a pink shirt, is sitting in a Western-style cafe in Jakarta, his cellphone at the ready, and his fried chicken growing cold as he explains how he tries to be a good Muslim by right action, not fighting.

That, he feels, is the best way of countering what he sees as the corrupting influence of American culture and morals on traditional Indonesian ways of life in the largest Muslim country in the world.

Until a few years ago, Zamzamy led a regular secular life, hanging out in bars and dating women. Then he met a Muslim teacher who became his spiritual guide. Now he follows Islamic teachings and donates most of his $1,300 monthly salary to his "guru" to be spent on building mosques and helping the poor.

He says he has made sure that none of the money goes to extremist groups that use violence in the name of Islam, such as the Laskar Jihad group, locked in bloody battle with Christians in the Maluku region of Indonesia.

Two years ago, in line with his growing religious beliefs, he quit the advertising agency he had worked for and set up his own company along Islamic lines: He

won't take banks or alcoholic-beverage producers as clients, for example, and he does no business on Friday, the Muslim holy day.

But he is relaxed about those who don't share his beliefs: He does not insist that his wife wear a headscarf, for example, and he is not uncomfortable sitting alongside the rich young Jakartans in the cafe who are flirting and drinking. They must make their own choices, he says.

And though he does not like the sexual overtones of American pop culture, he knows that "you can't hide from American culture." By living his life according to Islamic precepts, he says, "I am fighting America in my own way. But I don't agree with violence."

AMBIVALENCE ABOUT AMERICA

All over the Muslim world, young people like Zamzamy are juggling their sense of Islamic identity with the trappings of a globalized, secular society.

In a classroom of Al Khair University, set in a concrete office park in Islamabad, Nabil Ahmed, a business student, and his classmates are fuming over their president's betrayal of the Pakistani people by pledging to support what they fear will turn into a crusade against Muslims.

Ahmed and his friends are well-dressed, middle-class boys, and represent neither the old-money security of Pakistan's elite nor dirt-poor peasants who make up the bulk of Pakistan's angry conservative masses. They are the silent majority of Pakistan, with their feet firmly planted in both the East and the West. On weekdays, they listen to Whitney Houston and Michael Bolton, wear Dockers and Van Heusen shirts. On weekends, many switch to traditional salwar kameez outfits and go with their fathers to the mosque to pray.

'It is [the] double standard that creates hatred.'

—Nabil Ahmed, a business student at
Al Khair University in Islamabad

They have much to gain from a Western style of life, and most have plans to move to the United States for a few years to make some money before returning home to Pakistan. Yet despite their attraction to the West, they are wary of it too.

"Most of us here like it both ways, we like American fashion, American music, American movies, but in the end, we are Muslims," says Ahmed. "The Holy Prophet said that all Muslims are like one body, and if one part of the body gets injured, then all parts feel that pain. If one Muslim is injured by non-Muslims in Afghanistan, it is the duty of all Muslims of the world to help him."

Like his friends, Ahmed feels that America has double standards toward its friends and enemies. America attacks Iraq if it invades Kuwait, but allows Israel to bulldoze Palestinian homes in the West Bank and Gaza Strip.

ROBERT HARRISON—STAFF

AFGHAN REFUGEES: These boys are among some 60,000 displaced Afghans at Jalozai refugee camp near Peshawar, Pakistan, along its border with Afghanistan. The camp is crowded, and Pakistan has recently forbidden the UNHCR to register any more refugees.

It ostracizes a Muslim nation like Sudan for oppressing its Christian minority, but allows Russia to bomb its Muslim minority into submission in Chechnya.

And while the US supported many "freedom fighter" movements in the past few decades, including the contra movement in Nicaragua, America labels Pakistan and Afghanistan as terrorist states because they support militant Muslim groups fighting in the Indian state of Kashmir and elsewhere.

'The Americans wanted to fight the Russians with Muslim blood.'

—Sheikh Abu Hamza al-Masri,
a radical Muslim cleric who runs a mosque in London

"There is only one way for America to be a friend of Islam," says Ahmed. "And that is if they consider our lives to be as precious as their own. "If Americans are concerned about the 6,500 deaths in the World Trade Center, let them talk also about the deaths in Kashmir, in Palestine, in Chechnya, in Bosnia. It is this double standard that creates hatred."

Ahmed's ambivalence about America—his desire to live and work there, his admiration for its values, but his anger at its behavior around the world—is broadly shared across the Muslim world and Arab world.

"I think they hate us because of what we do, and it seems to contradict who we say we are," says Bruce Lawrence, a professor of religion at Duke University, referring to people in the Middle East. "The major issue is that our policy seems to contradict our own basic values."

That seems clear enough to Muslims who sympathize with the Palestinians, and who say that Washington should force Israel to abide by United Nations resolutions to withdraw from the occupied territories. "The Americans say September 11th was an attack on civilization," says Mr Hariri, the Lebanese prime minister. "But what does civilized society mean if not a society that lives according to the law?"

It also seems clear to citizens of monarchical states in the Gulf, where elections are unknown and women's rights severely restricted. "Since the Cold War ended, America has talked about promoting democracy," says John Esposito, head of the Center for Muslim-Christian Understanding at Georgetown University in Washington. "But we don't do anything about it in repressive regimes in the Middle East, so you can understand widespread anti-Americanism there."

At the same time, the state-run media—which is all the media there is across much of the Middle East—often fan the flames of anti-American and anti-Israel sentiment because that helps focus citizens' minds on something other than their own government's shortcomings.

In Sana, the Yemeni capital, where queues of visa-seekers line up daily outside the US embassy, the ambivalence about America is clear. "When you go there, you really

50 YEARS OF US POLICY IN THE MIDDLE EAST

1947–48

UN votes to partition Palestine into two states—one for Jews, one for Palestinian Arabs. Arab states invade; 300,000 Palestinians flee Jewish-controlled areas. Jewish forces prevail, declaring Israeli independence. US recognizes Israel.

1953

CIA helps Iran's military stage a coup, deposing elected PM Mohammad Mossadeq, whom US sees as communist threat. US oversees installation of Shah Mohammad Reza Pavlavi as ruler of Iran.

1956

Israel attacks Egypt for control of Suez Canal. Britain and France veto US-sponsored UN resolution calling for halt to military action. British forces attack Egypt.

1960

Iran, Iraq, Kuwait, Saudi Arabia, and Venezuela form Organization of Petroleum Exporting Nations (OPEC).

1966

US sells its firs jet bombers to Israel, breaking with a 1956 decision not to sell arms to the Jewish state.

1967

Six-Day War. Israel launches preemptive strike against Arab neighbors, capturing Jerusalem, the Sinai Peninsula, the Gaza Strip, and the Golan Heights. Kuwait and Iraq cut oil supplies to US, UN adopts Resolution 242, calling on Israel to withdraw from captured territory. Israel refuses.

1968

First major hijacking by Arab militants occurs on El Al flight from Rome to Tel Aviv, marking decades of hostage-takings, hijackings, and assassinations as a strategy by Arab militant groups.

1969

Mummar Qaddafi comes to power in Libyan coup and orders US Air Force to evacuate Tripoli.

1972

Eight Arab commandos of Palestinian group Black September kill 11 Israeli athletes at the Munich Olympic Games.

1973

Egypt and Syria attack Israel over its occupation of the Golan Heights and the Sinai Peninsula. US gives $2.2 billion in emergency aid to Israel, turning tide of battle to Israel's favor. Arab states cut US oil shipments.

1974

UN General Assembly recognizes right of Palestinians to independence.

1976

The UN votes on a resolution accusing Israel of war crimes in occupied Arab territories. US casts lone "no" vote. US Ambassador to Lebanon Francis Meloy and an adviser are shot to death in Beirut. US closes Embassy there.

1978

Egypt and Israel sign US-brokered Camp David peace treaty. Eighteen Arab countries impose an economic boycott on Egypt. Egyptian president Anwar Sadat and Israeli Prime Minister Menachem Begin receive Nobel Peace Prize.

1979

Ayatollah Ruhollah Khomeini leads grass-roots Islamic revolution in Iran, deriding the US as "the great Satan." Iranian students storm US Embassy in Tehran, taking 66 Americans hostage for next 15 months. US imposes sanctions. Protesters attack US Embassies in Libya and Pakistan.

1981

Israel bombs Iraqi nuclear reactor. Muslim militants opposed to Egypt's peace treaty with Israel assassinate Egyptian President Sadat.

1982

Israel invades Lebanon to expel the Palestine Liberation Organization, facilitate election of friendly government, and form 25-mile security zone along Israel's border. Defense Minister Ariel Sharon permits Lebanese Christian militiamen to enter the Sabra and Shatila refugee camps outside Beirut. The ensuing three-day massacre kills 600 or more civilian refugees. US and other nations deploy peacekeeping troops in Lebanon.

1983

A truck bomb explodes in US Marines' barracks in Beirut, Lebanon, killing 241 soldiers. US forces withdraw.

1986

Us bombs Libya in retaliation for the bombing of a Berlin nightclub frequented by US servicemen. The airstrike kills 15 people, including the infant daughter of leader Muammar Qaddafi. All Arab nations condemn the attack.

1987

Start of the Palestinian intifada, or uprising, in the West Bank and Gaza Strip.

1990

Iraq invades Kuwait. Saddam Hussein links pullout to Israel's withdrawal from occupied territories. UN imposes sanctions that continue to hobble Iraq's economy in effort to force Iraqi compliance with weapons resolutions.

(continued)

50 YEARS *(continued)*

1991

US and coalition launch attacks against Iraq from Saudi Arabia. Gulf War ends after some three months, but US deployment continues even now, with 17,000 to 24,000 US troops in region at any time.

1993

World Trade Center in New York is bombed, killing six. US Special Forces, deployed as peacekeepers in Somalia, attempt to capture warlord Mohamed Farah Aidid. Eighteen US servicemen are killed. Israeli PM Yitzhak Rabin and Palestinian leader Yasser Arafat sign historic peace declaration in White House ceremony with President Clinton.

1994

Jordan and Israel sign peace treaty. Yasser Arafat, Yitzhak Rabin, and Foreign Minister Shimon Peres receive Nobel Peace Prize for 1993 agreement.

1995

US announces trade ban against Iran, reinforcing sanctions in effect since 1979. Rabin is assassinated, two years after peace deal with Palestinians. In Riyadh, Saudi Arabia, a car bomb explodes outside an office housing US military personnel. Seven are killed, including five Americans. Three Islamist groups claim responsibility.

1996

A truck bomb explodes outside a US military barracks in Khobar, Saudi Arabia, killing 19 US airmen. UN reports that sanctions cause 4,500 Iraqi children under 5 to die each month.

1997

Egyptian Islamic Group massacres 62 people, mostly foreign tourists, in Luxor, Egypt. The group claims it is retaliation for US imprisonment of Sheikh Omar Abdel al-Rahman, who is later convicted in 1993 World Trade Center bombing.

1998

Bombs explode outside US Embassies in Kenya and Tanzania, killing 224 people. US launches cruise-missile attacks on sites in Sudan and Afghanistan allegedly linked to Osama bin Laden. US indicts bin Laden for committing acts of terrorism against Americans abroad.

1999

Islamic militants, traced to bin Laden, are arrested for plot to bomb tourist sites during millennium celebrations.

2000

Camp David negotiations fail. Sharon visits Temple Mount in Jerusalem, sparking current Palestinian uprising. USS Cole bombing in Yemen's Aden harbor kills 17 American sailors. Bin Laden denies responsibility, but applauds the act.

2001

Hijackers crash two planes into World Trade Center in New York, one into Pentagon, and one in Pennsylvania. More than 7,000 people are dead or missing.

Compiled by Julie Finnin Day

SOURCES: "THE MIDDLE EAST" (CONGRESSIONAL QUARTERLY), NEWS REPORTS.

love the United States," says Murad al-Murayri, a US-trained physicist. "You are treated like a human being, much better than in your own country. But when you go back home, you find the US applies justice and fairness to its own people, but not abroad. In this era of globalization, that cannot stand."

Nor has the mood that has gripped Washington over the past two weeks done much to reassure skeptics, says François Burgat, a French social scientist in Yemen.

"When Bush says 'crusade', or that he wants bin Laden 'dead or alive', that is a *fatwa* (religious edict) without any judicial review," he cautions. "It denies all the principles that America is supposed to be."

A *fatwa* is something Amirul Haq, a Pakistani shopkeeper whose son died two years ago in a jihad in Kashmir, understands better than judicial review. "When I heard that my son died, I was satisfied," he says.

It's a sentiment shared by Azad Khan, too. On a hot Sunday afternoon in Mardan, Pakistan, Mr. Khan and his family have laid out a feast in a small guesthouse next to the local mosque. They are celebrating because they have just heard that Mr. Khan's 20-year-old son, Saeed, has

been killed in a gun battle with Indian troops in the part of Jammu and Kashmir state that is under Indian control. With his death, Saeed has become another *shahid*, a martyr and heroic defender of the Muslims against the enemies of Islam. According to the Koran, *shahideen* are not actually dead; they are still alive, they just can't be seen. And through acts of bravery, a *shahid* guarantees that his whole family will go to heaven.

"It is not a thing to be mourned. We are happy," says Khan, sitting down to a meal of chicken and mutton, rice and bread, along with leaders of the group with which Saeed had fought. "I told him to take part in jihad [holy war] because he is the son of a Muslim," Khan says. "And just as we fight in Kashmir, if we need to fight against the United States in Afghanistan we are ready, because we are Muslims. It is our duty to fight against any infidels who are threatening our Muslim brothers."

It's not likely that many Pakistanis, or other Muslims, will actually go to Afghanistan to fight the Americans—assuming American soldiers land there. Khan's militant views are not shared by most of his countrymen.

But in a broader sense, and in the longer term, many people in the Middle East fear that the coming war against terrorism—unless it is waged with the utmost caution—could unleash new waves of anti-American sentiment.

Jamal al-Adimi, a US-educated Yemeni lawyer, speaks for many when he warns that "if violence escalates, you bring seeds and water for terrorism. You kill someone's brother or mother, and you will just get more crazy people."

Trying to root out terrorism without re-plowing the soil in which it grows—which means rethinking the policies that breed anti-American sentiment—is unlikely to succeed, say ordinary Middle Easterners and some of their leaders.

On the practical level, Hariri points out, "launching a war is in the hands of the Americans, but winning it needs everybody. And that means everybody should see that he has an interest in joining the coalition" that Washington is building.

On a higher level, argues Bassam Tibi, a professor of international relations at Gottingen University in Germany, and an expert on political Islam, "we need value consensus between the West and Islam on democracy and human rights to combat Islamic fundamentalism. We can't do it with bombs and shooting—that will only exacerbate the problem."

Reported by staff writers Scott Baldauf in Islamabad, Pakistan; Cameron W. Barr in Amman, Jordan; Peter Ford in London; Nicole Gaouette in Jerusalem; Robert Marquand in Beijing; Scott Peterson in Sana, Yemen; Ilene R. Prusher in Tokyo; as well as contributors Nicholas Blanford in Beirut, Lebanon; Sarah Gauch in Cairo; and Simon Montlake in Jakarta, Indonesia.

India, Pakistan and the Bomb

The Indian subcontinent is the most likely place in the world for a nuclear war

by M. V. Ramana and A. H. Nayyar

As the U.S. mobilized its armed forces in the aftermath of the terrorist attacks of September 11, the world's attention focused on Pakistan, so crucial to military operations in Afghanistan. When Pakistani president Pervez Musharraf pledged total support for a U.S.-led multinational force on September 14, many people's first thought was: What about Pakistan's nuclear weapons? Could they fall into the hands of extremists? In an address to his nation, Musharraf proclaimed that the "safety of nuclear missiles" was one of his priorities. The Bush administration began to consider providing Pakistan with perimeter security and other assistance to guard its nuclear facilities.

The renewed concern about nuclear weapons in South Asia comes a little more than three years after the events of May 1998: the five nuclear tests conducted by India at Pokharan in the northwestern desert state of Rajasthan, followed three weeks later by six nuclear explosions conducted by Pakistan in its southwestern region of Chaghai. These tit-for-tat responses mirrored the nuclear buildup by the U.S. and the former Soviet Union, with a crucial difference: the two cold war superpowers were separated by an ocean and never fought each other openly. Neighboring India and Pakistan have gone to war three times since British India was partitioned in 1947 into Muslim-majority and Hindu-majority states. Even now artillery guns regularly fire over the border (officially, a cease-fire line) in the disputed region of Kashmir.

In May 1999, just one year after the nuclear tests, bitter fighting broke out over the occupation of a mountain ledge near the Kashmiri town of Kargil. The two-month conflict took a toll of between 1,300 (according to the Indian government) and 1,750 (according to Pakistan) lives. For the first time since 1971, India deployed its air force to launch attacks. In response, Pakistani fighter planes were scrambled for fear they might be hit on the ground; air-raid sirens sounded in the capital city of Islamabad. High-level officials in both countries issued at least a dozen nuclear threats. The peace and stability that some historians and political scientists have ascribed to nuclear weapons— because nuclear nations are supposed to be afraid of mutually assured destruction—were nowhere in sight.

Wiser counsel eventually prevailed. The end of the Kargil clash, however, was not the end of the nuclear confrontation in South Asia. The planned deployment of nuclear weapons by the two countries heightens the risks. With political instability a real possibility in Pakistan, particularly given the conflict in Afghanistan, the dangers have never been so near.

Learning to Love the Bomb

BOTH COUNTRIES have been advancing their nuclear programs almost ever since they gained independence from Britain. Understanding this history is crucial in figuring out what to do now, as well as preventing the further proliferation of nuclear weapons. Although the standoff between Pakistan and India has distinct local characteristics, both countries owe much to other nuclear states. The materials used in their bombs were manufactured with Western technology; both countries' justifications for joining the nuclear club drew heavily on cold war thinking. The continued reliance of the U.S. and Russia on thousands of nuclear weapons on hair-trigger alert only adds to the perceived need for nuclear arsenals in India and Pakistan.

While setting up the Indian Atomic Energy Commission (IAEC) in 1948, Jawaharlal Nehru, India's first prime minister, laid out his desire that the country "develop [atomic energy] for peaceful purposes." But at the same time, he recognized that "if we are compelled as a nation to use it for other purposes, possibly no pious sentiments will stop the nation from using it that way." Such ambivalence remained a central feature of India's nuclear policy as it developed.

MAKING NUCLEAR WEAPONS MATERIAL

THE MOST DIFFICULT part of making nuclear weapons is manufacturing the fuel, either plutonium or highly enriched uranium. The starting point is natural uranium, which is 99.3 percent uranium 238 and 0.7 percent uranium 235. Only the latter can sustain a chain reaction. To build a uranium bomb, one needs to increase the uranium 235 content to 80 percent or more. Most modern enrichment facilities, including the ones in Pakistan and India, use high-speed centrifuges [see "The Gas Centrifuge," by Donald R. Olander; SCIENTIFIC AMERICAN, August 1978].

The alternative route involves plutonium. This element is not found in nature. It is produced by irradiating uranium fuel in nuclear reactors, then extracted through a chemical process called reprocessing [see "The Reprocessing of Nuclear Fuels," by William P. Bebbington; SCIENTIFIC AMERICAN, December 1976.] In the most commonly followed reprocessing scheme, the irradiated fuel is chopped up, dissolved in acid and exposed to a solvent called tributyl phosphate mixed with kerosene. The solvent separates out the plutonium and uranium from other fission products. Plutonium is then precipitated out by a reductant, a chemical that changes it to an insoluble form.

—M.V.R. and A.H.N.

1 URANIUM MINING The ore is extracted, crushed into fine particles and leached with acid or alkali to separate out the uranium

2 PROCESSING The uranium is converted to a chemical form suited to either enrichment or fuel fabrication

3a ENRICHMENT Uranium 235 is concentrated, converted to metallic form and fabricated into bomb cores

3b FUEL FABRICATION Natural or slightly enriched uranium is clad in metal casing to make fuel rods

4 REACTOR During a chain reaction, some of the uranium absorbs neutrons and transmutes into plutonium. Afterward the fuel rods cool in a water pool

5 REPROCESSING Plutonium is extracted through a chemical process, converted to metallic form and fabricated into bomb cores

INDIAN PLUTONIUM INVENTORY

Cumulative production (in reactors): 450–722 kg
Consumption (in tests and reactors): 165 kg
Net stock: 285–557 kg (equivalent to 55–110 bombs)

PAKISTANI ENRICHED-URANIUM INVENTORY

Cumulative production (by enrichment): 450–750 kg
Consumption (during tests): 120 kg
Net stock: 330–630 kg (equivalent to 20–40 bombs)

To Indian leaders, the program symbolized international political clout and technological modernity. Over the next two decades, India began to construct and operate nuclear reactors, mine uranium, fabricate fuel and extract plutonium. In terms of electricity produced, these activities often proved uneconomical—hardly, one might think, where a developing nation should be putting its resources. Politicians and scientists justified the nuclear program on the grounds that it promoted self-sufficiency, a popular theme in postcolonial India. Rhetoric aside, India solicited and received ample aid from Canada, the U.S. and other countries.

UNDERGROUND NUCLEAR EXPLOSIONS conducted by India on May 11, 1998, caused the surface immediately above to collapse. Seismic readings suggest that the total explosive yield was between 16 and 30 kilotons, about half of what India claimed.

After India's defeat in the 1962 border war with China, some right-wing politicians issued the first public calls for developing a nuclear arsenal. These appeals became louder after China's first nuclear test in 1964. Countering this bomb lobby were other prominent figures, who argued that the economic cost would be too high. Many leading scientists advocated the bomb. Homi Bhabha, the theoretical physicist who ran the IAEC, claimed that his organization could build nuclear weapons "within 18 months." Citing a Lawrence Livermore National Laboratory report, Bhabha predicted that nuclear bombs would be cheap. He also promised economic gain from "peaceful nuclear explosions," which many American nuclear researchers extolled for, say, digging canals.

In November 1964 Indian prime minister Lal Bahadur Shastri compromised, permitting the commission to explore the technology for such an explosion. It turned out that Bhabha had already been doing some exploring. In 1960 he reportedly sent Vasudev Iya, a young chemist, to France to absorb as much information as he possibly could about how polonium—a chemical element used to trigger a nuclear explosion—was prepared. Bhabha died in 1966, and design work on the "peaceful" device did not begin for another two years. But by the late 1960s, between 50 and 75 scientists and engineers were actively developing weapons. Their work culminated in India's first atomic test—the detonation on May 11, 1974, of a plutonium weapon with an explosive yield of five to 12 kilotons. For comparison, the bomb dropped on Hiroshima had a yield of about 13 kilotons.

Nuclear Tipping Point

THE 1974 TEST was greeted with enthusiasm within India and dismay elsewhere. Western countries cut off cooperative efforts on nuclear matters and formed the Nuclear Suppliers Group, which restricts the export of nuclear technologies and materials to nations that refuse to sign the 1968 Nuclear Non-Proliferation Treaty, including both India and Pakistan.

In the years that followed, the bomb lobby pushed for tests of more advanced weapons, such as a boosted-fission design and a hydrogen bomb. It appears that in late 1982 or early 1983, Prime Minister Indira Gandhi tentatively agreed to another test, only to change her mind within 24 hours. One of the causes for the volte-face is said to have been a conversation with the Indian foreign secretary, whom an American official had confronted with satellite evidence of preparations at the test site. The conversation seems to have convinced Gandhi that the U.S. reaction would create economic difficulties for India. Instead, it is reported, she wanted to "develop other things and keep them ready."

The "other things" she had in mind were ballistic missiles. In 1983 the Integrated Guided Missile Development Program was set up under the leadership of Abdul Kalam, a renowned rocket engineer. This followed an earlier, secret attempt to reverse-engineer a Soviet antiaircraft missile that India had purchased in the 1960s. Although that effort did not succeed, it led to the development of several critical technologies, in particular a rocket engine. Kalam adopted an open management style—as compared with the closed military research program—and involved academic institutions and private firms. Anticipating restrictions on imports, India went on a shopping spree for gyroscopes, accelerometers and motion simulators from suppliers in France, Sweden, the U.S. and Germany.

In 1988 India tested its first short-range surface-to-surface missile. A year later came a medium-range missile; in April 1999, a longer-range missile. The latter can fly 2,000 kilometers, well into the heart of China. Despite this ability, India is unlikely to achieve nuclear parity with China. According to various estimates, China has 400 warheads and an additional 200 to 575 weapons' worth of fissile material. If India's plutonium production reactors have been operating on average at 50 to 80 percent of full power, India has somewhere between 55 and 110 weapons' worth of plutonium. The stockpile could be much larger if commercial reactors earmarked for electricity generation have also been producing plutonium for weapons.

Eating Grass

PAKISTAN'S NUCLEAR PROGRAM drew on a general desire to match India in whatever it does. The country set up its Atomic Energy Commission in 1954, began operating its first nuclear research reactor in 1965 and opened its first commercial reactor in 1970. As scientific adviser to the government, physicist Abdus Salam, who later won the Nobel Prize in Physics, played an important role.

The program was severely handicapped by a shortage of manpower. In 1958 the commission had only 31 scientists and engineers; it was run by Nazir Ahmad, the former head of the Textile Committee. The commission pursued an active program of training personnel by sending more than 600 scientists and engineers to the U.S., Canada and western Europe. With generous help from these countries, some of which also aided India, Pakistan had a few nuclear research laboratories in place by the mid-1960s.

THROUGH THE STREETS OF KARACHI, a mock missile is paraded by Pasban, a youth wing of Pakistan's main fundamentalist party, Jamaat-e-Islami. The parade took place in February 1999 on a day of solidarity with Kasmiris in India-administered Kashmir. Such enthusiasm for nuclear weapons is widespread, though not universal, in both India and Pakistan.

After the 1965 war with India, many Pakistani politicians, journalists and scientists pressed for the development of nuclear weapons. The most prominent was Foreign Minister Zulfikhar Ali Bhutto, who famously declared that if India developed an atomic bomb, Pakistan would follow "even if we have to eat grass or leaves or to remain hungry." After Pakistan's defeat in the December 1971 war, Bhutto became prime minister. In January 1972 he convened a meeting of Pakistani scientists to discuss making bombs.

As the first prong of their two-pronged effort to obtain weapons material, researchers attempted to purchase plutonium reprocessing plants from France and Belgium. After initially agreeing to the sale, France backed down under American pressure. But a few Pakistani scientists did go to Belgium for training in reprocessing technology. Returning to Pakistan, they constructed a small-scale reprocessing laboratory in the early 1980s. Using spent fuel from a plutonium production reactor that opened in 1998, this lab is capable of producing two to four bombs' worth of plutonium annually.

As the second prong, researchers explored techniques for enriching uranium—that is, for concentrating the bomb-usable isotope uranium 235. In 1975 A. Q. Khan, a Pakistani metallurgist who had worked at an enrichment plant in the Netherlands, joined the group. With him came classified design information and lists of component suppliers

Nuclear Arena

For five decades, India and Pakistan have fought an incessant low-level war in Kashmir and engaged in a nuclear arms race. They now possess large and diverse nuclear weapons infastructures. Meanwhile hundreds of millions of people in the region remain impoverished.

Pakistan's Nuclear Establishment§	India's Nuclear Establishment§
REACTORS	
Research/Plutonium Production Reactor, 40–70 MW* **LOCATION:** Khushab **OPENED:** 1998 **FOREIGN PARTNER:** China **MODERATOR:** heavy water [?] **COOLANT:** heavy water **ANNUAL OUTPUT:** 6.6–18 kg of plutonium†	**CIRUS, 40 MW*** **LOCATION:** Mumbai **OPENED:** 1960 **FOREIGN PARTNER:** Canada **MODERATOR:** heavy water **COOLANT:** light water **ANNUAL OUTPUT:** 6.6–10.5 kg of plutonium **Dhruva, 100 MW*** **LOCATION:** Mumbai **OPENED:** 1985 **MODERATOR:** heavy water **COOLANT:** heavy water **ANNUAL OUTPUT:** 16–26 kg of plutonium† **Fast Breeder Test Reactor, 40 MW*** **LOCATION:** Kalpakkam **OPENED:** 1983 **FOREIGN PARTNER:** France **COOLANT:** liquid sodium **ANNUAL OUTPUT:** 4–6.4 kg of plutonium†
PLUTONIUM REPROCESSING	
New Labs **LOCATION:** Rawalpindi **OPENED:** early 1980s **ANNUAL OUTPUT:** 10–20 kg of plutonium	**Trombay** **LOCATION:** Mumbai **COMMISIONED:** 1964 **ANNUAL CAPACITY:** 30–50 tons of spent metallic fuel **PREFRE** **LOCATION:** Tarapur **COMMISSIONED:** 1977 **ANNUAL CAPACITY:** 100 tons of spent oxide fuel **KARP** **LOCATION:** Kalpakkam **COMMISSIONED:** 1997 **ANNUAL CAPACITY:** 100–125 tons of spent oxide fuel
URANIUM ENRICHMENT	
Khan Research Laboratories **LOCATION:** Kahuta **OPENED:** 1984 **ANNUAL OUTPUT:** 57–93 kg of highly enriched uranium	**Rattehalli††** **LOCATION:** Mysore **OPENED:** 1990 **ANNUAL PRODUCTION:** unknown
URANIUM MINE	
Dera Ghazi Khan **OPENED:** 1974 **ANNUAL PRODUCTION:** 23–30 tons	**Jadugoda** **OPENED:** 1968 **ANNUAL PRODUCTION:** 200 tons

*thermal power output
†running at 50%–90% of capacity
††said to produce fuel for a nuclear submarine
§bomb-related facilities; commercial power reactors omitted

in the West, many of which proved quite willing to violate export-control laws [see box "Secrets, What Secrets"]. Success came in 1979 with the enrichment of small quantities of uranium. Since then, Pakistan is estimated to have produced 20 to 40 bombs' worth of enriched uranium. Every year it produces another four to six bombs' worth.

By 1984 designs for aircraft-borne bombs were reportedly complete. Around this time, some American officials started alleging that China had given Pakistan the design for a missile-ready bomb. China and Pakistan have indeed exchanged technology and equipment in several areas, including those related to nuclear weapons and missiles. For example, it is believed that Pakistan has imported short-range missiles from China. But the accusation that China supplied Pakistan with a weapons design has never been substantiated. And understandably, Pakistan's nuclear scientists have denied it.

In spring 1990 events in Kashmir threatened to erupt into another full-scale war. According to a 1993 *New Yorker* article by American journalist Seymour M. Hersh, U.S. satellites detected a convoy of trucks moving out of Kahuta, Pakistan's uranium-enrichment facility, toward an air base where F-16 fighter jets stood ready. Hersh reported that American diplomats conveyed this information to India, which recalled the troops it had amassed at the border. But the overwhelming opinion among scholars who have analyzed these claims is that Pakistan never contemplated the use of nuclear weapons; experts are also skeptical that U.S. satellites ever detected the claimed movement. Nevertheless, the Pakistani bomb lobby has used the allegations to assert that nuclear weapons protect the country from Indian attack. In India, officials have never acknowledged Hersh's story; it would be an admission that Pakistan's nuclear capability had neutralized India's conventional military advantage.

"Now I Am Become Death"

FURTHER BUILDUP of nuclear capabilities in both countries took place against a background transformed by the end of the cold war. Superpower arsenals shrank, and the Comprehensive Test Ban Treaty, which prohibits explosive tests, was negotiated in 1996. But the five declared nuclear states—the U.S., Russia, Britain, France and China—made it clear that they intend to hold on to their arsenals. This ironic juxtaposition strengthened the bomb lobbies in India and Pakistan.

Domestic developments added to the pressure. India witnessed the rise of Hindu nationalism. For decades, parties subscribing to this ideology, such as the Bharatiya Janata Party (BJP), had espoused the acquisition of greater military capability—and nuclear weapons. It was therefore not surprising that the BJP ordered nuclear tests immediately after coming to power in March 1998.

The Indian tests, in turn, provided Pakistani nuclear advocates with the perfect excuse to test. Here again, religious extremists advocated the bomb. Qazi Hussain Ahmad of the Jamaat-e-Islami, one of the largest Islamist groups in Pakistan, had declared in 1993: "Let us wage jihad for Kashmir. A nuclear-armed Pakistan would deter India from a wider conflict." Meanwhile the military sought nuclear weapons to counter India's vastly larger armed forces.

This lobbying was partially offset by U.S. and Chinese diplomacy after India's tests. In addition, some analysts and activists enumerated the ill effects that would result from the economic sanctions that were sure to follow any test. They suggested that Pakistan not follow India's lead—leaving India to face international wrath alone—but to no avail. Three weeks after India's blasts, Pakistan went ahead with its own tests. Bombast notwithstanding, the small size of seismic signals from the tests of both countries has cast doubt on the declared explosive yields. The data released by the Indian weapons establishment to support its claims are seriously deficient; for example, a graph said to be of yields of radioactive by-products has no units on the axes. Independent scientists have not been able to verify that the countries set off as many devices as they profess.

Whatever the details, the tests have dramatically changed the military situation in South Asia. They have spurred the development of more advanced weapons, missiles, submarines, antiballistic missile systems, and command-and-control systems. In August 1999 the Indian Draft Nuclear Doctrine called for the deployment of a triad of "aircraft, mobile land-missiles and sea-based assets" to deliver nuclear weapons. Such a system would cost about $8 billion. This past January the Indian government declared that it would deploy its new long-range missile. A month later the Pakistani deputy chief of naval staff announced that Pakistan was thinking about equipping at least one of its submarines with nuclear missiles.

Critical Mass

DEPLOYMENT INCREASES the risk that nuclear weapons will be used in a crisis through accident or miscalculation. With missile flight times of three to five minutes between the two countries, early-warning systems are useless. Leaders may not learn of a launch until they look out their window and see a blinding flash of light. They will therefore keep their fingers close to the button or authorize others, geographically dispersed, to do so.

Broadly speaking, there are two scenarios. The first postulates that India crosses some threshold during a war—its troops reach the outskirts of Lahore or its ships impose a naval blockade on Karachi—and Pakistan responds with tactical nuclear weapons as a warning shot. The other scenario supposes that under the same circumstances, Pakistan decides that a warning shot would not work and instead attacks an Indian city directly. In 1998 one of us (Ramana) conducted the first scientific study of how much damage a modest, 15-kiloton bomb dropped on Bombay

INDIAN MISSILES

AGNI ("Fire") 1
TYPE: Solid-fueled, first stage; liquid-fueled, second stage
RANGE: 1,500–2,000 km
WARHEAD: 1,000 kg
DEVELOPMENT STAGE: Suspended

AGNI II
TYPE: Solid-fueled, two-stage
RANGE: 2,000 km
WARHEAD: 1,000 kg
DEVELOPMENT STAGE: Tested April 1999

PRITHVI ("Earth") I
TYPE: Liquid-fueled, single-stage; engine based on Russian SA-2 air defense missile
RANGE: 150 km
WARHEAD: 1,000 kg
DEVELOPMENT STAGE: Deployed

PRITHVI II
TYPE: Liquid-fueled, single-stage
RANGE: 250 km
WARHEAD: 500 kg
DEVELOPMENT STAGE: Tested January 1996

PRITHVI III
TYPE: Liquid-fueled, single-stage naval missile
RANGE: 350 km (?)
WARHEAD: Unknown
DEVELOPMENT STAGE: Under development

SAGARIKA ("Born on the Ocean")
TYPE: Submarine-launched cruise/ballistic missile
RANGE: 300 km (?)
WARHEAD: Unknown
DEVELOPMENT STAGE: Under development

PAKISTANI MISSILES

HATF ("Armor") I
TYPE: Solid-fueled, single-stage; based on French sounding rocket
RANGE: 60–80 km
WARHEAD: 500 kg
DEVELOPMENT STAGE: Tested January 1989

HATF II
TYPE: Solid-fueled, single-stage
RANGE: 280–300 km
WARHEAD: 500 kg
DEVELOPMENT STAGE: Tested January 1989

HATF III
TYPE: Solid-fueled, single-stage
RANGE: Up to 600 km
WARHEAD: 250 kg
DEVELOPMENT STAGE: Tested July 1997

M-11
TYPE: Solid-fueled, single-stage
RANGE: 290 km
WARHEAD: 500 kg
DEVELOPMENT STAGE: Allegedly imported from China; in storage?

GHAURI (name refers to 12th century Afghan king)
TYPE: Liquid-fueled, single-stage; similar to North Korean missile
RANGE: 1,500 km
WARHEAD: 700 kg
DEVELOPMENT STAGE: Tested April 1998; serial production started November 1998

SHAHEEN ("Eagle")
TYPE: Solid-fueled, single-stage
RANGE: 600-750 km
WARHEAD: 1,000 kg
DEVELOPMENT STAGE: Tested April 1999

SHAHEEN II
TYPE: Solid-fueled, two-stage
RANGE: 2,400 km
DEVELOPMENT STAGE: Under development

would cause: over the first few months, between 150,000 and 850,000 people would die.

The Indian military is already preparing for these eventualities. This past May it carried out its biggest exercises in more than a decade, called Operation Complete Victory. Tens of thousands of troops, backed by tanks, aircraft and attack helicopters, undertook drills close to the border with Pakistan. The stated aim was to train the armed forces to operate in an "environment of chemical, biological and nuclear assault" and "to teach the enemy a lesson once and for all." In one significant exercise, the military had to "handle a warlike situation wherein an enemy aircraft is encountered carrying a nuclear warhead." Abdul Kalam, head of India's missile program, said that India's nuclear weapons "are being tested for military operations … for training by our armed forces."

Even before September 11, South Asia had all the ingredients for a nuclear war: possession and continued development of bombs and missiles, imminent deployment of nuclear weapons, inadequate precautions to avoid unauthorized use of these weapons, geographical proximity, ongoing conflict in Kashmir, militaristic religious extremist movements, and leaders who seem sanguine about the dangers of nuclear war.

The responses of India and Pakistan to the events of September 11 and the U.S.-led attack on targets in Afghanistan reflect the strategic competition that has shaped much of their history. India was quick to offer air bases and logistical support to the U.S. military so as to isolate Pakistan. Attempting to tie its own problems in Kashmir with the global concern about terrorism, Indian officials even threatened to launch attacks on Pakistani supply lines and alleged train-

Secrets, What Secrets?

Terrorists might exploit Pakistan's cavalier attitude toward nuclear information

by David Albright

Over the years, successive Pakistani governments have assured the West that they had a secure grip on the country's nuclear weapons, materials and technology. But nuclear analysts have never been entirely comforted by these assertions. Many people in the Pakistani nuclear weapons program and the military could well be sympathetic to radical Islamist or anti-American causes. What is especially worrisome is that the historical development of Pakistan's program has heightened the risk of illegal assistance and other security violations.

From its inception, the program has relied on illicit procurement and deliberate deception. It has fostered extensive contacts with the world of shady middlemen and companies whose allegiance to Western export controls depends on the price one is willing to pay. In the organizational culture of such a program, disaffected individuals could find plenty of justifications and opportunities to transfer classified information or sensitive items. Others might be disinclined to report on the suspicious actions of colleagues. Some might even feel ownership over parts of the program and believe it is their right to sell their contributions for personal benefit.

Such problems affect India less, because it started its nuclear weapons program earlier than Pakistan did. India obtained much of its nuclear infrastructure from foreign suppliers before Western governments understood the extent to which developing countries were misusing civilian nuclear assistance to make nuclear explosives. To be sure, Pakistan is not alone in dealing with an organizational culture that scorns security guidelines. The German civilian gas-centrifuge program was notorious for its weak security. In the late 1980s German nuclear experts secretly assisted Iraq.

A key component of Pakistan's program, the production of highly enriched uranium for bombs, was born in an act of industrial espionage. In the mid-1970s the father of that effort, A. Q. Khan, worked at a Dutch engineering firm and was given the task of translating classified designs and specifications for gas centrifuges. He gained access to a wide variety of sensitive information. On his return to Pakistan, Khan founded the Engineering Research Laboratories, now known as the Dr. A. Q. Khan Research Laboratories, to transform this knowledge into a bomb factory.

According to a declassified 1983 U.S. State Department memorandum, the enrichment program disguised its activities by providing false statements about the final use of items imported from Western countries. Pakistan once described its gas-centrifuge plant as a synthetic butter factory. In a 1999 interview in the Egyptian newspaper *Al-Ahram,* Khan said that his program purchased items through offshore front companies in Japan, Singapore and elsewhere. Those companies took a cut of 15 to 25 percent of the purchase price.

Khan and his colleagues took a Robin Hood approach to classified information. In the late 1980s they published a series of technical articles in Western journals about gas centrifuges. The intention was to demonstrate Pakistan's self-sufficiency in centrifuges and thereby signal that the country was ready to make a bomb. One paper stated its purpose thus: to "provide useful and practical information, as technical information on balancing of centrifuge rotors is hardly available because most of the work is shrouded in the clouds of the so-called secrecy." These articles aided other countries, such as South Africa, in their own nuclear programs.

One Pakistani article is the only publicly available study on bellows built from maraging steel, a superstrong type of steel. For years, Urenco—a British, German and Dutch enrichment consortium—considered the mere mention of these bellows a violation of its secrecy rules.

How much further did the Pakistani nuclear scientists go in spreading the art of bomb making? The U.N. arms inspections in Iraq came across a one-page Iraqi intelligence document, marked TOP SECRET, that contained an offer of nuclear weapons assistance from the Pakistanis. According to the document, an intermediary approached Iraqi intelligence in October 1990—two months after the Iraqi invasion of Kuwait and three months before the U.S.-led counterattack—with the following proposition: Khan would give Iraq bomb designs, help to procure materials through a company in Dubai and provide other services. In return, Iraq would pay handsomely.

Arms inspectors were unable to find the middleman, and Pakistan and Khan have denied any involvement. Nerverthe-less, the Iraqis took this offer as genuine—and apparently rejected it. Khidhir Hamza, a former weapons scientists who left Iraq in 1994 and worked with me in the late 1990s, says he knew of this offer at the time and believes Iraq would not have pursued it, for fear that Khan would gain too much knowledge about, and control over, Iraq's nuclear programs. Khan already had a track record of misleading the Iraqis, having used a contract for a petrochemical facility as a cover to obtain maraging steel.

In March of this year the government of Pakistan removed Khan as head of the nuclear laboratory and offered him a position as a special science and technology adviser. The move is widely viewed as an attempt to rein him in. This past summer, however, reports emerged that the laboratory has kept up its ties with North Korea's ballistic-missile program, reviving fears of nuclear cooperation. Pakistani officials have denied any connection.

No evidence links elements in the Pakistani government with any terrorist group, but the Pakistani government has had extensive contact with the Taliban. It is conceivable that terrorists could exploit these connections to gain access to sensitive nuclear items. The culture within the nuclear program increases this risk.

David Albright is a physicist, president of the Institute for Science and International Security in Washington, D.C., and a former U.N. weapons inspector in Iraq.

ing camps for militants fighting in Kashmir. Pakistan, for its part, realizing both the geopolitical advantage it possessed and the dangers of civil instability, deliberated before agreeing to provide support to fight the Taliban. The diplomatic machinations, war in Afghanistan and violence in Kashmir may well have worsened the prospects for peace on the subcontinent. The lifting of American sanctions, which had been imposed in the 1990s, freed up resources to invest in weapons.

The limitations of Western nonproliferation policy are now painfully obvious. It has relied primarily on supply-side export controls to prevent access to nuclear technologies. But Pakistan's program reveals that these are inadequate. Any effective strategy for nonproliferation must also involve demand-side measures—policies to assure countries that the bomb is not a requisite for true security. The most important demand-side measure is progress toward global nuclear disarmament. Some people argue that global disarmament and nonproliferation are unrelated. But as George Perkovich of the W. Alton Jones Foundation in Charlottesville, Va., observed in his masterly study of the Indian nuclear program, that premise is "the grandest illusion of the nuclear age." It may also be the most dangerous.

MORE TO EXPLORE

Fissile Material Production Potential in South Asia. A.H. Nayyar, A.H. Toor and Zia Mian in *Science and Global Security*, Vol. 6, No. 2, pages 189–203; 1997.

The Making of the Indian Atomic Bomb: Science, Secrecy and the Postcolonial State. Itty Abraham. Zed Books, 1998.

India's Nuclear Bomb: The Impact on Global Proliferation. George Perkovich. University of California Press, 1999.

Out of the Nuclear Shadow. Edited by Smitu Kothari and Zia Mian. Zed Books, 2001.

For articles by M. V. Ramana, visit
www.geocities.com/CollegePark/5409/nuclear.html

For extensive information on both countries' nuclear weapons, visit:
www.isis-online.org/publications/southasia/index.html
www.ceip.org/files/Publications/trackingTOC.asp
www.fas.org/nuke/hew/

SCIENTIFIC AMERICAN ARTICLES ON NUCLEAR PROLIFERATION

Since the dawn of the nuclear age, SCIENTIFIC AMERICAN *has published articles on nuclear weapons policy. A sampling:*

The Hydrogen Bomb: II. Hans A. Bethe. April 1950.

The Proliferation of Nuclear Weapons. William Epstein. April 1975.

Stopping the Production of Fissile Materials for Weapons. Frank von Hippel, David H. Albright and Barbara G. Levi. September 1985.

The Future of American Defense. Philip Morrison, Kosta Tsipis and Jerome Wiesner. February 1994.

The Real Threat of Nuclear Smuggling. Phil Williams and Paul N. Woessner. January 1996.

Iran's Nuclear Puzzle. David A. Schwarzbach. June 1997.

M. V. RAMANA and *A. H. NAYYAR* are physicists and peace activists who have worked to bridge the divide between India and Pakistan. Ramana, a research staff member in Princeton University's Program on Science and Global Security (www.princeton.edu/~globsec), is a founding member of the Indian Coalition for Nuclear Disarmament and Peace. He was born and raised in southern India and has written extensively on the region's classical music. Nayyar, a physics professor at Quaid-e-Azam University in Islamabad, is co-founder of the Pakistan Peace Coalition. He also runs a project to provide education to underprivileged children.

China as Number One

"The inferred assumption in most American scenarios is one in which a dominant China is a threat to its neighbors and the United States. Yet what if China acts as a benevolent hegemon, or at least a benign one?"

SOONG-BUM AHN

Within the next few decades, East Asia will once again become sinocentric. The prospect of an East Asia centered around a more powerful China is not as disastrous as some in the United States have argued. Fears that American interests will be threatened if China dominates the region are largely unfounded and potentially dangerous. Indeed, it is in America's national interest to come to terms with China's historic leadership in the region and foster a positive relationship that will facilitate a constructive role for China. The alternative—a confrontation with China—would set a perilous course for the United States.

Many in the United States who call for global American primacy will attack this argument as counterintuitive—that it advocates a premature American capitulation to the Chinese, and is fundamentally un-American and antithetical to the concept of primacy. Some may even dismiss it as unrealistic naïveté. However, the argument is counterintuitive only to the extent that thinking on these issues is tainted by outmoded cold-war strategies. Political realism would also dictate that we look at China and its future based on how it actually is and not how we would like it to be. To expect that the United States could contain or engage China unilaterally and at whim without serious consequences is unrealistic. Americans must once and for all abandon the concept of a "pro-China" or "anti-China" policy.

The rapid reemergence of China as a global power will return East Asia to a system resembling the region's traditional order. From the American perspective, this will require a reevaluation of roles, interests, and policies in the coming decades. To lay the foundation for a positive United States role in East Asia into the twenty-first century and beyond, the United States needs to review and rethink not only its policy toward China, but also toward Taiwan, the Koreas, and Japan.

THE RETURN TO SINOCENTRISM

The Chinese civil war that established modern China in 1949 was a major challenge to United States foreign policy It represented the "communist threat" on a massive scale and marked the onset of the cold war in Asia, which would cost millions of Asian lives and more than a hundred thousand American lives. America's cold-war record in East Asia is marred by critical policy failures in China and Vietnam—where the United States backed what proved to be the wrong sides—and in Korea, where the United States agreed to the initial division of the peninsula and then failed to protect South Korea by not including it among America's security commitments in Asia. In retrospect, the earlier events and decisions that led to the devastating cold war in East Asia still form the basis for America's current role in the region. The United States is doomed to repeat the tragic aspects of its history in East Asia unless it changes its policies to reflect major events unfolding in the region. In East Asia, China will be preeminent and it is in the United States interest to learn to live with and prosper in that region. The traditional United States foreign policy approach based on bilateral alliances that isolate China will not work. Indeed, this approach may lead to catastrophe.

Significant contrasts mark the Western experience in international relations and the traditional mode of relations among the states of East Asia. Whereas great-power multipolarity characterizes the international structure in Europe, the traditional East Asian system was generally unipolar. China, as the Middle Kingdom, assumed the role of the unipole for lengthy periods of East Asian history. Another point of contrast is the different levels of "anarchy" in the two systems. Modern political realists have based their study of international relations on the dynamics among comparable European states vying for power under an "ordering principle of anarchy." In East Asia, China's mass and gravity brought a hierarchical order to regional dynamics. Along with its power, the Confucian influence on international politics and the tributary system of trade enabled China to manage the regional system in a relatively peaceful manner until the nineteenth century.

> *The idea that the best way to safeguard United States interests in East Asia is to portray China as a looming threat is short-sighted and dangerous.*

The United States is a status quo superpower while China is a revisionist power. Ravaged by modern history, mainly in the form of Western imperialism, modern China under the leadership of the Chinese Communist Party has made national restoration the country's number-one priority. It seeks to undo history and regain its rightful place as a great civilization and power. This has been the major theme and a source of legitimacy for the party. In essence, China must struggle for regional dominance. Its history and size have instilled a need to dominate that is impossible to renounce. At the same time, the Communist Party cannot allow the United States to dominate Asia because the United States government continues to pose the greatest threat to Communist rule in China in the form of its military alliances with the surrounding states and because of its regular condemnation of Beijing's policies. Amid the gap in perception and divergence in goals, the best way for the United States to make an enemy out of China is to portray it as such—a potentially devastating case of a self-fulfilling prophecy.

In international relations, the perception of intent is more important than the intent itself. A large gap in perception divides the United States and China. Although current United States policy toward China may very well be generally defensive and benign, the Chinese see it in more malignant terms. United States attempts to prod and cajole the Chinese into playing by America's rules and, in particular, America's duplicitous relationship with Taiwan, are seen in Beijing as meddling and an attempt to set up roadblocks to Chinese progress. The more the United States tries to persuade China of the legitimacy of these policies, the more the Chinese are convinced to the contrary. This gap in perception widened considerably as a result of the United States bombing of the Chinese embassy in Belgrade during NATO's Kosovo air war in 1999, the recent spy-plane incident off Hainan Island, and the diplomatic fallout that ensued after each incident.

Obviously, a return to the traditional sinocentric tributary system is impossible to envision under modern international conditions. However, the emergence of an East Asia that is centered on a modernized, powerful China is not difficult to imagine. China would be the center of gravity for the region in terms of population, landmass, economy, and military strength. In terms of economic activity, China, with its large and growing middle class of consumers, is already an attractive market for the exporting economies of East Asia as well as a good source of low-cost products and raw materials. Additionally, the cultural aspect of China's role in the region cannot be ignored. To varying degrees, Chinese culture has had an enduring impact on most of the societies of East Asia. This element in the relationships adds to China's influence.

Admittedly, the region's future security environment will depend largely on Chinese behavior and the perception its neighbors form of its intent.[1] America's evolving relationship with China will largely determine the course and nature of China's emerging role. If the Chinese achieve domination after a long hegemonic struggle with the United States, that supremacy is likely to be harsh, destructive, and based on military power. If, however, Chinese primacy comes about under a positive Sino-American relationship, then China is likely to be a benign or even a benevolent regional power. The regional states will once again learn to live within a China-dominated East Asia, and a United States role in the region will be affirmed. In that case, we are likely to witness a China-dominated regional system not susceptible to American influence and coercion, but open to American engagement and trade.

> *The coming Chinese challenge is in the realm of national wealth, not ideology. Very few in Beijing would place ideology before profits.*

AGAINST AMERICAN PESSIMISM

The inferred assumption in most American scenarios is one in which a dominant China is a threat to its neighbors and the United States. Yet what if China acts as a benevolent hegemon, or at least a benign one? Many, including prominent scholars such as the late Gerald Segal in his book *Defending China*, have argued that historically, Chinese military action has been defensive or punitive in nature and seldom imperialistic. And the People's Republic's "Five Principles of Peaceful Coexistence," as laid out by the late Prime Minister Zhou Enlai, are based on the concepts of noninterference and no stationing of troops outside one's own territory. Thus, from the Chinese perspective, military force is used only for domestic stability (as in the case of Tibet and Taiwan) or national defense (as in the case of Korea, India, and Vietnam).

As East Asia continues to emerge as a regional system, a form of "soft balance" may be achieved whereby China's power is checked by other states in the region and by the United States. This would not constitute an unstable balance of military power in the traditional sense. Rather, in response to pressures placed on it by its economic interdependence with the other states, China would seek to form consensus on regional issues with the other players rather than risk conflict and enmity through bullying. In the larger context of Asia, India and Russia will be factors in an overall structure for balance. The United States can play the crucial role of offshore balancer.

The core of future United States engagement in East Asia is economics. This does not mean that the United States must maintain dominance in East Asia to secure its economic interests there. If this were the case, every global economic power would seek to gain dominance and monopoly over regions key to its economic health. Even after its dominance is secured, China would gain little from disrupting the stability that enables economic exchange and prosperity in the region: a breakdown of the relationships across the Pacific would be disastrous for the Chinese. Roughly 35 percent of China's exports go to the United States. Moreover, two of its most important trading partners, Japan and South Korea, also depend on their ability to export to the American market. With China's forthcoming membership in the World Trade Organization, mutual reliance on this trans-Pacific trading network will continue to expand. Admittedly, East Asia and the Pacific are far behind Europe in terms of regional cooperation and integration. However, it is erroneous to think that a rising China will inevitably set off arms races and instability. The shared desire for peace and prosperity that undergird the European Union are at work in East Asia as well. The coming Chinese challenge is in the realm of national wealth, not ideology. Very few in Beijing would place ideology before profits and even fewer would risk stability and wealth to spread a defunct ideology. To argue otherwise is hyperbole and primordialist.

THE UNACCEPTABLE ALTERNATIVE

A change in the global distribution of power has usually been a traumatic event in world politics, with the emergence of a new dominant power often accompanied by war. Historically, war has caused the downfall of a superpower, enabling the emergence of a new power, or two rival states have fought each other for supremacy. In the latter case, war resulted because the reigning hegemon resisted the challenge of the emerging global power. The United States must avoid this trap.

The idea that the best way to safeguard United States interests in East Asia is to portray China as a looming threat is shortsighted and dangerous. Playing this new China card may arouse American concerns, and gather some domestic support for an aggressive United States security policy to deal with the "threat." The East Asian reaction, however, will be counter to United States interests. China will respond by reallocating resources. Japan and South Korea may decide that United States policy is flawed and pursue independent policies to accommodate China.

Historically China has been a continental power and will continue to maintain its strategic focus on the Asian landmass.

Given this continental outlook, any potential adversary must be prepared to engage in protracted land campaigns that would require troop-intensive formations and the expectation of large numbers of casualties. Fighting China in a protracted land war in Asia would be a disaster for the United States because American forces would not be able to threaten China's traditional centers of gravity—its vast landmass and population—without great loss of American life. In the absence of a direct threat to United States national security, it is difficult to imagine a scenario where the American public would support a conflict with China over dominance in East Asia, especially in light of the American aversion to casualties and mishaps, even minor ones.

For example, on Taiwan the United States opposes the use of military force and implies that it would risk a confrontation with China for the "security" of the island (President George W. Bush has made this more explicit by stating in a rather unstatesmanlike manner that the United States would indeed come to the aid of Taiwan). Yet a great asymmetry can be seen in the importance that China and the United States place on this issue. Rhetoric aside, the People's Republic is fully committed to preventing Taiwanese secession, even at the cost of a large-scale military conflict with the United States. Conversely, it is hard to imagine the United States sacrificing large numbers of Americans in a war with China for the sake of Taiwanese "security" (an assessment the Chinese also share).

Any future conflict scenario for East Asia must assume that China will fight for survival as a coherent state, whereas America will fight for a precarious dominance over a region that sees it as a distant foreign power. For the United States, a hegemonic war with China in East Asia would prove too costly, and in the end, unwinnable.

KOREA: BALANCING OR BANDWAGONING?

The Chinese strategic view of Korea was historically that of a reliable client state and buffer on China's eastern flank. Today the Chinese consider a divided Korea a way to maintain the regional balance of power and a check against the United States. Barring the continuation of the status quo on the peninsula, China's desired outcome is a peaceful reunification process resulting in a pro-Chinese Korean state without an alliance with the United States and without foreign troops. One could even envision a Finlandization process whereby Korea assumes a neutral status within the Chinese sphere. Some lesser-desired scenarios for China are Korean reunification resulting in the strengthening of the current South Korea–United States alliance, or worse yet, a truly independent Korean state with strategic capability.

According to the literature on international relations, states facing a threat generally have two options: balance against the threat, or "bandwagon" or align with the threat. In the case of Korea with regard to China, it appears that, if forced to make a choice, a reunified Korea will bandwagon with China. Besides the historical precedent for its close relations with China, compelling geopolitical reasons can be made for Korean bandwagoning, such as lack of a reliable long-term ally, geographic proximity to China, and the overwhelming nature of the potential threat. The preferred strategic option for Korea is to align with a distant power to balance against the local threat. The United States has been the ideal distant ally. However, as China continues to grow in importance and the rationale for a United States presence fades, Korea will lean toward China.

China also looms large for Korea in the context of south-north reconciliation and reunification. During a transition period on the peninsula, China will exert its considerable influence to shape the resulting political-military landscape. Considering China's potential impact on reunification for good or ill, it is clearly in Korea's interest to consider how to gain China's support for a reunified Korea. And China will leverage the Korean yearning for reunification to win political concessions.

Korea's strategic calculations are unique in that its disadvantageous geopolitical position requires it to be constantly mindful of the plans of the greater powers around it. For this reason, China is a major factor in Korean strategic calculations. When China does come to dominate East Asia, Korea will thus likely lean toward it to safeguard its national interests.

JAPAN'S DILEMMA

From the Chinese perspective, Japan has historically posed a threat. The failed Hideyoshi invasions of Korea in the sixteenth century were intended to provide a staging area for the conquest of China. At the end of the nineteenth century, after remaking itself in the image of the Western imperialist powers, Japan began to assert itself on the Asian continent, defeating a weakened China and Russia, annexing Korea, and establishing a puppet state in northeast China.

Since the end of World War II, Japan has served as the center of the United States presence in the region and host to numerous United States military bases. The end of the cold war and the diminished North Korean threat put into question the rationale for these bases and the United States–Japan alliance itself. The danger that the alliance will be perceived as an anti-China front is real and should concern all in the region. From the Chinese perspective, the purpose of such an alliance is to thwart the rise of China and maintain the status quo in East Asia. The Bush administration's abandonment of former President Bill Clinton's "strategic partnership" rhetoric, the identification of China as a potential future adversary by defense officials, and Deputy Secretary of State Richard Armitage's call for an increase in Japan's military capabilities add weight to China's perception.

Like Korea, Japan will probably choose to lean toward China, especially if it begins to question United States resolve and commitment. As in the Korean case, the perception of United States commitment to the region will be crucial to Japan's decision to continue to link its national security to the United States. Also as with Korea, the United States will not be able to offer a security commitment to Japan sufficient to convince Japan to maintain the alliance, at least in its current form. United States defense-treaty commitments would ring hollow without a military presence. However, that presence is already questioned in both countries, with organized calls for reduction or withdrawal. In the near future, absent a clear Chinese threat, the United States will find itself in the difficult position of advocating a forward presence to show resolve while opposition in the host countries gains increasing support.

Japan will be the most likely battleground in a China–United States rivalry in East Asia. It cannot afford to bear the brunt of the negative fallout from a Sino-American confrontation. If Japan concludes that China's hegemony is inevitable, that it will not be malevolent, and that the United States commitment is less than assured, it may disengage itself from the United States, seeking an equidistant relationship between the United States and China or even an accommodation with China. China and Japan share many common, long-term interests that can lead to compromise and cooperation.

Absent an alliance with the United States, Japan does have more options than Korea but they will be difficult and dangerous to bring to fruition. For Japan, balancing against China through an independent defense buildup will not be easy. Any sizable Japanese remilitarization is likely to make the region more unstable. And it would be dangerous if Japan were to pursue such a policy, since it would likely force China (and Korea) to adopt an aggressive posture toward Japan, thus making Japan more vulnerable.

For China, Japan will continue to be a crucial source of technology and capital as it modernizes its economy. Likewise, as its economy and its middle class continue to grow, China will become an important market for Japan's consumer goods. It is already a large market for low-cost goods and an important source of raw materials. Even in the age of rapid globalization, the attractiveness of such a huge regional market has the potential to overshadow the economic linkages across the Pacific Ocean. China's regional dominance will afford both China and Japan easier and more reliable access to the assets crucial to their long-term development.

Finally, China's fear of a resurgent Japan (and hence its implicit support of continued United States military presence) is much exaggerated. This "cork-in-the-bottle" argument is useful as a rhetorical tool but lacks power to be considered seriously. All factors, including Japan and the Japanese, are fundamentally different today than they were before World War II. As discussed earlier, without a clear and direct Chinese threat to Japan, any attempts at Japanese remilitarization would prove counterproductive, and hence be a poor policy option. Also, China's desire to supplant the United States as the dominant power in East Asia overshadows any benefit China currently derives from the United States presence in the region.

AMERICAN IMPERATIVES

Since the end of the cold war, the American foreign policy community has been searching for a new paradigm and a new focus. As America looks at China, its perceptions are distorted by the lingering effects of obsolete cold-war strategies. The concept of a threatening China, as reflected in a number of Chinese "threat" articles, is an example of this. American policymakers cannot adopt a single ideology or theory as a guide for the future. The United States must abandon zero-sum calculations that dictate an either/or dichotomy that ultimately leads to a China–United States conflict. Likewise, it should be wary of overly sanguine perspectives that predict peace and prosperity for all. The actual consequences of China's rise in East Asia will be more a function of the nature of the China–United States relationship and the environment in which the transition takes place rather than a predetermined chain of inescapable events based on theory. America should reject policies based on short-term interests and view its East Asia role through a long-term perspective based on enlightened self-interest.

Looking to the midterm future in East Asia, the United States should assess its interests within the context of a regional order centered on China. America's main source of influence will come from its regional economic linkages. Therefore, it should intensify economic cooperation with the states of East Asia, especially China and its key trading partners, Japan and Korea. In the near term, the United States should help restructure the Sino-American relationship into an equal partnership in East Asia to aid in the transition of power from the United States to China over the next several decades. This will enable the United States to play a guiding role as China continues to reform and ascend, allowing for a smooth transition to Chinese dominance and minimizing the possibility for conflict during this unstable period. The United States should abandon its policy of engaging China only if it fits into its own design for the region. No real progress in the Sino-American relationship can occur if the United States insists that everyone must play only by its rules. Historical precedent shows that the Chinese will not engage the United States under those conditions.

Based on their partnership, China and the United States should promote multilateralism. This would facilitate the development of a formalized regional framework of military, economic, and political ties. Building on existing relationships and influence, China and the United States could facilitate confidence building, gradually producing frameworks for stability and cooperation in East Asia. In addition, Americans must recognize that values are in fact relative across different levels of political and economic development and across different cultures, and that the gap in values between most of East Asia and the United States is still quite large. The United States should thus stop promoting American values regarding human rights and democracy to the societies of East Asia, which resent this intrusion. Although this will be difficult, the United States government must not overreact to the insecure belligerence of the Chinese Communist Party and the Chinese military. American policymakers must see beyond Communist rule and seek to establish a relationship with the Chinese people that is based on mutual respect and cooperation; the system in Beijing is bound to change eventually. Finally, the United States should be consistent in its dealing with the states in the region.

For Taiwan, the current United States policy is ill conceived and dangerous and must be brought in line with its commitment to the one-China policy. American allies in the region have already expressed their concern over what they see as a dangerous United States tendency to entangle them in a conflict with China. A war over Taiwan will rupture the United States-centered bilateral alliance structure in Asia. An American abandonment of or setback in Taiwan, at a time of crisis, also would be a major blow to United States prestige and credibility since it would confirm suspicions regarding the "hollowness" of United States might.

The continued United States support of Taiwan also militates against its interest in promoting democratization in China. Although the Chinese are split on many issues, they are uniformly supportive of the Communist Party's stance that Taiwan is an integral part of China and that the issue is strictly an internal matter. Taiwan is a good way for the party to muster popular support and firm up national solidarity, especially in times of domestic turmoil. And the United States in effect helps the Communists maintain a monopoly on power in Beijing by continuing to raise Taiwan as a banner for the party to wave. The best way to secure Taiwanese security and encourage peaceful Beijing–Taipei interaction is to reaffirm America's commitment to Chinese sovereignty by cutting all military ties to Taiwan and insisting that the island abide by the one-China policy.

For Korea, the United States must recognize the unique geostrategic position that Korea occupies. The United States must also recognize that a reunified Korea will once again lean toward China as it seeks to secure its interests. A key factor is that the Sino-American relationship will greatly determine the nature of the United States–Korea relationship. For now, China and the United States should foster close relations with both South and North Korea so as to play constructive roles during the reunification process.

For Japan, the United States must help it normalize relations with its neighbors. East Asian security and prosperity and Japan's place in the region cannot be fully secured as long as Japan's relations with its neighbors are constrained by history. Japan's failure to fully reconcile with its past is a lasting obstacle to better relations with its two most immediate neighbors, China and Korea. Correspondingly, the focus of the United States–Japan alliance should shift from security concerns to economic and global issues. This will demilitarize United States–Japan ties and enhance the ability of both countries to deal with global issues. It will also minimize Chinese anxiety and mistrust.

The best outcome for the entire region would be a network of densely interconnected relationships regionally and across the Pacific that would balance all actors, moderate their behavior, and promote cooperation. This would be a long-term stable arrangement that accommodates China's overall regional primacy while safeguarding American economic interests and establishing a regional balance of power by nonmilitary means.

This vision of East Asia's future bears out the dangers of current United States policy and its probable negative outcome. It also acknowledges China as the long-term regional dominant power and maintains that peace and prosperity can be achieved only through mutual compromise by the two powers. It seeks to prevent the emergence of a new bipolar confrontation in the region and recognizes that the realist-based tendencies of China will require the United States to come to terms with it sooner rather than later. And it recognizes that the United States can work with China now to ensure that East Asia develops into a peaceful and prosperous region from which the United States will benefit.

A NEW OUTLOOK

United States interests in East Asia have changed since the cold war, and the United States needs to build a new foundation for its role and policies in the region. China's rise to displace the United States regionally in the coming decades will test America's capacity to respond to a change of this magnitude. United States interests lie in a stable and open region in which it continues to play a significant role while China assumes the mantle of dominant regional power and police officer. Taking on the number-two role in East Asia would allow the United States to continue to benefit from a dynamic region and at the same time contribute to the maintenance of peace and prosperity. A reduced United States role would mean a significant trimming of its massive array of military assets in the region; the current costs incurred by United States primacy would be greatly reduced across the Pacific. Another important outcome is that the United States and China will be better able to form a durable relationship. As a committed revisionist power, China will most likely not reconcile with a United States that is dominant in a region that China sees as its own sphere. The United States as number two will no longer represent the Western imperialism that the Chinese consider a major cause of their historic downfall and current second-rate status. A China that is number one is less likely to behave aggressively or threaten its neighbors to secure its position and interests. Rather, it will find it in its interest to promote regional cooperation and global stability. The United States can assure a smooth transition to this new structure in East Asia by engaging China in a constructive partnership today.

According to Samuel Huntington, in a multipolar world "there is no reason why Americans should take responsibility for maintaining order if it can be done locally." For years, the United States has provided the common good of stability in East Asia, and all have benefited. It is time to let the Chinese provide this common good so the United States can benefit.

Note

1. Currently, China's interests are served by a United States military presence in East Asia because it promotes stability. However, once the North Korean threat is substantially reduced or removed, an American military presence will no longer be in China's interest. A precondition for continued United States military presence in East Asia is American leadership in the region, a role in direct opposition to the Chinese view of China's rightful place in the region. A comprehensive regional framework for security cooperation is one scenario where a limited United States military presence may coexist with a regionally dominant China. This would require a formal multilateral structure where a United States military presence would clearly be for regional collective security and not part of an anti-China alliance.

LIEUTENANT COLONEL SOONG-BUM AHN *wrote this essay while serving as a United States Army Research Fellow at RAND. He is currently the United States Army Senior Service College Fellow at the Korean Institute for Defense Analysis in Seoul. The views expressed here are solely those of the author.*

Reprinted from *Current History*, September 2001, pp. 250-256. © 2001 by Current History, Inc. Reprinted by permission.

Battlefield: Space

Space-based warfare used to seem pure fantasy. Now,
to the delight of war planners, and to the dismay of many civilians,
it's closer to reality than you'd think.

By Jack Hitt

The Defense Department's newest satellite technology, War-fighter I, sits inside a protected clean room in Germantown, Md. To enter, you must run your shoes through a cleaning device and then don a "bunny suit," a layered hooded outfit that covers every part of your body except your eyes.

"Human skin sloughs off as many 30,000 particles a second," says the program manager, Michael Lembeck, as we step onto a tacky mat, essentially an enormous piece of fly-paper. "If one speck of skin got on the Warfighter's lens," he adds with friendly hyperbole, "it would set us back 20 years."

The satellite, which is not much bigger than a college soph-omore's dorm refrigerator, is undergoing final tests. Several different machines—producing an artificial magnetic field, dig-itally created blinking stars, phony sunshine and computer-gen-erated Global Positioning System signals—are fooling the satellite into acting as if it were in real orbit. Several lights click on and motors grind. "It must think it just cleared the North Pole," Lembeck says, "and is reorienting itself toward the sun."

After a few more tests confirm the on-board systems are working, Lembeck says, "We'll get all the graybeards in the room, tell them what we've done here and they will bless us and say, 'Go fly.'"

The Next Generation

The U.S. military has plans for a diverse arsenal in space. Microsatellites could sabotage enemy satellites with high-power microwave beams. Flechettes are rods that could theoretically be dropped from space to smash into targets on earth. And laser cannons could, the Pentagon hopes, zap targets anywhere on the planet.

In fact, Warfighter I is an extremely powerful camera, one that will give the Pentagon revolutionary new powers of sur-veillance. But its importance goes beyond its technological wiz-ardry. The launch of Warfighter—scheduled for early September—will mark the latest effort by the Pentagon to end a new threat to American security. According to the nation's war planners, America has had a free ride in space during the last 40 years, when the only country capable of even getting there was Russia. Now there is a satellite rush in the final frontier, with both countries and companies entering space. Commercial space launches started to outnumber military ones in 1998. Of the 1,000 active satellites currently in orbit, about an eighth be-long to the U.S. military, and that percentage will diminish by the end of the decade, when experts estimate that operating sat-ellites in space will reach 2,000. (Warfighter is being launched by a private company called Orbital Imaging, itself a sign of the times.)

America's war planners fear that we could soon lose our ad-vantage in space. As a result, the military has commissioned nu-merous studies and long-range plans, all of them coming to the same conclusion. Space, the Pentagon believes, is the ultimate military "high ground"—the tower from which to pour boiling oil. Therefore, America's goal there should be, in the felicitous phrase used in an early study, "Global Battlespace Dominance."

Perhaps that term sounded a little too Strangelove, for the Pentagon's preferred phrase has since become "Full Spectrum Dominance." Last year, the Air Force developed its Strategic Master Plan for space, which states our goal bluntly: "To main-tain space superiority, we must have the ability to control the 'high ground' of space. To do so, we must be able to operate freely in space, deny the use of space to our adversaries, protect ourselves from attack in and through space and develop and de-ploy a N.M.D. capability."

N.M.D. stands for national missile defense, the controversial $8.3 billion missile shield that President Bush and his secretary of Defense, Donald Rumsfeld, have championed. (Last month, the Pentagon announced that it was ready to pour concrete on the first missile-defense test site, in Alaska.) And yet the political attention devoted to national missile defense, which is an updated version of President Reagan's Strategic Defensive Initiative, has obscured its larger purpose. According to the Strategic Master Plan, N.M.D. is but one part of a triad of technologies—along with improved space surveillance and antisatellite offensive weaponry—that, the Air Force hopes, will lead to total "space control." George Friedman, an intelligence consultant and the author of "The Future of War," calls the national missile defense plan a "Trojan horse" for the real issue: the coming weaponization of space.

The cost of expanding our space assets is only now beginning to show itself. Many of the specific systems for space have had their budgets increased in President Bush's first defense-spending proposal, which has been otherwise criticized for being stingy. A new system of space sensors went from $239 million to $420 million. (By comparison, the Air Force's new F-22 Raptor fighter plane has a price tag of $180 million.) A previously unfunded space-based radar program is budgeted at $50 million. And a line for "space control technology"—a euphemism for antisatellite weaponry—was expanded from $8 million to $33 million. Carefully budgeted space technologies like the Warfighter will cost only $42 million, but the more exotic ideas face a long climb up the technological curve and will cost billions.

Warfighter's camera features a new form of imaging called hyperspectral. Space is already home to multispectral cameras, which can take a picture of an ecosystem and discern conifer from deciduous trees. But hyperspectral goes much further, distinguishing the subtle "light signatures" that separate a field of oats from barley and telling you the precise *species* of oats. And then whether the field contains natural or genetically altered oats. And then whether the field is infested with insects or damaged by nitrogen depletion.

The eventual commercial potential of such a technology is obvious. But if you talk to enough colonels and experience what old Pentagon hands call "death by briefing,"—and I have—you will hear mentions of hyperspectral quickly followed by the new mantra of contemporary war planners: *tanks under trees.* To put it briefly: as with oats, so with tanks. Warfighter I will be able to discern the unique light signatures of extremely specific things—like tanks hiding under trees or tanks covered in camouflage or tanks painted with a paint meant to make them not look like tanks.

Consider what such space-assisted technology would have meant to a commander in, say, Kosovo two years ago. He could have swept the contested area with Warfighter I and zeroed in on every enemy tank, missile, ammo dump or plane, almost no matter how hard the Serbs tried to conceal them. Then the commander could have called in a cruise missile to blast each one. In theory, the entire conflict could have been finished off in time for lunch. It's a nice, sweet, hammock-tempting image if you're a war planner.

In preparation, space planners have already engaged in some feverish brainstorming. They envision a high-tech arsenal that will take full advantage of the military potential of space, ranging from the near-term possible to long-term notional: ki-

netic energy rods, microwave guns, space-based lasers, pyrotechnic electromagnetic pulses, holographic decoys, robo-bugs, suppression clouds, 360-degree helmet-mounted displays, cluster satellites, oxygen suckers, microsatellites, destructo swarmbots, to name a few.

Some civilians find these plans deeply troubling. "If you start talking about putting actual weapons in space, you can take the unhappiness that our allies, Russia and China already have with the missile shield and multiply it by 10," warns Lisbeth Gronlund, a physicist with the Union of Concerned Scientists. Such critics see the Pentagon's effort to weaponize space as profoundly dangerous for national security—not to mention expensive and potentionally unfeasible.

"Once you start spinning this baby out," says Dan Smith, an analyst with the Center for Defense Information, "it becomes more complex, more expensive and more impossible to protect ourselves. After the next country introduces space weaponry, *then* what do we do? Live with a new, unpredictable threat orbiting right above us? Or commit an act of war by pre-emptively removing their weapons from space? The basis of security is that it never works for just one. You have to have security for everyone or it fails."

Not surprisingly, the *Realpolitik* leadership at the Pentagon disagrees. Just before taking over Defense, Rumsfeld led a space commission that was established not long after Congressional Republicans grew enraged that Clinton had line-item-vetoed funds for a space plane, antisatellite weapons and a missile-defense technology. The commission issued its report nine days before Bush was sworn in as president, and it concluded: "Every medium—air, land and sea—has seen conflict. Reality indicates that space will be no different." And Warfighter I, it turns out, is the beginning of a many-splendored arsenal to ensure we're ready for battle when it does.

MUCH OF THE MILITARY'S RESEARCH INTO SPACE TECHNOLOGY takes place at the Space Research Lab. It is divided into 10 missions scattered across the country, ranging from the Propulsion Directorate to the Munitions Directorate. On a blazing hot afternoon in June, I arrive at Kirtland Air Force Base in Albuquerque to get cleared into the Space Vehicles Directorate, which specializes in satellite technology. Many outposts of the emerging bureaucracy of space distill their enthusiasms into a shoulder patch. The First Space Operations's patch shows stars and a plane above the words "Always in Control." The 50th Space Wing's logo is an image of Pegasus above the claim "Master of Space." Some divisions have more informal slogans. The motto of the Space Warfare Center is "In Your Face From Outer Space."

I first meet with Alok Das, the head of the Space Vehicle Directorate's innovative concepts group. His latest work has been perfecting the microsatellite. Unlike traditional satellites, which can weigh tons, microsatellites are the size of a suitcase and weigh about 200 pounds. Since it costs "a bar of gold to launch a can of Coke," as Das put it, lightweight microsatellites will be much cheaper to launch than their obese precursors. The idea is to send microsatellites into space in flocks. In this cluster, they would be reprogrammable, able to switch to new tasks when the

Pentagon required it. They might be set in linear formation to conduct ground reconnaissance or grouped in a circle to serve as a communications satellite. "It's like going from a mainframe computer to a network of PC's," Das says brightly. "Together, they'd form a larger virtual satellite."

Yet a flock could also be launched with separate missions. One microsatellite might refuel a larger satellite or upgrade its software. Others might scoot about with small on-board cameras to provide live video feeds from space—a capability no nation currently has.

A microsatellite could do both surveillance and sabotage. As one official explains, 'It could go right up to an enemy satellite and look at it real close—maybe even bump it.'

As I am escorted into a clean room to the see the first microsatellite under construction, one officer offhandedly confides, "It could also go right up to an enemy satellite and look at it real close—maybe even *bump* it."

That's how easy it is to go from peaceful mission to offensive weapon. A suitcase-size microsatellite would just have to put a little shoulder and some thrust into an S.U.V.-size satellite to push it off its proper orbit and render it temporarily unable to communicate with the ground. Another idea is to mount a microwave gun on board so that once the microsatellite maneuvered right beside an enemy satellite, it could emit a pulse of microwaves and fry the electronics permanently. Space planners call this application a high-power microwave pill. Better yet, this microsatellite's sabotage operations would be covert, undetectable from earth. It would give a nation complete deniability: that Chinese satellite that Saddam Hussein has been using doesn't work? Must have been a solar storm.

The first microsatellite launch is planned for this fall.

Later, I talk with the lab's experts in hyperspectral imaging. How, I ask them, will the Warfighter learn the precise "light signature" of, say, a tank hiding beneath a pine-forest canopy?

"Think of them as fingerprints," says Tom Cooley, one of the lab's top researchers. "The wavelength of any kind of camouflage, regardless of composition, can be distinguished—by the dyes, cotton, different lignants from plants. If you look at black-and-white images of camouflage next to scrub brush, they look the same. But a leaf from the scrub brush does not look at all like camouflage to hyperspectral. It would be sharply different."

Before hyperspectral can work, it will require some novel research and testing, says Col. Jack Anthony, chief of space experiments. "Take a tank under a tree," he says, explaining some coming tests. "We'll take some panels made of wood and paint them with different paints, government paint, some paint you might buy at a store. Then we'll take some images with the Warfighter I, and that will give us what's called 'truth.'"

To build what Anthony calls "a library of light signatures," a lot of truth will have to be collected. All possible contingencies—tank under trees, tank under branches, tank under government paint—must be cataloged, one by one. "So if the bad guys are hiding tanks under trees," Anthony explains, "and you have a good idea what the bad guy's tank is made out of and you know what the local trees look like, then you can screen out the trees' wavelength and just see the tank's signature. Then you're going to know there's something bad under that tree. And we can arm our soldiers accordingly."

Cooley adds that "anything from Somalia to Bosnia to Haiti would have dramatically different backgrounds," making it necessary to bank in a library the differences among, say, Honduran swamps and Libyan deserts. "And by the way, water vapor is terribly opaque and will cause the special signature to be completely invisible." However, Cooley continues, another project will be to gather data in order to "correct for water vapor that may blur some of those special features."

To a civilian, hyperspectral surveillance can sound amazing and then—once you hear about light-signature libraries and water-vapor snafus—it can seem a bit iffy, about as dependable as launching a Xerox machine into the stress of low-earth orbit and then counting on it to work during a war.

That's how the Pentagon's critics see it. "There are *already* countermeasures for this kind of technology," Lisbeth Gronlund says. She describes a new kind of camouflage that entails bundling, say, two dozen Mylar balloons beside a nuclear warhead. After launch and the boost phase, the balloons and the warhead are scattered into space. Each has a slightly different light signature. So which target do you shoot down? "The military is *very* sensitive about this problem," Gronlund says.

Yet Anthony is doggedly optimistic. He believes that hyperspectral could be working successfully in the battlefield before the end of the decade. And he thinks the technology will help save lives: "It makes me feel good if I can help a soldier, sailor, airman, marine to know there is something bad hiding on the other side of that hill. We're just putting another arrow in our quiver."

Anthony's robust enthusiasm for space is shared among the research scientists. This enthusiasm is extraordinary. The Nasdaq bubble that burst around election time last year has not affected the military. Space-wise, war planners are prebubble techno-enthusiasts. (And their visions of space warfare are as cinematic as a summer blockbuster. Just look at the language: "Full Spectrum Dominance," "destructo swarmbots," "robobugs." It's hard to imagine the Pentagon's idea of space without Hollywood's.)

Inside the military, all technological setbacks—like the fact that two out of the four major missile-defense simulations conducted so far have failed—are set aside as part of the natural arc of any technological testing. Failure is just proof that there needs to be more research. But the real reason the military is so excited by space is that so much that is already up there, both civilian and military, works splendidly. Nearly all the emblems of our technologically quotidian life—the A.T.M., credit-card transactions, cell phones, the Internet—rely upon satellites.

150

When space technology has catastrophically failed, the public's reaction has not been greater skepticism but mere annoyance. In May 1998, the Galaxy IV satellite malfunctioned, causing 45 million pagers to shut down and credit-card transactions to cease. The public did not decide to return to making house calls, paying cash and reading by candlelight: it simply expected it to be fixed because it has so internalized the presumption that such technology works, and works wonders. And so has the military.

If the A.T.M. is the shorthand symbol of how easy modern space-based technology has made our lives, then the precision-guided munition is that symbol for the average grunt. The invention of a missile that can be aimed after it has been fired has fundamentally changed modern warfare. It is why arguments about the possible failure of new technologies bounce off space researchers as if off a force field.

Back in World War II, it took, on average, 5,000 bombs to take out one target like a bridge. By the time of the Vietnam War, the ratio had dropped to 500. But in all those wars, bombs were dumb, meaning once you let go of them, they fell in the general direction in which they were pitched.

Then came the gulf war. During this conflict, the U.S. military used space to conduct nearly all of its secret communications, reconnaissance missions and bombing raids. And space-based technology guided new "smart bombs" with such accuracy that the hit ratio plummeted to 1 in 10. "The 500-year history of ballistic warfare has come to an end," George Friedman says. "The gulf war was the first space war."

Although not of the same scale, one notable fact of the Kosovo conflict of 1999 is that no Americans died in combat. Military planners credit that result in part to munitions directed by the Global Positioning System, a constellation of 24 satellites orbiting the earth that is capable of precisely geo-locating any object equipped with the proper receiver. Couple such technological progress with the ultimate lesson of Vietnam—no body bags on TV—and you begin to understand the military's profound enthusiasm about space and why there has been so much blue-sky planning to maintain "Full Spectrum Dominance."

INSIDE THE LAB OF THE DIRECTED ENERGY DIRECTORATE, where research on everything from microwave beams to lasers begins, the machines thrum to a start. A long pipe of fuzzy purple light in a large tube seems to vibrate like a plucked string. In an adjacent chamber that has had most of the air removed to mimic the high altitude of a missile trajectory, a piece of carbonized steel like that which might clad a rocket fuel tank is set in a grip. It begins to spin rapidly to simulate a missile in its ascent. Visual access to the vacuum room is supplied by a closed-circuit television. Technicians call out from one system to another that they are ready. The machines screech into action. On the TV screen, the piece of spinning metal is suddenly blasted with bursts of columnated light that scorch it, back-splashing in a dramatic laser fan.

"We're testing the laser's effect on what would be the body of a rocket spinning in flight," says Capt. Eric Moomey, the chief of this facility. (His insignia reads "Peace Through Light.") In effect, what I am seeing is a small part of what might one day become the national missile-defense shield.

At one point, Moomey clamps a four-inch-square piece of thick plexiglass in a C-clamp and orders the crew to fire up the laser. We all put on safety goggles as the laser shouts for a portion of a second. Burned neatly in the center is an indentation, just big enough, the captain tells me as he hands me the square, to hold a coffee cup. It is holding mine right now. I suspect that my souvenir coaster is not the first of its kind.

Such laser parlor tricks suggest just how far we've come since President Reagan first suggested this idea. Back then, the technology was far off and impossible. The Strategic Defense Initiative amounted to a bluff against the Soviets, and in the end it collapsed amid political ridicule. Back in the early 80's, the idea of shooting down a missile with another missile was widely scoffed at as trying to "shoot a bullet with a bullet." The Star Wars program specifically designed to do this was called Brilliant Pebbles. Besides being technologically complex, it frightened many people with its inherent idea: ringing the planet with thousands of space-borne projectiles, each of which could drop down into the atmosphere to collide with an enemy's missile.

Brilliant Pebbles is now being revived by President Bush, but given the instantaneous speed of lasers, it may soon be joined by a companion technology. With the ability to lock onto the trajectory of a missile, Moomey explains, you might be able to aim an air-based laser at an enemy missile's fuel tank and rapidly heat up the cladding so that "the liquid propulsion vents out and it rips open like a tin can." Moomey says that this kind of laser defense weapon, budgeted at $11 billion, should be operational sometime around 2010.

I next speak with Doug Beason, another expert on laser weaponry. Colonel Beason is a thin, amiable man and a widely read scientist. His magazine rack has well-thumbed editions of Sky and Telescope, Science and Wired. He is the author (sometimes co-author) of 10 novels, including "Virtual Destruction," "Assemblers of Infinity" and "Assault on Alpha Base." A few of his works have just been issued in paperback. When I casually use the word "sci-fi" in a sentence, Beason stops me politely to say that "techno-thriller" is the genre in which he labors. Sci-fi is a "50's expression," he says, trying to be cordial, even though it's clear that I've committed a faux pas on the order of asking Jane Campion about her next chick flick. There are bright lines in Beason's world—between techno-thriller and science fiction, but also between research that looks great on paper and technology he can help put in the hands of an American space warrior.

"The time between invention and mass use of the fluorescent lamp was 79 years," he says. "For the jet engine, 14 years; for the wireless, 8 years."

> The Warfighter I camera, riding on [a] satellite, uses "hyperspectral" technology to identify enemy targets on earth—even those hidden beneath camouflage. The results of such surveillance could be beamed to an Air Force bomber pilot.

This lag time is shrinking rapidly, he says. "We have the tools to exploit the technology, and that's why I'm so excited. Lasers, for example, are no longer used just for CD's and light pointers."

As a result, the Pentagon has its hopes set on a space-based laser. President Bush doubled the research budget this year to $165 million. The estimated cost for a working space laser test is about $5 billion. Actual testing in space is expected to take place as early as 2008.

"This is the technology that can provide the next revolution in military affairs," Beason says, "the Buck Rogers kind of thing."

He adds that lasers have many warfare applications besides outright weaponry. "We've also been working on a flexible-membrane mirror," Beason says, one that would be deployed in space. Then, from earth, a commander could fire a certain frequency of laser, bounce it off the mirror and "onto the battlefield to light up the night only to people with certain types of goggles."

Whenever I express any sense that these technologies sound a bit too, um, sci-fi, Beason responds the same way all his colleagues do. "These are all *concepts*," he explains, "and like any weaponry in a mature technological arsenal, it all depends on how much money you want to spend." Men like Beason are supremely confident in the technology; it's the political will to have space-based weapons that's the problem.

The peculiar thing about space warfare is that many of the innovations that sound the most far-fetched—like illuminating a battlefield at night with light that only one side can see or the deployment of high-power microwave pills—are actually much closer to existence, technologically, than some items that might seem more logically in line for development. Consider the spaceplane. It would be a tremendous tool for the military, since it could get to any point on the globe in a few hours. But building a manned craft that can quickly glide in and out of low orbits has proved incredibly daunting. Earlier this year, the X-33, NASA's big experiment in flying into space, ended in failure. The image that most people have of "Star Wars"-style combat—manned spaceplanes engaging in dogfights near the moon—is very far off. But the use of space for weaponry directed back at earth or guided from space is pretty much at hand.

"I'm particularly excited about high-power microwaves," Beason tells me. Lacking the thousand-mile reach of lasers, H.P.M.'s, as they are called, can be projected only about a half-mile. But were an unmanned plane guided from space able to transport a high-powered microwave device close to a battlefield, the possibilities could push the Pentagon's bomb-to-target ratio even closer to perfection. To an invading army of modern soldiers, a massive hit by high-powered microwave could ground their high-tech weapons, leaving them to wage modern warfare with their fists.

The time lag between the current R.&D. on microwaves and its application in the battlefield may be a while. Beason himself estimates 15 years, although one use is on the verge of showing up in battlefields soon. On the ground, a microwave weapon could be used to drive back an invading squadron. "It'll feel like opening the door of an oven," Beason says.

"We're testing it on humans now." He pauses and worries that he is bumping up against classified information. "If you want to know more," he adds, "you'll have to contact the Human Effectiveness Directorate."

THE PENTAGON'S PASSION FOR SPACE ALSO DERIVES FROM the thrill of discovering the medium's own peculiar disadvantages and advantages. True, you have to worry about new problems—space debris traveling at 16,000 miles per hour, solar flares, the Van Allen radiation belt. But it is never overcast in space, the field of vision is planetary and the speed of light is really, really fast. For the far term, war planners have conceived scores of new and exciting weapons. Talking about them is not a conversation the military wants to have in public, given the gnarly debate over the missile shield, but it is one they have been having in private for some time.

Among the internal reports generated by the war colleges and service branches are a half-dozen that imagine how space will be integrated into the U.S. military: The Strategic Master Plan, New World Vistas, Long Range Plan, Guardians of the High Frontier, Almanac 2000, Joint Vision 2010, Spacecast 2020 and Air Force 2025. Taken together, they form an encyclopedia of our war planners' dreams.

Any military response in the future would rely heavily on technologies aloft in space or directed from there. As a result, the U.S. Air Force will little resemble the service as we now romantically conceive it. According to a study entitled Counterair: The Cutting Edge, "uninhabited aerial vehicles will be widespread in 2025." Our new fleet of pilot-free planes would be directed from space and would range from small devices permitting a squadron leader to see over a hill to much larger craft that could deliver powerful weapons to a distant battlefield with tremendous speed. For example, one notion for an unmanned space-directed vehicle—called Strike Star—could "loiter over an area of operations for 24 hours" to deliver "'stun bombs' producing overbearing noise and light effects to disrupt and disorient groups of individuals."

Weapons like the Strike Star would exist on earth but be orchestrated from space. If we can get used to the idea of weapons actually in space, though, then a new arsenal would emerge. For example, if a laser cannon were to be inserted in space, its potential as an offensive weapon would make a cruise missile look like a firecracker. Why? Because, according to one study on directed energy, "a full-power beam can successfully attack ground or airborne targets by melting or cracking cockpit canopies, burning through control cables, exploding fuel tanks, melting or burning sensor assemblies and antenna arrays, exploding or melting munitions pods, destroying ground communications and power grids and melting or burning a large variety of strategic targets (e.g., dams, industrial and defense facilities and munitions factories)—all in a fraction of a second."

Just as the sea and the air presented different advantages in maneuverability, so will space. Having a weapon up there means being at the top of the "gravity well" so that the force that frustrates rocketeering is suddenly your friend. "Kinetic energy weapons" are the subject of a study included in Air Force 2025, with one application being rods, or "flechettes," designed to be tossed down to earth from space. Like the legendary penny tossed off the Empire State Building boring 10 feet into the sidewalk, flechettes could travel at supersonic speed (by aiming a laser just in front of them to create an "air spike," eliminating most of the effects of shock and heat). At such a speed, they could pierce the earth's surface to a depth of one-half mile and obliterate a hidden underground bunker.

Another idea is to set into orbit a number of "giant mirrors" that would take a boy's notion of burning ants with a magnifying lens and loft it into space. "This concept constructs a 10-kilometer magnifying glass or focusing element in space to illuminate targets on the ground or in space," reads one report touting it. "This illumination can turn night to day on the ground, scorch facilities or overheat satellite components." There is a database of such ideas at the Air War College in Alabama. This "solar energy weapon" is colloquially known as "concept No. 900163."

What precisely some of these concepts do is not known, but their names can be tantalizingly glimpsed in footnotes throughout the reports that reference the space database. For example: No. 901178, "space debris repulsion field"; No. 900168, "meteors as a weapon"; No. 900231, "gnat robot threat detectors"; No. 900288, "swarms of micromachines"; No. 900390, "holographic battlefield deception"; No. 900522, "space-based AI-driven intelligence master mind."

In these internal documents, real-world constraints like political will are postponed and the enormous issue of cost is finessed. The one roadblock that is seriously addressed is the bureaucratic resistance from pilots upset at the very concept of unmanned warcraft. In such moments, the tone of the language is melancholic—the problem referred to sorrowfully as "pro-pilot bias"—and suggests that listening to such woes is akin to hearing out the complaints of old sergeants a century ago harrumphing about all that crazy talk of a horseless cavalry.

On a clear blue Colorado afternoon, a bus with high-security officers, civilian engineers and computer techies rumbles into the entrance tunnel to Cheyenne Mountain, the underground cold-war city built on giant springs to withstand a Soviet ICBM attack. I have come here to try to see the emerging space bureaucracy, the elements that may one day make up a new branch of the military, the United States Space Force. At the first checkpoint, we set out on foot. A cool persistent wind practically pushes us through the 30-ton blast doors. For most of the last 40 years, Cheyenne was famous for being the home of Norad, the North American Aerospace Defense Command— the U.S.-Canadian early-warning system that scanned the globe looking for the telltale launch plume of an intercontinental ballistic missile. In fact, it is a Canadian officer from Norad who escorts me into the command room and to the chair where a commanding general would make the decision to launch a nuclear weapon.

"Don't mash the distress button under the desk there," the Canadian warns me, "or armed guards will storm the room." Before me are a wall of television screens reporting global data.

(On account of my presence, several are draped with blankets marked Top Secret.) And right away, the shift toward space is obvious. The main screen reads "Combined Command Center for NORAD/USSPACECOM."

The U.S. Space Command is the proto-bureaucracy of our emerging space force. Its current commander, a four-star general named Ralph E. Eberhardt, was given more prominence last May when Rumsfeld reorganized the space command structure. Eberhardt is being touted as the possible next chairman of the Joint Chiefs. Should he be appointed, it will be the most powerful signal yet that President Bush's campaign promise to "leapfrog" to the next generation of weaponry will mean the militarization of space.

The clearest evidence is across town from Cheyenne at Schriever Air Force Base. The Space Warfare Center was established there in 1993. It has three branches, the Space Battle Lab (patch: "Above All Others"); the Space Warfare School (patch: image of missile shooting off lightning bolts); and, as of last October, the 527th Space Aggressor Squadron (patch: image of cartoon bird standing on a cloud tossing a missile to earth). A good deal of the theory about how space can assist our troops during wartime on earth—today—is being developed here. It is the Space Battle Lab that will soon be figuring out how to take a reading from the hyperspectral camera aboard Warfighter I and make that information meaningful to a pilot flying an Air Force bomber.

"We are trying to bring the utility of space *directly* to the fighter," says the battle lab commander, Col. Ron Oholendt, "by either increasing lethality or mission effectiveness." Another project under way is to make better use of space for "bomb-impact assessment."

"As a cruise missile is heading for its target," Oholendt says, "it would transmit a data burst into space just before impact. It might tell us, 'I'm armed; here's where I am; the scene I see matches the target I was given.' So we'd have a confidence it was successful. Or it might say, 'I'm here; I don't see anything familiar so I'm going to blow up some dirt.' After we downloaded the information from the satellite, we'd be fairly confident that site would have to be retargeted."

Rumsfeld has said that the military must prepare itself to avoid a "space Pearl Harbor." This is where such preparations are being made. The commander of the space aggressor squadron, Col. Conrad Widman, spends his days envisioning how an enemy might exploit space—in order to train our forces how to react.

"The one thing you don't want to do is go to war and encounter the enemy's capability for the first time," Widman says. In one simulation exercise, he and the 527th posed as an Iranian terrorist cell set against some real U.S. troops stationed in South Asia. During the exercise, Widman hired a French satellite to take a picture, which can be paid for with a credit card.

"The guys on the Iranian team were able to count airplanes and see entry control points," Widman explains. "They could even see the tent-city area and figure out how many people they had deployed. They could also tell there was some kind of air-defense batteries. They knew that Patriot missiles often played that role, so they went to the Raytheon home page and learned that Patriot batteries are normally laid out in a format with the radar in the center." By the time the 527th had finished the simulation, they had learned the surrounding landscape, the best approach path and the entry points into the concertina-wire-protected camp.

"Is this how the terrorists in Yemen figured where the U.S.S. Cole was?" Widman says chillingly. Widman's work repeatedly reveals that technologies once carefully held as national-security secrets are now commonplace because of satellite proliferation and the Internet. "More and more," Widman's colleague Col. James Rogers says, "the problem is not another superpower, but a guy with a credit card."

As a sign of space's growing importance to the military, the first large-scale war game devoted to space issues was held for five days in January. The hypothetical conflict was set in the year 2017 and involved fighting a space battle with a "near-peer competitor" country named Red that resembled China. During the simulation exercise, which involved 250 people, the two main weapons used to duke it out were laser cannons and microsatellites. Even though select journalists were invited to "watch," the Pentagon did not provide many details of the fighting, except to say that the conflict hinged on attempts to blind each other's satellites as a first step toward waging war. The message of the demonstration, however, was clear: whoever doesn't control space in the next conflict will lose.

The future of space depends a great deal on how we describe it, a struggle that is largely metaphorical. Is space merely an extension of the air or an entirely separate arena?

THE FUTURE OF SPACE DEPENDS A GREAT DEAL ON HOW WE describe it, a struggle that is largely metaphorical. Is space merely an extension of the air and therefore the province of the Air Force? Or is it an entirely separate medium for power, like the land or sea, in need of a new doctrine? The first comparison more easily allows a militarization of space as just more of what we already have, while the second challenges us to debate space as the frontier it still is.

Rumsfeld leans toward the first comparison. His reorganization of the space command structure two months ago put Eberhardt and the Air Force in charge. The changes are even linguistic; the Air Force has revived the antique word "aerospace" to remarry the two domains. The Strategic Master Plan, for example, describes the current Air Force as being engaged in a "transition from a cold-war garrison force to an expeditionary aerospace force" in order to train "21st-century aerospace warriors."

At every stop, I was reminded of the incremental militarization of air after World War I. The Air Force began as a wing of the Army, flying over enemy territory and providing surveillance. Then the pilots began shooting one another down; later they started to drop bombs. Space can be seen as undergoing the

same process, progressing out of its current stage as an arena of surveillance to microsatellites attacking other satellites to, finally, space-based lasers aiming down at fighter jets to blast them from the sky.

Yet at some point the future of space will emerge as a great American debate. Over and over, as I interviewed military scientists and generals assigned to space, I was reminded that the decision to move into space will, at the end of the day, be made in Washington. Already, a few politicians have foreseen this conversation and staked out positions.

"Space is our next manifest destiny," explains Senator Bob Smith, Republican of New Hampshire, "because it's a dangerous world out there." Smith says that we have to weaponize space before somebody else does or face the consequences: "I don't want to see a president in the position where he has to step up to the microphones and say that the next Iraq has threatened us with a full-scale attack tomorrow, and we've either got to surrender or nuke them."

On the other side is Representative Dennis Kucinich, Democrat of Ohio. This fall, he intends to introduce a bill to ban completely the weaponization of space. "It's bad enough that we've turned space into a junkyard, but they want to turn space into a place of death," he says. "Think about the metaphysics. For all of human history, space was a place of wonder, of dreams, of aspirations—an almost visual portrayal of Browning's poem: 'Ah, but a man's reach should exceed his grasp,/Or what's a heaven for?'"

Ugh. Maybe this is how the debate must begin—duck-and-cover fear-mongering versus mawkish piety. Yet both positions are really built around the same fear: *weaponizing space is terrifying*. Smith resolves his fear by weaponizing first; Kucinich by appealing to a pristine notion of space that hasn't existed for 40 years. But this fear is real precisely because space weapons, unlike those at sea or on land, would orbit invisibly above us all. That fear would be irresolvable, like the nuclear nightmares of the last century, with their bomb shelters, gas masks and decades of mass-destruction anxiety. It is bad enough that space-surveillance technology has conspiracy theorists convinced the government can see them stepping out of the shower. Can you imagine the global neuroses if deadly lasers could be fired from space?

There is, however, a middle ground between hang-nukes-from-every-star and leave-space-the-inky-domain-of-magi, one that is occupied by some civilian theorists and military war planners.

"If we aggressively move weaponry into space," warns Michael Krepon of the Stimson Center, a Washington think tank, "then we will start an arms race." By inspiring nations to compete directly and immediately with our space-based assets, we will almost certainly guarantee the loss of the very advantages we seek to protect. Krepon supports a doctrine called "space sanctuary," a woolly phrase that sounds more feel-good than it is. His position is really that of a space pragmatist.

Pragmatists like Krepon want the military to continue research into space technologies; it would be foolish not to do so. But instead of testing or deploying a space-based arsenal, pragmatists would hold up a threat: if any rival country goes into space to test armaments, then America will go up with its own devices immediately. In the meantime, pragmatists believe, the United States should be promoting efforts to create rules of the road for space. As a model, Krepon suggests the bilateral agreements that currently regulate behavior among blue-water ships on the oceans. They are informally negotiated navy to navy, rather than through the more potentially hostile venues of governments and treaty arrangements.

Space pragmatists also believe there is great danger in abandoning the treaties that so far have guided behavior in space: the 1967 Outer Space Treaty, which forbids putting weapons of mass destruction in space, and the 1972 Antiballistic Missile Treaty, which created the surveillance system to prevent nuclear conflict (and forbids most antimissile testing). President Bush has roundly condemned the ABM treaty as a "relic" and has said that he will test antimissile technology no matter what—prompting precisely the kind of reaction Krepon fears. Even our allies have expressed "concern."

"If the ABM treaty is trashed, its protections of satellites also go by the boards," Krepon cautions. "The ABM treaty contains the most explicit protections of satellites on the books. They pertain only to those satellites that monitor treaty provisions, but when you kill the treaty, you also remove the protections." Indeed, if the U.S. abandons the treaty, a rogue nation might well respond by tossing into orbit what experts call a "keg of nails"—that is, putting thousands of metal shards into a 16,000-mile-per-hour counterorbit against our low-orbit satellites.

Kaboom.

The Pentagon's certainty that "Full Spectrum Dominance" is the only answer is curious because its own actions undercut the theory. Throughout all the conversations I had, I was perplexed by one glaring paradox. The linchpin of our precision-guided munitions is the Global Positioning System. After making the system public in the 90's, we opened it up further two years ago so that anyone on the earth can use its efficacy down to one meter of accuracy. This is an amazing gift to the world. Why did we make it? I kept asking the officers this question and heard an answer that didn't quite satisfy: "American businessmen could make some money off it."

But there is one other theory that is not stated so publicly: if we permitted everyone to use it, then no one would feel driven to build a competing system. Rather, everyone would become dependent on it. And, in fact, everyone has. The world has incorporated our G.P.S. into its daily life as rapidly as Americans took up the A.T.M. banking network, and the rules of the G.P.S.

road are getting written. The entire military forces of Australia now rely upon our G.P.S., and the new generation of cell phones will automatically locate a 911 caller.

By sharing G.P.S., no one feels so threatened to compete with it. And its use is so ubiquitous internationally that any country that damaged it would provoke a global fury. There is a sense of transparency on our part by giving away access to the G.P.S., even a feeling of generosity. Naturally, there are encryption devices on our satellites. In a crisis, we could block a bellicose nation's access to G.P.S. What was done with G.P.S. is a kind of space pragmatism.

A similar protocol could be done for introducing direct video access to space. Once it is developed, the U.S. military could make technology that allows us to see and confirm exactly what is happening up in space publicly available. This would, once again, be viewed as American generosity. It would ease competitive tensions since there would be mutually assured awareness in space. A nation with a defunct satellite would be able to confirm that it was not sabotage but the usual wear and tear of, say, subatomic bombardment (another new space hazard) that caused a breakdown. The benefit for us would be that when the crunch time of a crisis came, the visual infrastructure to see precisely what's going on in space, like G.P.S., could be made unavailable to a hostile force.

The strength of the pragmatic position is that it seeks neither to march into space while locking and loading nor does it naïvely strive for a purity that no longer exists. Space pragmatism doesn't pretend to keep space unsullied, because it can't. Without a doubt, more and more satellites will go up. More businesses will operate there, new uses will be discovered and quarrels will occur. And gradually, a military presence that is already there will get expanded. But the pragmatist intent is to hold the line at surveillance.

Can we? Can we hold the line without necessarily filling space with weaponry? The pragmatist position holds out the hope that by writing rules now—and by sharing technology—the United States could make it much harder for anyone to ever breach that line. On the other hand, if we plan, test and deploy aggressively as the lone superpower, we make certain that after a brief respite from the cold war's nuclear competition, we will once again embark on a fresh and costly arms race. And with it, assume the dark burden of policing a rapid evolution in battlespace.

Jack Hitt is a contributing writer for the magazine.

UNIT 6
Cooperation

Unit Selections

Key Points to Consider

- Itemize the products you own that were manufactured in another country.

- What recent contacts have you had with people from other countries? How was it possible for you to have these contacts?

- Do you use the Internet to access people or Web sites in other countries?

- Identify nongovernmental organizations in your community that are involved in international cooperation (e.g., Rotary International).

- What are the prospects for international governance? How do trends in this direction enhance or threaten American values and constitutional rights?

- What new strategies for cooperation can be developed to fight terrorism, international narcotics trafficking, and other criminal threats?

- How can conflict and rivalry be transformed into meaningful cooperation?

 **Links: www.dushkin.com/online/**
These sites are annotated in the World Wide Web pages.

American Foreign Service Association
http://www.afsa.org/related.html

Carnegie Endowment for International Peace
http://www.ceip.org

Commission on Global Governance
http://www.sovereignty.net/p/gov/gganalysis.htm

OECD/FDI Statistics
http://www.oecd.org/statistics/

U.S. Institute of Peace
http://www.usip.org

An individual can write a letter to another person who is located just about anywhere in the world, and, assuming it is properly addressed, the sender can be relatively certain that the letter will be delivered. This is true even though the sender pays for postage only in the country of origin and not in the country where it is delivered. A similar pattern of international cooperation is true when an individual boards an airplane in one country and never gives a second thought to the issues of potential language and technical barriers, even though the flight's destination is halfway around the world.

Many of the most basic activities of our lives are the direct result of multinational cooperation. International organizational structures, for example, have been created to monitor threats to public health and to scientifically evaluate changing weather conditions. The flow of mail, the safety of airlines, and the monitoring of changing international conditions are just some of the examples of individual governments recognizing that their self-interest directly benefits from cooperation (in some cases by giving up some of their sovereignty through the creation of international governmental organizations, or IGOs).

Transnational activities are not limited to the governmental level. There are now tens of thousands of international nongovernmental organizations (INGOs). The activities of INGOs range from staging the Olympic Games to organizing scientific meetings to actively discouraging the hunting of seals. The number of INGOs along with their influence has grown tremendously in the past 50 years.

During the same period in which the growth in importance of IGOs and INGOs has taken place, there also has been a parallel expansion of corporate activity across international borders. Most U.S. consumers are as familiar with Japanese or German brand-name products as they are with items made in their own country. The multinational corporation (MNC) is an important nonstate actor. The value of goods and services produced by the biggest MNCs is far greater than the gross domestic product (GDP) of many countries. The international structures that make it possible to buy a Swedish automobile in Sacramento or a Swiss watch in Singapore have been developed over many years. They are the result of governments negotiating treaties that create IGOs to implement the agreements (e.g., the World Trade Organization). As a result, corporations engaged in international trade and manufacturing have created complex transnational networks of sales, distribution, and service that employ millions of people.

To some observers these trends indicate that the era of the nation-state as the dominant player in international politics is passing. Other experts have observed these same trends and have concluded that the state system has a monopoly of power and that the diverse variety of transnational organizations depends on the state system and, in significant ways, perpetuates it.

In many of the articles that appear elsewhere in this book, the authors have concluded their analysis by calling for greater international cooperation to solve the world's most pressing problems. The articles in this section provide examples of successful cooperation. In the midst of a lot of bad news, it is easy to overlook the fact that we are surrounded by international cooperation and that basic day-to-day activities in our lives often directly benefit from it.

STRATEGIES FOR WORLD PEACE:

THE VIEW OF THE UN SECRETARY-GENERAL

By Kofi A. Annan

Today, in Afghanistan, a girl will be born. Her mother will hold her and feed her, comfort her and care for her, just as any mother would anywhere in the world. In these most basic acts of human nature, humanity knows no divisions.

But to be born a girl in today's Afghanistan is to begin life centuries away from the prosperity that one small part of humanity has achieved. It is to live under conditions that many of us would consider inhuman. Truly, it is as if it were a tale of two planets.

I speak of a girl in Afghanistan, but I might equally well have mentioned a baby boy or girl in Sierra Leone. No one today is unaware of this divide between the world's rich and poor. No one today can claim ignorance of the cost that this divide imposes on the poor and dispossessed who are no less deserving of human dignity, fundamental freedoms, security, food, and education than any of us. The cost, however, is not borne by them alone. Ultimately, it is borne by all of us—North and South, rich and poor, men and women of all races and religions.

Today's real borders are not between nations, but between powerful and powerless, free and fettered, privileged and humiliated. Today, no walls can separate humanitarian or human-rights crises in one part of the world from national-security crises in another.

Scientists tell us that the world of nature is so small and interdependent that a butterfly flapping its wings in the Amazon rain forest can generate a violent storm on the other side of the earth. This principle is known as the "Butterfly Effect." Today, we realize, perhaps more than ever, that the world of human activity also has its own "Butterfly Effect"—for better or for worse.

The Universal Bond of Tragedy

We have entered the third millennium through a gate of fire. If today, after the horror of September 11, we see better and we see further, we will realize that humanity is indivisible. New threats make no distinction among races, nations, or regions. A new insecurity has entered every mind, regardless of wealth or status. A deeper awareness of the ties that bind us all—in pain as in prosperity—has gripped young and old.

In the early beginnings of the twenty-first century—a century already violently disabused of any hopes that progress toward global peace and prosperity is inevitable—this new reality can no longer be ignored. It must be confronted.

The twentieth century was perhaps the deadliest in human history, devastated by innumerable conflicts, untold suffering, and unimaginable crimes. Time after time, a group or a nation inflicted extreme violence on others, often driven by irrational hatred and suspicion or by unbounded arrogance and thirst for power and resources. In response to these cataclysms, the leaders of the world came together at mid-century to unite the nations as never before.

A forum was created—the United Nations—where all nations could join forces to affirm the dignity and worth of every person and to secure peace and development for all peoples. Here, states could unite to strengthen the rule of law, recognize and address the needs of the poor, restrain man's brutality and greed, conserve the resources and beauty of nature, sustain the equal rights of men and women, and provide for the safety of future generations.

We thus inherit from the twentieth century the political, scientific, and technological power that—if only we have the will to use them—give us the chance to vanquish poverty, ignorance, and disease.

Redefining the United Nations

In the twenty-first century, I believe the mission of the United Nations will be defined by a new, more profound awareness of the sanctity and dignity of every human life, regardless of race or religion. This will require us to look beyond the framework of states and beneath the surface of nations or communities. We must focus, as never before, on improving the conditions of the individual men and women who give the state or nation its richness and character.

Over the past five years, I have often recalled that the United Nations' Charter begins with the words, "We the peoples." What is not always recognized is that "We the peoples" are made up of individuals whose claims to the most fundamental rights have too often been sacrificed in the supposed interests of the state.

The United Nations' priorities for the future are to promote democracy, prevent conflict, and lessen the burden of global poverty.

A genocide begins with the killing of one man not for what he has done but because of who he is. A campaign of "ethnic cleansing" begins with one neighbor turning on another. Poverty begins when even one child is denied his or her fundamental right to education. What begins with the failure to uphold the dignity of one life all too often ends with a calamity for entire nations.

In this new century, we must start from the understanding that peace belongs not only to states or peoples but to each member of those communities. The sovereignty of states must no longer be used as a shield for gross violations of human rights. Peace must be made real and tangible in the daily existence of every individual in need. Peace must be sought, above all, because it is the condition that enables every member of the human family to live a life of dignity and security.

Focusing on Diversity And Dialogue

From this vision of the role of the United Nations in the next century flow three key priorities for the future: eradicating poverty, preventing conflict, and promoting democracy.

Only in a world that is rid of poverty can all men and women make the most of their abilities. Only where individual rights are respected can differences be channeled politically and resolved peacefully. Only in a democratic environment, based on respect for diversity and dialogue, can individual self-expression and self-government be secured and freedom of association be upheld.

The idea that there is one people in possession of the truth, one answer to the world's ills, or one solution to humanity's needs has done untold harm throughout history—especially in the last century. Today, however, even amid continuing ethnic conflict around the world, there is a growing understanding that human diversity is both the reality that makes dialogue necessary and the basis for that dialogue.

We understand, as never before, that each of us is fully worthy of the respect and dignity essential to our common humanity. We recognize that we are the products of many cultures, traditions, and memories; that mutual respect allows us to study and learn from other cultures; and that we gain strength by combining the foreign with the familiar.

In every great faith and tradition one can find the values of tolerance and mutual understanding. The Qur'an, for example, tells us, "We created you from a single pair of male and female and made you into nations and tribes, that you may know each other." Confucius urged his followers, "When the good way prevails in the State, speak boldly and act boldly. When the State has lost the way, act boldly and speak softly." In the Jewish tradition, the injunction to "love thy neighbor as thyself" is considered to be the very essence of the Torah.

This thought is reflected in the Christian Gospel, which also teaches us to love our enemies and pray for those who wish to persecute us. Hindus are taught, "Truth is one, the sages give it various names." And in the Buddhist tradition, individuals are urged to act with compassion in every facet of life.

Each of us has the right to take pride in our particular faith or heritage. But the notion that what is "ours" is necessarily in conflict with what is "theirs" is both false and dangerous. It has resulted in endless enmity and conflict, leading men to commit the greatest of crimes in the name of a higher power.

It need not be so. People of different religions and cultures live side by side in almost every part of the world, and most of us have overlapping identities that unite us with very different groups. We can love what we are without hating what—and who—we are not. We can thrive in our own tradition, even as we learn from others, and come to respect their teachings.

This will not be possible, however, without freedom of religion, of expression, of assembly, and basic equality under the law. Indeed, the lesson of the past century has been that where the dignity of the individual has been trampled or threatened—where citizens have not enjoyed the basic right to choose their government or the right to change it regularly—conflict has too often followed, with innocent civilians paying the price in lives cut short and communities destroyed.

PEACE HAS NO PARADE: KOFI A. ANNAN ON THE NOBEL PEACE PRIZE

In 1960, the Nobel Peace Prize was awarded for the first time to an African—Albert Luthuli, one of the earliest leaders of the struggle against apartheid in South Africa. In 1961, the Prize was first awarded to a secretary-general of the United Nations—posthumously, because Dag Hammarskjöld had already given his life for peace in Central Africa. For me, as a young African beginning his career in the United Nations a few months later, those two men set a standard that I have sought to follow throughout my working life.

This award belongs not just to me. My own path to service at the United Nations was made possible by the sacrifice and commitment of my family and many friends from all continents—some of whom have passed away—who taught me and guided me. To them, I offer my most profound gratitude.

In a world filled with weapons of war and all too often words of war, the Nobel Committee has become a vital agent for peace. Sadly, a prize for peace is a rarity in this world. Most nations have monuments or memorials to war, bronze salutations to heroic battles, archways of triumph. But peace has no parade, no pantheon of victory.

Only by understanding and addressing the needs of individuals for peace, for dignity, and for security can we at the United Nations hope to live up to fulfill the vision of our founders. This is the broad mission of peace that United Nations staff members carry out every day in every part of the world.

—*Kofi A. Annan*

The obstacles to democracy have less to do with culture or religion than with the desire of those in power to maintain their position at any cost. This is neither a new phenomenon nor one confined to any particular part of the world. People of all cultures value their freedom of choice and feel the need to have a say in decisions affecting their lives.

The United Nations, whose membership comprises almost all the states in the world, is founded on the principle of the equal worth of every human being. It is the nearest thing we have to a representative institution that can address the interests of all states and all peoples. Through this universal, indispensable instrument of human progress, states can serve the interests of their citizens by recognizing common goals and pursuing them in unity. No doubt, that is why the Nobel Committee says that it "wishes, in its centenary year, to proclaim that the only negotiable route to global peace and cooperation goes by way of the United Nations."

Our era of global challenges leaves us no choice but to cooperate at the global level. When states undermine the rule of law and violate the rights of their individual citizens, they become a menace not only to their own people but also to their neighbors and the world. What we need today is better governance—legitimate, democratic governance that allows each individual to flourish and each state to thrive.

A Test of Common Humanity

I began with a reference to the girl born in Afghanistan today. Even though her mother will do all in her power to protect and sustain her, there is a one-in-four risk that she will not live to see her fifth birthday. Whether she does is just one test of our common humanity—of our belief in our individual responsibility for our fellow men and women. But it is the only test that matters.

If we remember this girl, then our larger aims—to fight poverty, prevent conflict, or cure disease—will not seem distant or impossible. Indeed, those aims will seem very near and very achievable—as they should. Beneath the surface of states and nations, ideas and language, lies the fate of individual human beings in need. Answering their needs will be the mission of the United Nations in the century to come.

Kofi A. Annan is secretary-general of the United Nations, #S-8000, New York, New York 10017. Web site www.un.org. This article is drawn from his Nobel Lecture delivered on December 10, 2001, in Oslo, Norway.

Justice Goes Global

(International Criminal Court Is Created)

More than 160 nations voted to establish the International Criminal Court, which will be located in The Hague, Netherlands. The U.S. refused to sign the treaty to create the global tribunal to judge war crimes because of reservations about sovereignty and jurisdiction.

Despite U.S. dissent the world community finally creates a new court to judge the crimes of war.

The spectre of the century's slaughtered millions haunted Rome as the world's nations struggled for five weeks to create the first permanent international body dedicated to punishing the crimes of war. "Victims of past crimes and potential victims are watching us," said U.N. Secretary-General Kofi Annan. "They will not forgive us if we fail."

They did not fail. Cheers and applause echoed as representatives of some 160 nations, assisted by more than 200 non-governmental organizations, gathered last week in the plush maze of the U.N. Food and Agriculture Organization's building, voted overwhelmingly to create the International Criminal Court (I.C.C.). But success came only after frantic last-minute negotiations to bridge philosophical divides that left the U.S. in opposition to the treaty and at odds with most of its major allies. Just how viable the court will be if the world's superpower carries out its threat to "actively oppose" the new institution remains to be seen, but 18 judges will gather in The Hague within the next few years, ready to try cases of genocide, war crimes and crimes against humanity.

The duality of mankind's urges to both wage war and curb its own bellicosity is virtually as old as warfare itself. But it was only in the 19th century that refinements in the technology of battle concentrated minds on serious attempts to find judicial ways to combat their brutality. The laws and customs of war were codified at Conventions in The Hague in 1899 and 1907, and efforts continued between the two World Wars. But those laudable agreements were impotent in face of the unprecedented carnage of the 20th century's first half, when an estimated 58 million died in Europe alone. After World War II, international tribunals at Nuremburg and Tokyo tried and convicted the conflict's instigators for war crimes, crimes against peace and against humanity itself. But these judgments were carried out within a temporary judicial framework imposed by the victors.

The Geneva Conventions of 1949 continued to build a body of international law governing the conduct of war, but the problem of applying the provisions remained. The newly formed U.N. had commissioned a study in 1948 to look into establishing a permanent tribunal, but the cold war prevented any real progress. The topic surfaced again only in 1989, when the International Law Commission began preparing a draft statute for an International Criminal Court. But what really galvanized the international community was the chaotic disintegration of Yugoslavia and the atrocities that accompanied it.

The U.N. eventually moved to create an ad hoc criminal tribunal on the crimes committed during the Bosnian war in 1993, followed a year later by another one-off body for Rwanda. The distinguished South African jurist Richard Goldstone, the original chief prosecutor for both tribunals, says that those courts represented "the first real international attempt to enforce international humanitarian law." But

establishing those bodies took up to two years of preparatory work and negotiation. "The thing is to avoid having to spend six months looking for a prosecutor," notes Theodor Meron, professor of international law at New York University Law School, "and a year looking for a building."

Even though Bosnia's most notorious accused war criminals have not yet been brought before The Hague tribunal, it has indicted some 60 people, holds 27 men in custody and has handed down two judgments. This month the court launched the first genocide prosecution in Europe. Deputy Prosecutor Graham Blewitt says that "We have been a model for the creation of the new court."

International law has always involved an inherent tension between national sovereignty and accountability. But the continuing carnage since 1945—another 18 million dead and the likes of Idi Amin, Pol Pot and Saddam Hussein reigning in terror—reinforced the U.N.'s determination to act. As it did so, Washington began to fret. Michael Scharf, currently professor of law at New England Law School, was the State Department's point man on the court under President George Bush. "One of my jobs, which I did not enjoy," recalls Scharf, "was to find ways to stall it forever." President Clinton has been far more supportive, but his administration, too, developed serious qualms. Recalling the invasions of Grenada and Panama, and the bombing of Libya, the U.S. worried that similar actions in the future could involve officials all the way up the chain of command being hauled before the I.C.C.

As the conference convened in Rome on June 15, it was beset by disagreement. The most divisive questions revolved around the precise definition of the crimes to be within the court's jurisdiction, the breadth of that jurisdiction and just who would determine which cases should be brought. The U.S. went in with goals that allied it uncomfortably with China, Russia and India, as well as Libya and Algeria, but put it at odds with most of its usual friends who gathered among the so-called Like-Minded Nations seeking a strong and independent Court. "We are not here," said Washington's U.N. ambassador, Bill Richardson, "to create a court that sits in judgment on national systems." The U.S. is concerned that its many soldiers serving overseas could become involved in confrontations that would make them vulnerable to what an Administration official called "frivolous claims by politically motivated governments."

The Washington negotiators—who rejected universal jurisdiction, subjecting any state, signatory or not, to the court's remit—agreed that the court should have automatic jurisdiction in the case of genocide, giving it the ability to prosecute individuals of any country that had signed the treaty. But they sought a clause allowing countries to opt out of the court's jurisdiction on war crimes and crimes against humanity for 10 years. The agreed statute allows states to opt out of the court's jurisdictions only on war crimes and only for seven years. It also includes the crime of "aggression" within the court's jurisdiction, subject to a precise definition of aggression. Washington had also wanted to give only the Security Council and states party to the agreement the right to bring cases to the court. The statute, however, also empowers the prosecutor to initiate cases. The U.S. did manage to get a compromise, promoted by Singapore, allowing the Security Council to call a 12-month renewable halt to investigations and prosecutions included in the text. "If states can simply opt in or out when they want, the court will be unworkable," said a senior official in the German delegation. Without an independent prosecutor, he added, "crimes will be passed over for political reasons."

Although conference chairman Philippe Kirsch of Canada had already successfully chaired at least eight international conferences—brokering agreements on issues such as terrorism and the protection of war victims—all his undoubted mediation skills failed to resolve the disputes. As Washington became increasingly isolated, a copy of U.S. "talking points" circulated among the delegations, suggesting that if the court did not meet U.S. requirements Washington might retaliate by withdrawing its troops overseas, including those in Europe. Although few believed in that possibility and the Administration downplayed it, State Department spokesman Jamie Rubin explained that "The U.S. has a special responsibility that other governments do not have."

After all the wrangling, what emerged was a court to be located in The Hague—where the International Court of Justice already deals with cases brought on a civil basis by states against other states. It is to contain four elements: a Presidency with three judges; a section encompassing an appeals division, trial and pre-trial divisions; a Prosecutor's office; and a Registry to handle administration. The court, which will act only when national courts are "unwilling or unable genuinely" to proceed, will confine its maximum penalty to life imprisonment.

How the court will fare without the support of the U.S. is unclear. Washington has provided vital political backing for the Yugoslav and Rwanda tribunals and continues to be their leading financier. "We have shown that the only way to get war criminals to trial is for the U.S. to take a prominent role," said one Administration official last week. "If the U.S. is not a lead player in the creation of this court, it doesn't happen."

Nevertheless, the fact that a court with teeth has actually been created was an unprecedented move by the world community to make the rule of law finally prevail over brute force—a step towards fulfilling Secretary-General Annan's pledge that "At long last we aim to prove we mean it when we say 'Never Again.'"

Meet the World's Top Cop

Interpol's Raymond Kendall explains why today's world has him worried.

What is a city cop to do? In the same way that businesses have relocated to profit from a shrinking world, crime networks have stretched thousands of miles to penetrate new markets, find new sources of revenue and influence, or get an edge on the compensation at home. Governments have been less adroit, especially at building effective multilateral mechanisms to meet this more sophisticated threat to their citizens. For the last 15 years, Raymond Kendall has been secretary general of Interpol, a kind of United Nations for the world's police forces. As he prepared to step down in November of last year, Kendall met for several hours with *FP* Editor Moisés Naím in New York City for an exclusive interview on the state of global crime. The world needs to change the way it fights back, he argues. Legalize drug use? Privatize police intelligence? Globalize the courts? Everything is on the table.

FOREIGN POLICY: In your 15 years at the helm of Interpol, what have you learned about global crime that the world doesn't seem to understand?

Raymond Kendall: Well, I'm not sure whether the world is conscious of how rapidly the magnitude of crime has grown. Take the film *The French Connection*, which was a hit almost 30 years ago when I was starting at Interpol. The criminals in *The French Connection* are trying to move 100 kilos of heroin from Marseilles to New York City. Now in those days, there were a few registered heroin addicts in the United Kingdom, but there was no real drug problem in Europe. By today's standards, 100 kilos of heroin is nothing.

Take Operation Icicle, which involved Interpol offices in Athens, London, Ljubljana, Rome, and Vienna, plus Interpol's General Secretariat Analytical Criminal Intelligence Unit, or ACIU. In July 1999, British police found out that a suspect container was due to arrive in Italy. When the container arrived, Italian police searched it and discovered 1,400 kilograms of cocaine. The police decided to leave 500 kilograms of cocaine in the container and follow it to its final destination. It went from Greece, to Macedonia, back to Italy, and then to Vienna, where the cocaine was seized, and its recipients arrested. People read about it in the newspapers and are concerned. But what the world does not quite understand is how much net-works have penetrated all countries, how entrenched and dangerous they are.

FP: So complacency reigns?

RK: It really frightens me to see how, in this period of 25 years, we have gone from a situation where there were virtually no drugs to a situation where, quite frankly, today the European continent is pretty well flooded in drugs. And everything else that goes with them. Public outrage has not grown in the same proportion.

FP: The scale of crime, then, has increased globally. Have the kinds of crimes you encounter changed as much? What new crimes have emerged since you've been involved in law enforcement?

RK: I don't think there have been what we might call new, really new, types of crime. There have been changes in intensity, changes in methodology, changes in the way criminals use technology—really, a more businesslike approach to what they do. Local organized crime still exists. But at the same time, criminal groups on the European continent, and extending around the world, instead of specializing in particular types of criminal activity—whether it's drug trafficking, trafficking weapons, trafficking stolen art objects, and more recently trafficking people as a commodity—they have diversified, created networks through which, at any given time, they can adapt themselves to evade capture and profit from many types of criminal activity.

FP: Your job is huge. The world needs to fight a U.S. $400-billion drug trade. The United Nations estimates that criminal syndicates worldwide are making U.S. $1.5 trillion a year. The International Monetary Fund (IMF) estimates that U.S. $600 billion is laundered every year. More than 60,000 works of art are missing. The trade in people is soaring. Terrorism is on the rise. Then you have the stolen cars, the pirated software, and the cybercrime. Recently, you had to deal with more political matters—the extradition of Gen. Augusto Pinochet of Chile and the arrest warrant against Slobodan Milosevic of Yugoslavia. And for all of that, Interpol has a budget of U.S. $23 million?

RK: Yes.

Once Upon a Crime

The idea for Interpol was born in 1914 when Monaco's Prince Albert convened more than 20 countries to discuss international crime. The outbreak of World War I killed their plans for a new international police organization. But in 1923, Vienna's chief of police, Johann Schober, resuscitated the idea, joining up with 138 delegates from 20 countries to form the International Criminal Police Commission (ICPC). Based in Vienna, the ICPC aimed to facilitate international police cooperation.

During World War II, the Nazis took over the ICPC and moved the headquarters to a town near Berlin. Under the command of Reinhold Heydrich, the Nazis used ICPC files, which recorded a suspect's religion and sexual orientation, to track down European Jews and homosexuals.

After the war, Belgian Inspector-General Florent Louwage formed a committee that rebuilt the ICPC as it was intended. In 1946, the organization, then with delegates from 17 countries, chose ICPC's telegraphic address, "Interpol," as its new name and relocated to Paris—then to Saint-Cloud, France, in 1966, and finally, in 1989, to Lyon. Today, Interpol boasts 178 member states.

Interpol has never had policing powers of its own. Rather, it acts as a global clearinghouse for information on crime and maintains vast databases of fingerprints, mug shots, reproductions of missing art, license plate numbers of stolen cars, and more. A system of notices, ranging from red notices for criminals wanted for extradition to green notices for informational purposes, keeps its National Central Bureaus up-to-date. Interpol also helps police bureaus counter a wide range of crimes, from trafficking in human beings and drugs to terrorism. At the request of Interpol Moscow, for instance, U.S. immigration officials located, arrested, and in June 1999 deported a Russian murder fugitive.

In October 1999, Interpol assisted the South African Endangered Species Protection Unit and the Portuguese police in arresting four suspects who were smuggling 150 African elephant tusks and 1,000 kilograms of cannabis.

Interpol's 373-person bureaucracy of about 120 police experts and 200-plus support staffers runs on a budget financed by the annual contributions of member states. When a new state joins Interpol, it agrees to pay fixed annual dues ranging from 2 to 100 "units," according to its resources, G7 (Group of Seven) countries pay the maximum contribution. The General Assembly, Interpol's supreme governing body—made up of approximately 400 delegates appointed by the member states—determines the budget at annual meetings by deciding how much money each unit will be worth. In 2000, those units added up to 177 million French francs (U.S. $23 million). The General Assembly also elects officials, such as the secretary general, and decides how to allocate money to Interpol's various programs and initiatives.

In addition to helping countries connect, Interpol now cooperates with agencies within states, as well as non-governmental and international organizations. In October 2000, Interpol agreed to help the Tequila Regulating Council in Mexico combat the export of adulterated tequila by affixing the Interpol hologram on export permits. The United Nations is a frequent partner as well. In 1996, the United Nations granted Interpol "Permanent Observer" status to the United Nations General Assembly, which entitles Interpol to participate in U.N.-sponsored international conferences. And last year, the United Nations Educational, Scientific, and Cultural Organization collaborated with Interpol to create a CD-ROM that reproduces and profiles more than 14,000 stolen works of art.

—FP

FP: Didn't you feel at times that perhaps it was pointless? That there is no point in trying to fight so much and so many different kinds of crimes with $23 million a year?

RK: We are a small-budget organization if you compare us with the challenges we face. One of my biggest frustrations, and maybe I have a certain responsibility for this, I don't know, has been how to get the people and politicians to recognize this. In the last few years, for example, at the G8 [Group of Eight] meetings, they say their number-one priority is to deal with organized crime, drug trafficking, and terrorism—and then some of the people at those meetings go back home and immediately reduce their local law-enforcement budgets by 10 percent or something like that. I have seen that happen many times.

FP: Which countries?

RK: It's not a matter of one particular country. It is a general trend—spending lots of time talking about international crime, but acting only when something happens. For instance, in June of last year, 58 Chinese people suffocated in a truck that was going from Belgium to England. And suddenly French politicians decided that the illegal movement of people is a problem, that they must do something. But in fact, trafficking in people is something we've been concerned about, and asking governments to do something about, for at least 10 years.

FP: Sure. But isn't there a long tradition of the police bashing politicians, accusing politicians of getting interested in their problems only when a high-visibility crime generates a public reaction?

RK: You're absolutely right. Politicians make decisions on the basis of good information. It seems that effective communication between police and governments is not there.

FP: Could it be that communication is breaking down because of Interpol? That governments do not trust Interpol, and therefore they would rather bypass you and develop relationships with specific foreign police departments? For instance, when the case of money laundering at the Bank of New York broke, you said you were reluctant to share information with your Russian counterparts because you had a sense that the information could end up in the wrong hands.

RK: I think that is a fact of international political life. We have to manage as best we can. We cooperate with nations that have so many different types of regimes, so many different types of police systems, so many possibilities for corruption. The misconception on the part of the people who have this mistrust is that they think that any information that comes into an organization like ours will be automatically distributed to all the member countries simply because they are members. This is absolutely untrue. We don't own any information, we are entrusted with it. The country that gives us information sets the conditions under which we can share it.

FP: Is Interpol really that secure? A former senior U.S. intelligence official told me with certainty that Interpol had been compromised.

RK: Interpol is secure. We may have, from time to time, like in any organization, some corruption. But I can remember only one such incident at the General Secretariat in almost 30 years.

FP: Tell me, what can Interpol do well and what can it not do well?

RK: There are some basic things that only Interpol can do. For example, being able to circulate throughout the world, throughout 178 countries, information about a wanted person, a stolen work of art, or a missing child. What it cannot do is undertake street action. Interpol cannot do that, and I don't believe it will ever be able to do that in the foreseeable future.

FP: Our sources tell us that recently, Interpol has launched a new project, a new kind of information database. That you have moved from sharing basic criminal information tied to specific events to sharing criminal intelligence and analysis. We are told that a core of 50 or so Interpol member states, including the United States, is collaborating on a project designed to track the activities of Eurasian organized crime in and outside of the former Soviet sphere. Is this true?

RK: It is true.

FP: How is this project an innovation for Interpol?

RK: Before, a lot of information was passed to us that then went into a database that was never exploited. Now, all that data is being examined globally from the intelligence point of view. The key role of strategic intelligence is not to identify specific operational targets, but rather to focus attention on new threats, identifying changing situations and providing the basis for forecasting future trends. We look at such things as the movement of people or goods to find if there is any way we can predict where the next organized activity is going, where we should be looking next.

FP: What is Interpol's Millennium Project?

RK: A majority of European countries are participating in the Millennium Project, providing intelligence information on the newest and most dynamic of the major international organized criminal groups—gangs from the former Soviet Union. The latest estimates are that about 1,000 Russian organized-crime groups are operating internationally, plus 8,000 to 10,000 operating in the area of the former Soviet Union. Each group ranges in size from 50 to 1,000 members. They are best understood as loose networks; they may work together in specific cases and cooperate against common threats, but there is little top-down control and coordination. Their activities include cigarette and other contraband smuggling, drug smuggling, illegal immigration, extortion, prostitution, vehicle theft, and arms dealing. The Millennium Project has demonstrated the difficulties in investigating these organizations in one country or even bilaterally. They operate from several countries and are diversified enough to make it extremely difficult to investigate using traditional methods. Unless approached in a truly multilateral way, with countries really collaborating efficiently, these criminal groups will always have the upper hand.

FP: The millennium Project has been quite effective in tracking property that Russian crime bosses are buying in Columbia.

RK: Yes, that's correct.

CRIME'S NEW POLITICS

FP: What was the most controversial decision you made in your 15-year tenure?

RK: Perhaps the Rainbow Warrior case. As you may recall, in 1985 a ship belonging to Greenpeace was sunk in the harbor in northern New Zealand. The people who claimed to be responsible for the attack were carrying, of course, Swiss passports. To cut a long story short, it turned out that they were members of the French Secret Service. The New Zealand president's position was that the crime has been committed, people should be punished, this is a criminal matter. The French authorities said, "You should not be dealing with this case. This is a political case." Well, I said, "No, this is a criminal case."

FP: What other controversial decision or what other decision looms large in your memory?

RK: Several cases led to controversy when Interpol cooperation was requested. When Interpol refused to help in an investigation of the Chinese religious group Faun Gong, for instance, or an investigation of the instigators of a coup in Qatar. And of course, there was Interpol's involvement in the Pinochet case.

FP: How did the decision to go after General Pinochet come about?

RK: It was initiated by a Spanish judge. If the British had said they didn't agree, that this action shouldn't involve Interpol, then I would have never made the decision to say, "Well, I think it should." But I didn't have to make the decision. Nobody objected.

FP: What was your role, Interpol's role?

RK: Interpol supplied the official channel to enable the Spanish magistrate to contact the British police and provided the legal cover to Spain's extradition request.

FP: And now you're charged with the capture of Slobodan Milosevic.

RK: When I heard there were plans to create this international tribunal for dealing with war crimes in the former Yugoslavia, I said, well, that's fine to have a tribunal, but how are they going to bring people before it? So I wrote to [then United Nations Secretary-General] Boutros Boutros-Ghali and said we will place at the U.N.'s disposal Interpol's system for circulating warrants on these people. At the moment I think there are something like 40 or 50 of these warrants in existence, including one for Milosevic.

FP: And of those that were issued, how many have been captured?

RK: I think about half a dozen. It's kind of interesting that, in fact, there can be real international legal action. And the plan now to create an international criminal tribunal, which unfortunately is being opposed by a certain number of countries, including the United Kingdom and the United States, is a good sign that we will see more of this.

FP: We are increasingly witnessing how countries cede some specific aspects of their sovereignty to an international body such as the World Trade Organization or the IMF. We are also beginning to see more frequent instances in the area of international law enforcement. Do you think that the world is heading toward more and more global tribunals and, perhaps one day, a global justice systems?

RK: The movement started with the war-crimes tribunal that was created for Yugoslavia and was then extended to Rwanda. I think you will find that inevitably it will become a global thing and not restricted to certain countries. The more crime becomes global, the more justice and law enforcement in general will have to become global.

THE CASE FOR DECRIMINALIZATION

FP: You've repeatedly made governments furious by deciding early on and publicly stating that the criminalization of drug use is wrong.

RK: Part of the confusion in attitudes comes from terminology. Words like "depenalization," "liberalization," "decriminalization," and so on can be interpreted a bit

freely. I have always strongly believed that a drug abuser is not the same as a criminal. He is not victimizing somebody else, he's not stealing anything. He has a problem, a personal problem that leads him to take drugs. Over the years, I have advocated not reducing the amount we spend on law enforcement, but balancing whatever we spend on repression with at least an equal sum dealing with treatment, education, and so on. That way, we reduce the demand, so producing countries can't say it's not their problem. Because they have a point. It's not really fair to accuse them all the time of being responsible for our problems.

FP: The Clinton administration's drug czar, Gen. Barry McCaffrey, has said that your way of thinking would result in "significantly higher rates of drug abuse particularly among young people and exponential increase in human and social costs to society." How would you respond?

RK: I met General McCaffrey in Brussels a couple of years ago, and he was sympathetic, shall I say, maybe not agreeing of course, but he was sympathetic to the way I view things—making clear what we mean by "decriminalization," as I just did.

FP: Bottom line, your position is that if somebody is captured with cocaine, and there's no evidence that this person has been trafficking or reselling cocaine, this person should not go to jail.

RK: There should be an alternative that enables him to go for treatment. And the only reason, I repeat, the only reason I think there has to be some administrative way of sending people for treatment is because it has been shown that the drug addict, unless you force him, will not voluntarily seek treatment.

FP: A generation or so from now, do you think your successors will look back at the way we fought the drug war these last two or three decades as a big mistake?

RK: Yes.

FP: A big misallocation of resources and—

RK: Now let's be clear about this. I'm not saying that you should take resources away from what the law-enforcement people are doing and put them elsewhere. I say that the mistake is in not giving equal resources to deal with treatment.

FP: How does the balance fall now, globally?

RK: At the global level something like 80 percent is spent on law enforcement and only 20 percent on treatment.

FP: What are the most powerful tools to deal, not with drug consumers, but with the people who are actively engaged in drug production, funding, trafficking, and distribution?

RK: In terms of producers, a good, if limited example, is what the U.N. has tried to do with crop substitution for Bolivia's coca fields. There are a number of approaches like this. If our job in law enforcement is to interrupt the supply of drugs, then our most powerful tool for attack would be international cooperation, good international intelligence, and so on. We produce at the moment a weekly drug-intelligence bulletin. Even our U.S. critics would have to recognize that quite a number of the big recent seizures they've made were based on information coming from our bulletin.

THE CHALLENGE OF CYBERCRIME

FP: You have said that governments were caught unaware by Internet crime and were unprepared to deal with it. But even if they had been aware, what could they have done? What can they do?

RK: I don't think that governments can deal with this issue unless they do it with the private sector. Governments may say they are concerned about cybercrime, that they have created a special group to deal with it, and what does that mean? They just take somebody from the drug squad, somebody from the money-laundering squad, and pull them into a group and call it a cybercrime squad. The kind of expertise we need in this area, the kind of research and development we need, are all in the private sector. And even if you want to recruit the appropriate, qualified people, you can't, because you can't compete with the private sector in terms of salaries and incentives.

FP: Do you have a cybercrime unit at Interpol?

RK: I have a unit, which I would not claim to be better than those government units I spoke about. My people don't have the expertise of those in the private sector. I'm willing to listen to anybody in the private sector who can come up with something we can do.

FP: Have you sought the help of the private sector?

RK: We are exploring a collaboration with a firm called Atomic Tangerine, an independent consulting firm based in Menlo Park, California. We are exploring plans to provide relevant Interpol information to private firms, and, in return, to have those private firms, through Interpol, advise member nations of Internet threats. In the future, more private companies will have access to global intelligence, without cost, to assist them in defending their Internet activities from cyberterrorism and "hactivism." At the same time, law enforcement agencies will be able to benefit from sophisticated technology intelligence gathered by leading Internet firms.

TERRORISM'S CHANGING FACE

FP: What percentage of Interpol's time and resources is spent combating terrorism?

RK: The General Secretariat dedicates 8 percent of its police staff resources—mainly police experts seconded by countries that were, or are, victims of terrorism, such as Italy, Germany, and Spain—to fighting terrorism.

FP: How are the terrorists you encounter today different from the ones you encountered when you took your position at Interpol 15 years ago? And before that, when you were a British police officer?

RK: From exclusively politically motivated or funded terrorism in the 1970s and 1980s, we see a new era of terrorism motivated by pseudo-religious purposes. In fact, these "religious" terrorists are simply criminals. Osama bin Laden comes to mind. Extortion of funds is among the most usual motivations of terrorists, but hidden by ideological cover—the Irish Republican Army, Basque ETA, or Corsican FNLC.

FP: Do you think your successor will have to deal with terrorist organizations that have nuclear weapons at their disposal?

RK: I personally don't think this is a serious threat. Making a nuclear bomb is not an easy thing to do. We have worked with atomic-energy authorities in a group of countries to try and assess the nature of the threat. The general conclusion up until now has always been that for a terrorist organization to do something like that, it would need a great deal of—There are other ways they can do things without going that far.

FP: Are you thinking about terrorism with biological or chemical weapons of mass destruction?

RK: Yes, that is a possibility, though still unlikely. It would be more likely in that area than in the atomic area, in my view.

FP: Would you say the world is doing a better job fighting terrorism than fighting drug trafficking?

RK: Yes, I think so.

FP: Why:

RK: Well, the nature of the problem is such that people other than police become involved to a much greater extent, even at the highest political level. Politicians seem more interested in combating terrorism that drug cartels.

FP: Is there anything that can be learned from the fight against terrorism and applied to the fight against drugs?

RK: Intelligence. A greater application of the methods used to deal with terrorist organizations.

FP: For instance?

RK: Intelligence collection and intelligence analysis again and again, undercover operations, satellite observation, phone and mobile-phone tapping, e-mail interception, and use of information technology.

FP: Which countries have the most effective approach to counterterrorism?

RK: I think the British have always had an approach to subversion that puts them in a good position, in the same ways the United States is in a good position. Also maybe the French. Any country that has a history of battling subversion is better able to handle terrorism than others.

FP: And the failures? Which countries frustrate you in their inability to fight terrorism effectively?

RK: Those countries where politics have prevented them from eradicating terrorism within their borders, and those countries that have used or are still using terrorism as a political tool. I won't give you any names, but open your newspaper on any given day and you will recognize them.

ENDGAME

FP: You said that you are relieved to be leaving your job. Was it a mistake for you to run last time, five years ago?

RK: I don't think it was a mistake. I thought the last time I hadn't really gotten far enough into completing all there was to be done. For instance, I hadn't had time to adapt Interpol's finances to future missions. Member countries request more and more from Interpol, but they refuse to increase our budget. Another term didn't help, though. A solution will have to be found by my successor. Interpol needs more money.

FP: So now it's time to retire?

RK: I think that it's time for somebody else to take over. I don't know why, but there are clear indications that the moment has come to turn the page. One of them, purely coincidental, was the death of Jean Nepote, my predecessor, a man who recreated and built up this organization immediately after World War II. He was a Frenchman, a farsighted person and, in a way, a little bit my mentor.

FP: What kind of advice did he give you?

RK: Mr. Nepote gave me a lot of advice, even after his retirement. The advice that has been most important to the organization was that Interpol needed to adapt to the information technology era.

FP: What have you done in the last 15 years at Interpol that makes you very proud?

RK: Well, I think the fact that the organization has reached the status that it has, in the United Nations General Assembly for instance. That puts you in a stronger position to deal with things than before. I mean, I personally have a kind of status that I did not have 10 years ago. When I started, if I visited a country, I never got to see the interior minister or somebody at that upper level. Recognition for Interpol is something I've struggled for, to get some recognition at the political level.

FP: But I'm interested in specifics. What were Interpol's greatest moments during your tenure? The ones where it showed what it can do, or hinted at its true potential?

RK: The G7 (Group of Seven] summer meetings held in Lyon, France, in 1996, when we gained official recognition of Interpol's prominent role in the fight against transnational terrorism; a day in June 1998 when I delivered my first speech to the United Nations General Assembly in New York; my numerous meetings with U.S. presidents in Washington, D.C., and Davos, Switzerland—these were among the greatest moments in my career.

FP: I am surprised by that answer. I would have thought that you would have singled out specific achievements in your fight against criminals.

RK: There were many of those, but without the political support for Interpol I was fighting for, our efforts to fight criminals would have been doomed.

FP: Is your successor, the U.S. lawyer Ronald Noble, the best person available? Was that a good pick?

RK: Time will tell. But I don't think I should comment on that.

FP: How does one get elected to head Interpol?

RK: You need a two-thirds majority, two thirds of the country representatives present to vote. That's usually about 140 to 150 members. Of course, it's not as if people are being asked to vote from a selection of candidates, there's a preelection procedure.

FP: And how does that work?

RK: There are 13 countries on our executive board. They choose one person out of a number of candidates presented by their member countries for the job. Then that one has to go before Interpol's General Assembly.

FP: So 13 people meet in a small room and decide who the next head of Interpol will be, and then they take it to the General Assembly for rubber stamping. Would that be an unfair characterization?

RK: I think it would be unfair. Those men sitting around in a small group are all individually elected by groups of countries. There are three members from each continent. So it's a bit like countries that have a system where their president is elected by their parliament as opposed to being elected directly by all the citizens of the country.

FP: Throughout the multilateral world, every time there is a need to pick a leader of a multilateral organization, be it the leader of the IMF, the World Trade Organization, or the U.N., there is always a highly political process. Many critics say that the process is so political that very often, the merits of the candidates are not as central to the selection as they should be. Is Interpol different?

RK: When I was appointed, I came from within the organization. I was head of the criminal division, I was seen as the natural successor. The organization had to invent the process for my successor. What is tricky is that now, Interpol has reached the point where political considerations have become very important.

FP: Among Interpol's staff, as well as in the secretary general's office, no? One of the criticisms that has been leveled against Interpol is that the organization gets stuck with either second-rate police officers who countries can afford to do without and send off to your headquarters in Lyon, or officers who are good at milking the system and have their eyes on the easy life of an Interpol functionary.

RK: I can agree with that as a general criticism. Once, perhaps it was true. I think it's no longer valid in the same way. There are ways of fighting that sort of thing, but it remains an issue. Still, so long as the organization doesn't pay its staff directly, totally finance its staff, there will be a problem. The situation has changed in the last decade, and now countries are recognizing the fact that sending highly qualified staff to Lyon contributes to their profile in global police cooperation. But if we want to have a good staff, we need to pay them accordingly.

FP: So, you have drug trafficking that cannot be contained. You have cybercrime that cannot be fought without recruiting the help of the private sector. You have international trafficking in individuals, including children. You have frail democratic regimes that are prone to corruption and capture by criminal organizations. You have international networks that are capable of matching governments technology for technology. And facing all of that, you have a multilateral organization, Interpol, that is neither staffed with the best people nor funded sufficiently to keep up with fast-moving crime. Are you pessimistic?

RK: If I were totally pessimistic, then I would not have remained in office. The feeling I have is that we have been doing a lot of experimentation, without a true strategy to deal with crime globally. Part of it has to do with the major disparity between developed and underdeveloped countries. But then again, there are sufficient indications of successes to go with the failures. For instance, although the world production of counterfeit currency has probably increased by something like 30 percent in the last few years, if we look at the counterfeiting of U.S. dollars, that seems to have gone down. We do have bits and pieces of good news here and there. But the global picture is certainly not cause for celebration.

FP: The European Union has established Europol, and the countries of the Associations of Southeast Asian Nations have launched Aseanpol. These groups of countries all seem to be setting up their own shop. What does that say about the performance of Interpol?

RK: All it says to me is that people have decided, all for different reasons, to reinforce their crime-fighting capabilities. Let's take the European Union. Here we have 15 like-minded countries that came together for political or economic reasons. Inevitably, they will become interested in their internal security, which is very laudable. And if they are willing to put resources into it, that's fine. But they have to understand that there are no drugs, other than synthetic drugs, produced in the EU. So to confront even a basic problem like drug trafficking, they are going to have to look outside the union. And then there is a need for cooperation with European countries outside the EU.

FP: And that's where Interpol comes in?

RK: No group of countries can act in isolation, period. The value and the power of an organization like ours is to be able to create a global frame within which these other people can work together. Unless the world learns how to do that quickly and effectively, international crime will continue to grow.

Want to Know More?

For broad surveys of global crime, see *Law Enforcement in a New Century and a Changing World: Improving the Administration of Federal Law Enforcement* (Commission on the Advancement of Federal Law Enforcement, January 2000); Claire Sterling's *Crime without Frontiers: The Worldwide Expansion of Organized Crime and the Pax Mafiosa* (London: Little, Brown & Co., 1994) and *Thieves' World: The Threat of the New Global Network of Organized Crime* (New York: Simon & Schuster, 1994); and H. Richard Friman and Peter Andreas, eds. *The Illicit Global Economy and State Power* (Lanham: Rowman & Littlefield Publishers, 1999).

On the decriminalization of drug use, read Ethan A. Nadelmann's *"Commonsense Drug Policy"* (*Foreign Affairs*, January/February 1998). For studies of regional crime, see Raimondo Catanzaro's *Men of Respect; A Social History of the Sicilian Mafia* (New York: Free Press, 1992); Gerard P. Burke and Frank J. Cilluffo, eds. *Russian Organized Crime* (Washington: Center for Strategic and International Studies, 1997); Martin Booth's *The Dragon Syndicates: The Global Phenomenon of the Triads* (Chicago: Carroll & Graf Publishers, 2000); Paul Klebnikov's *Godfather of the Kremlin* (New York: Harcourt, 2000); and a special double issue of *Transnational Organized Crime* (nos. 2.2 and 2.3, 1997), edited by Phil Williams. On terrorist organizations operating around the world, read the U.S. State Department's *Patterns of Global Terrorism Index*. On nuclear terrorism, see Brian Jenkins's valuable article **"Will Terrorists Go Nuclear?"** (*Orbis*, Fall 1985). FOREIGN POLICY has provided regular analysis of terrorist threats, including: **"The Great Superterrorism Scare"** (Fall 1998) and **"Rational Fanatics"** (September/October 2000), both by Ehud Sprinzak; **"Think Again: Terrorism"** (Fall 1997) by John Deutch; and **"Is Europe Soft on Terrorism?"** (Summer 1999) by Bruce Hoffman.

For links to relevant Web sites, as well as a comprehensive index of related FOREIGN POLICY articles, access **www.foreignpolicy.com.**

The New Containment

An Alliance Against Nuclear Terrorism

Graham Allison & Andrei Kokoshin

DURING THE Cold War, American and Russian policy-makers and citizens thought long and hard about the possibility of nuclear attacks on their respective homelands. But with the fall of the Berlin Wall and the disappearance of the Soviet Union, the threat of nuclear weapons catastrophe faded away from most minds. This is both ironic and potentially tragic, since the threat of a nuclear attack on the United States or Russia is certainly greater today than it was in 1989.

In the aftermath of Osama bin Laden's September 11 assault, which awakened the world to the reality of global terrorism, it is incumbent upon serious national security analysts to think again about the unthinkable. Could a nuclear terrorist attack happen today? Our considered answer is: yes, unquestionably, without any doubt. It is not only a possibility, but in fact the most urgent unaddressed national security threat to both the United States and Russia.[1]

Consider this hypothetical: A crude nuclear weapon constructed from stolen materials explodes in Red Square in Moscow. A 15-kiloton blast would instantaneously destroy the Kremlin, Saint Basil's Cathedral, the ministries of foreign affairs and defense, the Tretyakov Gallery, and tens of thousands of individual lives. In Washington, an equivalent explosion near the White House would completely destroy that building, the Old Executive Office Building and everything within a one-mile radius, including the Departments of State, Treasury, the Federal Reserve and all of their occupants—as well as damaging the Potomac-facing side of the Pentagon.

Psychologically, such a hypothetical is as difficult to internalize as are the plot lines of a writer like Tom Clancy (whose novel *Debt of Honor* ends with terrorists crashing a jumbo jet into the U.S. Capitol on Inauguration Day, and whose *The Sum of All Fears* contemplates the very scenario we discuss—the detonation of a nuclear device in a major American metropolis by terrorists). That these kinds of scenarios are physically possible, however, is an undeniable, brute fact.

After the first nuclear terrorist attack, the Duma, Congress—or what little is left of them—and the press will investigate: Who knew what, when? They will ask what could have been done to prevent the attack. Most officials will no doubt seek cover behind the claim that "no one could have imagined" this happening. But that defense should ring hollow. We have unambiguous strategic warning today that a nuclear terrorist attack could occur at any moment. Responsible leaders should be asking hard questions now. Nothing prevents the governments of Russia, America and other countries from taking effective action immediately—nothing, that is, but a lack of determination.

The argument made here can be summarized in two propositions: first, nuclear terrorism poses a clear and present danger to the United States, Russia and other nations; second, nuclear terrorism is a largely *preventable* disaster. Preventing nuclear terrorism is a large, complex, but ultimately finite challenge that can be met by a bold, determined, but nonetheless finite response. The current mismatch between the seriousness of the threat on the one hand, and the actions governments are now taking to meet it on the other, is unacceptable. Below we assess the threat and outline a solution that begins with a U.S.-Russian led Alliance Against Nuclear Terrorism.

Assessing the Threat

A COMPREHENSIVE threat assessment must consider both the likelihood of an event and the magnitude of its anticipated consequences. As described above, the impact of even a crude nuclear explosion in a city would produce devastation in a class by itself.[2] A half dozen nuclear explosions across the United States or Russia would shift the course of history. The question is: how likely is such an event?

Security studies offer no well-developed methodology for estimating the probabilities of unprecedented events. Contemplating the possibility of a criminal act, Sherlock Holmes investigated three factors: motive, means and opportunity. That framework can be useful for analyzing the question at hand. If no actor simultaneously has motive, means and opportunity, no nuclear terrorist act will occur. Where these three factors are abundant and widespread, the likelihood of a nuclear terrorist attack increases. The questions become: Is anyone *motivated* to instigate a nuclear attack? Could terrorist groups acquire the *means* to attack the United States or Russia with nuclear weapons? Could these groups find or create an *opportunity* to act?

I. Motive

There is no doubt that Osama bin Laden and his associates have serious nuclear ambitions. For almost a decade they have been actively seeking nuclear weapons, and, as President Bush has noted, they would use such weapons against the United States or its allies "in a heartbeat." In 2000, the CIA intercepted a message in which a member of Al-Qaeda boasted of plans for a "Hiroshima" against America. According to the Justice Department indictment for the 1998 bombings of the American embassies in Kenya and Tanzania, "At various times from at least as early as 1993, Osama bin Laden and others, known and unknown, made efforts to obtain the components of nuclear weapons." Additional evidence from a former Al-Qaeda member describes attempts to buy uranium of South African origin, repeated travels to three Central Asian states to try to buy a complete warhead or weapons-usable material, and discussions with Chechen criminal groups in which money and drugs were offered for nuclear weapons.

Bin Laden himself has declared that acquiring nuclear weapons is a religious duty. "If I have indeed acquired [nuclear] weapons," he once said, "then I thank God for enabling me to do so." When forging an alliance of terrorist organizations in 1998, he issued a statement entitled "The Nuclear Bomb of Islam." Characterized by Bernard Lewis as "a magnificent piece of eloquent, at times even poetic Arabic prose," it states: "It is the duty of Muslims to prepare as much force as possible to terrorize the enemies of God." If anything, the ongoing American-led war on global terrorism is heightening our adversary's incentive to obtain and use a nuclear weapon. Al-Qaeda has discovered that it can no longer attack the United States with impunity. Faced with an assertive, determined opponent now doing everything it can to destroy this terrorist network, Al-Qaeda has every incentive to take its best shot.

Russia also faces adversaries whose objectives could be advanced by using nuclear weapons. Chechen terrorist groups, for example, have demonstrated little if any restraint on their willingness to kill civilians and may be tempted to strike a definitive blow to assert independence from Russia. They have already issued, in effect, a radioactive warning by planting a package containing cesium-137 at Izmailovsky Park in Moscow and then tipping off a Russian reporter. Particularly as the remaining Chechen terrorists have been marginalized over the course of the second Chechen war, they could well imagine that by destroying one Russian city and credibly threatening Moscow, they could persuade Russia to halt its campaign against them.

All of Russia's national security documents—its *National Security Concept*, its military doctrine and the recently-updated *Foreign Policy Concept*—have clearly identified international terrorism as the greatest threat to Russia's national security. As President Putin noted in reviewing Russian security priorities with senior members of the Foreign Ministry in January 2001, "I would like to stress the danger of international terrorism and fundamentalism of any, absolutely any stripe." The illegal drug trade and the diffusion of religious extremism throughout Central Asia, relating directly to the rise of the Taliban in Afghanistan, threaten Russia's borders and weaken the Commonwealth of Independent States. The civil war in Tajikistan, tensions in Georgia's Pankisi Gorge, and the conflicts in South Ossetia, Abkhazia and Nagorno-Karabakh—all close to the borders of the Russian Federation—provide feeding grounds for the extremism that fuels terrorism. Additionally, Russia's geographical proximity to South Asia and the Middle East increases concerns over terrorist fallout from those regions. President Putin has consistently identified the dark hue that weapons of mass destruction (WMD) give to the threat of terrorism. In a December 2001 interview in which he named international terrorism the "plague of the 21st century," Putin stated: "We all know exactly how New York and Washington were hit.... Was it ICBMs? What threat are we talking about? We are talking about the use of mass destruction weapons terrorists may obtain."

Separatist militants (in Kashmir, the Balkans and elsewhere) and messianic terrorists (like Aum Shinrikyo, which attacked the Tokyo subway with chemical weapons in 1995) could have similar motives to commit nuclear terrorism. As Palestinians look to uncertain prospects for independent statehood—and never mind whose leadership actually increased that uncertainty in recent years—Israel becomes an ever more attractive target for a nuclear terrorist attack. Since a nuclear detonation in any part of the world would be extremely destabilizing, it threatens American and Russian interests even if few or no Russians or Americans are killed. Policymakers would therefore be foolish to ignore any group with a motive to use a nuclear weapon against any target.

II. Means

To the best of our knowledge, no terrorist group can now detonate a nuclear weapon. But as Secretary of Defense Donald Rumsfeld has stated, "the absence of evidence is not evidence of absence." Are the means beyond terrorists' reach, even that of relatively sophisticated groups like Al-Qaeda?

Over four decades of Cold War competition, the superpowers spent trillions of dollars assembling mass arsenals, stockpiles, nuclear complexes and enterprises that engaged hundreds of thousands of accomplished scientists and engineers. Technical know-how cannot be un-invented. Reducing arsenals that include some 40,000 nuclear weapons and the equivalents of more than 100,000 nuclear weapons in the form of highly enriched uranium (HEU) and plutonium to manageable levels is a gargantuan challenge.

Terrorists could seek to buy an assembled nuclear weapon from insiders or criminals. Nuclear weapons are known to exist in eight states: the United States, Russia, Great Britain, France, China, Israel, India and Pakistan. Security measures, such as "permissive action links" designed to prevent unauthorized use, are most reliable in the United States, Russia, France and the United Kingdom. These safeguards, as well as command-and-control systems, are much less reliable in the two newest nuclear states—India and Pakistan. But even where good systems are in place, maintaining high levels of security requires constant attention from high-level government officials.

Alternatively, terrorists could try to build a weapon. The only component that is especially difficult to obtain is the nuclear fissile material—HEU or plutonium. Although the largest stockpiles of weapons-grade material are predominantly found in the nuclear weapons programs of the United States and Russia, fissile material in sufficient quantities to make a crude nuclear weapon can also be found in many civilian settings around the globe. Some 345 research reactors in 58 states together contain twenty metric tons of HEU, many in quantities sufficient to build a bomb.[3] Other civilian reactors produce enough weapons-grade nuclear material to pose a proliferation threat; several European states, Japan, Russia and India reprocess spent fuel to separate out plutonium for use as new fuel. The United States has actually facilitated the spread of fissile material in the past—over three decades of the Atoms for Peace program, the United States exported 749 kg of plutonium and 26.6 metric tons of HEU to 39 countries.[4]

Terrorist groups could obtain these materials by theft, illicit purchase or voluntary transfer from state control. There is ample evidence that attempts to steal or sell nuclear weapons or weapons-usable material are not hypothetical, but a recurring fact.[5] Just last fall, the chief of the directorate of the Russian Defense Ministry responsible for nuclear weapons reported two recent incidents in which terrorist groups attempted to perform reconnaissance at Russian nuclear storage sites. The past decade has seen repeated incidents in which individuals and groups have successfully stolen weapons material from sites in Russia and sought to export them—but were caught trying to do so. In one highly publicized case, a group of insiders at a Russian nuclear weapons facility in Chelyabinsk plotted to steal 18.5 kg (40.7 lbs.) of HEU, which would have been enough to construct a bomb, but were thwarted by Russian Federal Security Service agents.

In the mid-1990s, material sufficient to allow terrorists to build more than twenty nuclear weapons—more than 1,000 pounds of highly enriched uranium—sat unprotected in Kazakhstan. Iranian and possibly Al-Qaeda operatives with nuclear ambitions were widely reported to be in Kazakhstan. Recognizing the danger, the American government itself purchased the material and removed it to Oak Ridge, Tennessee. In February 2002, the U.S. National Intelligence Council reported to Congress that "undetected smuggling [of weapons-usable nuclear materials from Russia] has occurred, although we do not know the extent of such thefts." Each assertion invariably provokes blanket denials from Russian officials. Russian Atomic Energy Minister Aleksandr Rumyantsev has claimed categorically: "Fissile materials have not disappeared." President Putin has stated that he is "absolutely confident" that terrorists in Afghanistan do not have weapons of mass destruction of Soviet or Russian origin.

For perspective on claims of the inviolable security of nuclear weapons or material, it is worth considering the issue of "lost nukes." Is it possible that the United States or Soviet Union lost assembled nuclear weapons? At least on the American side the evidence is clear. In 1981, the U.S. Department of Defense published a list of 32 accidents involving nuclear weapons, *many of which resulted in lost bombs*.[6] One involved a submarine that sank along with two nuclear torpedoes. In other cases, nuclear bombs were lost from aircraft. Though on the Soviet/Russian side there is no official information, we do know that four Soviet submarines carrying nuclear weapons have sunk since 1968, resulting in an estimated 43 lost nuclear warheads.[7] These accidents suggest the complexity of controlling and accounting for vast nuclear arsenals and stockpiles.

Nuclear materials have also been stolen from stockpiles housed at research reactors. In 1999, Italian police seized a bar of enriched uranium from an organized crime group trying to sell it to an agent posing as a Middle Eastern businessman with presumed ties to terrorists. On investigation, the Italians found that the uranium originated from a U.S.-supplied research reactor in the former Zaire, where it presumably had been stolen or purchased *sub rosa*.

Finally, as President Bush has stressed, terrorists could obtain nuclear weapons or material from states hostile to the United States. In his now-infamous phrase, Bush called hostile regimes developing WMD and their terrorist allies an "axis of evil." He argued that states such as Iraq, Iran and North Korea, if allowed to realize their nuclear ambitions, "could provide these arms to terrorists, giving them the means to match their hatred." The fear that a hostile regime might transfer a nuclear weapon to terrorists has contributed to the Bush Administration's development of a new doctrine of preemption against such regimes, with Iraq as the likeliest test case. It also adds to American concerns about Russian transfer of nuclear technologies to Iran. While Washington and Moscow continue to disagree over whether any safeguarded civilian nuclear cooperation with Iran is justified, both agree on the dangers a nuclear-armed Iran would pose. Russia is more than willing to agree that there should be no transfers of technology that could help Iran make nuclear weapons.

III. Opportunity

Security analysts have long focused on ballistic missiles as the preferred means by which nuclear weapons would be delivered. But today this is actually the least likely vehicle by which a nuclear weapon will be delivered against Russia or the United States. Ballistic weapons are hard to produce, costly and difficult to hide. A nuclear weapon delivered by a missile also leaves an unambiguous return address, inviting devastating retaliation. As Robert Walpole, a National Intelligence Officer, told a Senate subcommittee in March, "Nonmissile delivery means are less costly, easier to acquire, and more reliable and accurate."[8] Despite this assessment, the U.S. government continues to invest much more heavily in developing and deploying missile defenses than in addressing more likely trajectories by which weapons could arrive.

Terrorists would not find it very difficult to sneak a nuclear device or nuclear fissile material into the United States via shipping containers, trucks, ships or aircraft. Recall that the nuclear material required is smaller than a football. Even an assembled device, like a suitcase nuclear weapon, could be shipped in a container, in the hull of a ship or in a trunk carried by an aircraft. After this past September 11, the number of containers that are X-rayed has increased, to about 500 of the 5,000 containers currently arriving daily at the port of New York/New Jersey—approximately 10 percent. But as the chief executive of CSX Lines, one of the foremost container-shipping companies, put it: "If you can smuggle heroin in containers, you may be able to smuggle in a nuclear bomb."

Effectively countering missile attacks will require technological breakthroughs well beyond current systems. Success in countering covert delivery of weapons will require not just technical advances but a conceptual breakthrough. Recent efforts to bolster border security are laudable, but they only begin to scratch the surface. More than 500 million people, 11 million trucks and 2 million rail cars cross into the United States each year, while 7,500 foreign-flag ships make 51,000 calls in U.S. ports. That's not counting the tens of thousands of people, hundreds of aircraft and numerous boats that enter illegally and uncounted. Given this volume and the lengthy land and sea borders of the United States, even a radically renovated and reorganized system cannot aspire to be airtight.

The opportunities for terrorists to smuggle a nuclear weapon into Russia or another state are even greater. Russia's land borders are nearly twice as long as America's, connecting it to more than a dozen other states. In many places, in part because borders between republics were less significant in the time of the Soviet Union, these borders are not closely monitored. Corruption has been a major problem among border patrols. Visa-free travel between Russia and several of its neighbors creates additional opportunities for weapons smugglers and terrorists. The "homeland security" challenge for Russia is truly monumental.

In sum: even a conservative estimate must conclude that dozens of terrorist groups have sufficient motive to use a nuclear weapon, several could potentially obtain nuclear means, and hundreds of opportunities exist for a group with means and motive to make the United States or Russia a victim of nuclear terrorism. The mystery before us is not how a nuclear terrorist attack could possibly occur, but rather why no terrorist group has yet combined motive, means and opportunity to commit a nuclear attack. We have been lucky so far, but who among us trusts luck to protect us in the future?

Chto Delat?[9]

THE GOOD NEWS about nuclear terrorism can be summarized in one line: no highly enriched uranium or plutonium, no nuclear explosion, no nuclear terrorism. Though the world's stockpiles of nuclear weapons and weapons-usable materials are vast, they are finite. The prerequisites for manufacturing fissile material are many and require the resources of a modern state. Technologies for locking up super-dangerous or valuable items—from gold in Fort Knox to treasures in the Kremlin Armory—are well developed and tested. While challenging, a specific program of actions to keep nuclear materials out of the hands of the most dangerous groups is not beyond reach, *if* leaders give this objective highest priority and hold subordinates accountable for achieving this result.

The starting points for such a program are already in place. In his major foreign policy campaign address at the Ronald Reagan Library, then-presidential candidate George W. Bush called for "Congress to increase substantially our assistance to dismantle as many Russian weapons as possible, as quickly as possible." In his September 2000 address to the United Nations Millennium Summit, Russian President Putin proposed to "find ways to block the spread of nuclear weapons by excluding use of enriched uranium and plutonium in global atomic energy production." The Joint Declaration on the New Strategic Relationship between the United States and Russia, signed by the two presidents at the May 2002 summit, stated that the two partners would combat the "closely linked threats of international terrorism and the proliferation of weapons of mass destruction." Another important result yielded by the summit was the upgrading of the Armitage/Trubnikov-led U.S.-Russia Working Group on Afghanistan to the U.S.-Russia Working Group on Counterterrorism, whose agenda is to thwart nuclear biological and chemical terrorism.

Operationally, however, priority is measured not by words, but by deeds. A decade of Nunn-Lugar Cooperative Threat Reduction Programs has accomplished much in safeguarding nuclear materials. Unfortunately, the job of upgrading security to minimum basic standards is mostly unfinished: according to Department of Energy reports, two-thirds of the nuclear material in Russia remains to be adequately secured.[10] Bureaucratic inertia, bolstered by mistrust and misperception on both sides, leaves these joint programs bogged down on timetables that extend to 2008. Unless implementation improves significantly, they will probably fail to meet even this unacceptably distant target. What is required on both sides is personal, presidential priority measured in commensurate energy, specific orders, funding and accountability. This should be embodied in a new U.S.-Russian led Alliance Against Nuclear Terrorism.

Five Pillars of Wisdom

WHEN IT COMES to the threat of nuclear terrorism, many Americans judge Russia to be part of the problem, not the solution. But if Russia is welcomed and supported as a fully responsible non-proliferation partner, the United States stands to accomplish far more toward minimizing the risk of nuclear terrorism than if it treats Russia as an unreconstructed pariah. As the first step in establishing this alliance, the two presidents should pledge to each other that his government will do everything technically possible to prevent criminals or terrorists from stealing nuclear weapons or weapons-usable material, and to do so on the fastest possible timetable. Each should make clear that he will personally hold accountable the entire chain of command within his own government to assure this result. Understanding that each country bears responsibility for the security of its own nuclear materials, the United States should nonetheless offer Russia any assistance required to make this happen. Each nation—and each leader—should provide the other sufficient transparency to monitor performance.

Archy: "An optimist is a guy that has never had much experience."

—Don Marquis, *Archy and Mehitabel* (1927)

To ensure that this is done on an expedited schedule, both governments should name specific individuals, directly answerable to their respective presidents, to co-chair a group tasked with developing a joint Russian-American strategy within one month. In developing a joint strategy and program of action, the nuclear superpowers would establish a new world-class "international security standard" based on President Putin's Millennium proposal for new technologies that allow production of electricity with low-enriched, non-weapons-usable nuclear fuel.

A second pillar of this alliance would reach out to all other nuclear weapons states—beginning with Pakistan. Each should be invited to join the alliance and offered assistance, if necessary, in assuring that all weapons and weapons-usable material are secured to the new established international standard in a manner sufficiently transparent to reassure all others. Invitations should be diplomatic in tone but nonetheless clear that this is an offer that cannot be refused. China should become an early ally in this effort, one that could help Pakistan understand the advantages of willing compliance.

A third pillar of this alliance calls for global outreach along the lines proposed by Senator Richard Lugar in what has been called the Lugar Doctrine.[11] All states that possess weapons-usable nuclear materials—even those without nuclear weapons capabilities—must enlist in an international effort to guarantee the security of such materials from theft by terrorists or criminals groups. In effect, each would be required to meet the new international security standard and to do so in a transparent fashion. Pakistan is particularly important given its location and relationship with Al-Qaeda, but beyond nuclear weapons states, several dozen additional countries hosting research reactors—such as Serbia, Libya and Ghana—should be persuaded to surrender such material (almost all of it either American or Soviet in origin), or have the material secured to acceptable international standards.

A fourth pillar of this effort should include Russian-American led cooperation in preventing any further spread of nuclear weapons to additional states, focusing sharply on North Korea, Iraq and Iran. The historical record demonstrates that when the United States and Russia have cooperated intensely, nuclear wannabes have been largely stymied. It was only during periods of competition or distraction, for example in the mid-1990s, that new nuclear weapons states realized their ambitions. India and Pakistan provide two vivid case studies. Recent Russian-American-Chinese cooperation in nudging India and Pakistan back from the nuclear brink suggests a good course of action. The failure and subsequent freeze of North Korean nuclear programs offers complementary lessons about the consequences of competition and distraction. The new alliance should reinvent a robust non-proliferation regime of controls on the sale and export of weapons of mass destruction, nuclear material and missile technologies, recognizing the threat to each of the major states that would be posed by a nuclear-armed Iran, North Korea or Iraq.

Finally, adapting lessons learned in U.S.-Russian cooperation in the campaign against bin Laden and the Taliban, this new alliance should be heavy on intelligence sharing and affirmative counter-proliferation, including disruption and pre-emption to prevent acquisition of materials and know-how by nuclear wannabes. Beyond joint intelligence sharing, joint training for pre-emptive actions against terrorists, criminal groups or rogue states attempting to acquire weapons of mass destruction would provide a fitting enforcement mechanism for alliance commitments.

AS FORMER Senator Sam Nunn has noted: "At the dawn of a new century, we find ourselves in a new arms race. Terrorists are racing to get weapons of mass destruction; we ought to be racing to stop them."[12] Preventing nuclear terrorism will require no less imagination, energy and persistence than did avoiding nuclear war between the superpowers over four decades of Cold War. But absent deep, sustained cooperation between the United States, Russia and other nuclear states, such an effort is doomed to failure. In the context of the qualitatively new relationship Presidents Putin and Bush have established in the aftermath of last September 11, success in such a bold effort is within the reach of determined Russian-American leadership. Succeed we must.

Notes

1. This judgment echoes that of a Department of Energy task force on nonproliferation programs with Russia led by Howard Baker and Lloyd Cutler: "The most urgent unmet national security threat to the United States today is the danger that weapons of mass destruction or weapons-usable material in Russia could be stolen and sold to terrorists or hostile nation states and used against American troops abroad or citizens at home." *A Report Card on the Department of Energy's Nonproliferation Programs with Russia*, January 10, 2001.

2. Although biological and chemical weapons can cause huge devastation as well, "the massive, assured, instantaneous, and comprehensive destruction of life and property" of a nuclear weapon is unique. See Matthew Bunn, John P. Holdren and Anthony Wier, "Securing Nuclear Weapons and Materials: Seven Steps for Immediate Action," *Nuclear*

Threat Initiative and the Managing the Atom Project, May 20, 2002, p. 2. This report provides extensive, but not-too-technical detail on many of the points in this essay.

3. See U.S. Department of Energy, *FY 2003 Budget Request: Detailed Budget Justifications—Defense Nuclear Nonproliferation* (Washington, DC: DOE, 2002), p. 172.

4. Summarized in NIS Nuclear Trafficking Database, available at www.nti.org.

5. The Nuclear Threat Initiative maintains a database of cases and reported incidents of trafficking in nuclear and radioactive materials in and from the former Soviet Union. Available at www.nti.org.

6. U.S. Department of Defense, "Narrative Summaries of Accidents involving U.S. Nuclear Weapons: 1950-1980" (April 1981).

7. Joshua Handler, Amy Wickenheiser and William M. Arkin, "Naval Safety 1989: The Year of the Accident," *Neptune Paper No. 4* (April 1989).

8. U.S Senate Subcommittee on International Security, Proliferation, and Federal Services, "Statement of Robert Walpole before the Senate Subcommittee on International Security, Proliferation, and Federal Services," 107th Cong., 1st sess., March 11, 2002.

9. A proverbial Russian refrain: "What is to be done?"

10. Bunn, Holdren and Wier, "Securing Nuclear Weapons and Materials."

11. Speech by Senator Richard Lugar, May 27, 2002, at the Moscow Nuclear Threat Initiative Conference.

12. Sam Nunn, "Our New Security Framework," *Washington Post*, October 8, 2001.

Graham Allison is director of the Belfer Center for Science and International Affairs at Harvard's John F. Kennedy School of Government. Andrei Kokoshin is director of the Institute for International Security Studies of the Russian Academy of Sciences and a former secretary of the Security Council of Russia.

countdown TO ERADICATION

Rotary and its partners move ever closer to a polio-free world

by Anne Stein

In 1985, Rotary International committed to help immunize the world's children against polio, with a target date of a polio-free world by 2005, Rotary's centennial year. Since 1988, when the Global Polio Eradication Initiative was launched, millions of Rotary volunteers have raised funds and assisted in oral polio vaccine delivery, social mobilization, and garnering cooperation from national governments and health ministries—making Rotary the leading private-sector partner in the effort.

The program began with donations from Rotarians of more than US $240 million. By 2005, Rotary's financial contribution will reach an estimated $500 million. Along with its global partners—the World Health Organization (WHO), UNICEF, and the U.S. Centers for Disease Control and Prevention (CDC)—Rotary can take pride in the largest public health initiative in history.

ENDING POLIO Ten polio-endemic nations and a $275 million shortfall stand between failure and success.

According to the most recent data, only 537 cases of polio have been reported worldwide for 2001, although that number will likely increase somewhat as final reports are filed. However, it is sure to be down considerably from 2000, when 2,979 cases were logged. The program partners report an astounding 99 percent decrease in polio cases since 1988, the year WHO resolved to eradicate the disease. But despite the remarkable progress in reaching children and the unprecedented outpouring of time and resources from Rotarians and their partners, there's still much work to do and $1 billion in donor contributions needed to reach the final goal.

So far, $725 million has been pledged or projected, leaving a gap of $275 million. In response, the RI Board of Directors and The Rotary Foundation Trustees have announced a new, $80 million fundraising campaign—Rotary's contribution to closing the gap—that will run through the 2002-03 Rotary year. The campaign, Fulfilling Our Promise: Eradicate Polio, was officially launched in June during the RI Convention in Barcelona by 2001-02 Trustees Chairman Luis V. Giay.

"The very principles of Rotary are on the line," says Past RI Vice President Louis Piconi, chairman of the North American Polio Eradication Fundraising Campaign Committee. "We should never make a promise without fulfilling it; it's our word. To have Rotary be a major factor in eradicating polio will be a very significant milestone for the organization and its history."

"Each of these countries has made tremendous progress, but each has its own unique set of challenges."

About 60 percent of today's Rotarians—including all women members—have joined since the launch of the PolioPlus campaign, observes former RI General Secretary Herbert Pigman, who is now the director of the polio eradication fundraising campaign. "This new campaign gives them an opportunity to play an important part in our great humanitarian cause."

Up to $25 million of the contributions to this campaign can be matched by an equal amount from the Bill & Melinda Gates Foundation. The resulting $50 million can be further augmented by $75 million in assistance from the World Bank. "If we raise this money now, we will be saving all of humankind from this disease for all time, and that's priceless," Giay said in April during a joint news conference with representatives of the CDC, UNICEF, and WHO.

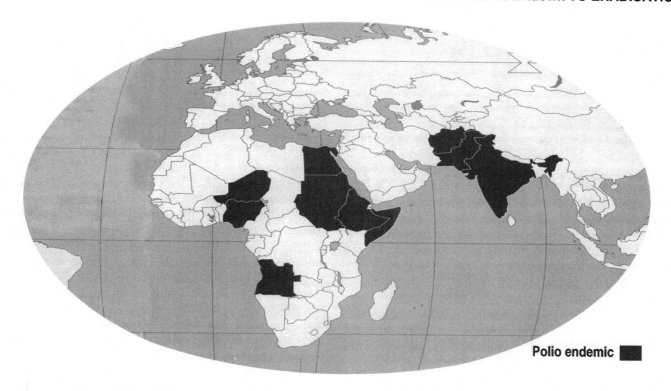

Countdown to 2005: The wild poliovirus remains endemic in 10 countries: Afghanistan, Angola, Egypt, Ethiopia, India, Niger, Nigeria, Pakistan, Somalia and Sudan.

What remains to be done? Quite simply, a lot. Intensified immunization activities must take place in the 10 remaining polio-endemic nations (down from 20 nations in 2000 and 127 in 1985) and in several more countries that are considered high risk for outbreaks. Surveillance must also be maintained to ensure polio-free status. "We need to immunize more than 600 million children over the next four years," explains Piconi. "Rotary's campaign will help to buy and get the vaccine to all of those children, no matter where they are."

Areas of greatest focus are the remaining polio-endemic nations: Afghanistan, India, Niger, Nigeria, and Pakistan, which collectively accounted for more than 85 percent of the 2001 polio cases and are considered high-intensity transmission areas, and Angola, Egypt, Ethiopia, Somalia, and Sudan, which are low-intensity transmission areas, accounting for less than 15 percent of new polio cases last year.

The high risk areas have been identified and are receiving special attention during the immunization drives.

"Each of these countries has made tremendous progress, but each has its own unique set of challenges," said Carol Bellamy, executive director of UNICEF, during the April news conference. "Through the battle to rid the world of polio, we have managed to reach children living in some of the most remote and challenging circumstances imaginable. Over the coming days and months, we must continue this unprecedented effort, using all of our resources to reach the very last child with polio vaccine. We have a unique opportunity to deliver a truly global victory in an uncertain world."

Deepak Kapur, India's National PolioPlus Committee Chairman and a member of the Rotary Club of Delhi South, says: "The job is almost done [in India]. But the most dangerous word in the dictionary is *almost*. We rejoiced when the thousands of cases fell to a few hundred, and today we have cause to celebrate when only a few cases are left. But this new chapter begins with those few cases."

The good news in India is that the geographic distribution of the virus has decreased. In 2001, India reported 268 cases of polio (a slight increase from 2000), about 95 percent of them in the northern states of Bihar and Uttar Pradesh. Kapur says that a range of projects will be undertaken this year, including mobilizing village volunteers and school students to spread awareness of the campaign; continuing the "video van" project and musical street dramas, which are especially useful in explaining PolioPlus to illiterate populations; and updating local media to enlist their editorial support.

According to Kapur, challenges include generally poor immunization rates for routine childhood diseases, over-worked public health officials, public indifference because of repeated immunization drives, and outright distrust of immunization efforts by members of certain populations.

"[PolioPlus India] Executive Director Rajiv Tandon and I went to Moradabad on a field visit. It's considered the 'polio capital of the world since more than half the cases have been genetically traced to the virus prevalent in this area," recounts Kapur. "One of the reasons for proliferation is the significant resistance of members of a particular minority community." The two visited a hut in a slum to persuade a mother to let them administer the polio drops to her infant son. Finally, after making the pair promise not to tell her husband, she allowed her baby to be immunized.

"Her husband was convinced that the immunization program was aimed at curtailing the population of their community by making children impotent," Kapur explains. "As we stepped out of the hut, we saw a polio-crippled girl playing in the mud. The neighbors told us that the little girl had contracted polio a few months ago and was that very mother's daughter." Polio workers in India face this kind of resistance because of incomplete information, misinformation, and rumors, says Kapur.

Afghanistan and Pakistan also pose challenges, but progress has been good. Despite the war and refugee crises in the wake of 11 September, Afghanistan reported only 11 cases of polio in 2001 (down from 120 in 2000), clustered in two main areas. Two rounds of National Immunization Days (NIDs) were held in September and November 2001, reaching 35 million children; this year, NIDs began anew in mid-April, coordinated with Pakistan to ensure coverage of cross-border areas.

"The most urgent need is to bring stability to the country [Afghanistan] and start reconstruction work without delay," says Rotarian Abdul Haiy Khan, chairman of Pakistan's National PolioPlus Committee and a member of the Rotary Club of Karachi. "The Afghans are resilient, and during my visits to Afghanistan to monitor their NIDs, I was impressed with their commitment. I think that once they get the opportunity to freely move around the country and have the assurance of obtaining the needed resources, the polio-eradication job will be done in a short time."

In Pakistan, Khan reports, "we have to deal with a large population, 140 million people with about 30 million of them children under age 5, nearly 30 percent living in densely populated or slum areas with poor sanitary conditions." However, the high-risk areas have been identified and are receiving special attention during the immunization drives. The high-transmission season of wild polio-virus runs from May to September.

"This will be the testing period to prove that we are in control of the situation and can hope to come close to the zero polio-case reporting status," says Khan. "The real challenge includes training and retraining health workers, volunteers and monitors, improving the surveillance system, and providing better transport for health workers and volunteers so they can reach the farthest, most isolated areas in the hot summer season."

But ingenuity and creativity bring success. Khan cites one story of an Afghan refugee camp on the outskirts of Karachi, where health workers reported that camp residents refused to let their children be immunized. "We spent some time investigating the matter and were pleasantly surprised to find that the refusal was based on a lack of communication. Our health workers didn't speak [their] language and couldn't convey the purpose of house-to-house visits. And the camp housed a large number of unregistered refugees, so camp elders were reluctant to let government officials (health workers) go door-to-door."

The solution? "We simply located two Afghan girls studying in a local medical college, explained the problem, and they came to the camp to talk to the elders. The next day, the two girls were accompanied by more than 100 fellow students, and in two days, working eight hours a day, they immunized 10,000 Afghan children under the age of 5 whom we had been missing for a couple of years."

The funding gap, along with the continuing threat of conflict worldwide and particularly within the 10 polio-endemic nations, brought a recent warning from experts: While polio has been pushed to the lowest level in history, efforts to complete eradication and immunization must be intensified now before global instability closes the window of opportunity.

"I urge the world to finish the job. Eradicate polio while we still have the opportunity."
– Dr. Gro Harlem Brundtland, WHO director-general

"When we began the eradication effort in 1988, polio paralyzed more than 1,000 children each day. In 2001, there were far fewer than 1,000 cases the entire year," said Dr. Gro Harlem Brundtland, director-general of WHO, during the joint April news conference. "But we're not finished yet, and the past year has reminded us that we live in a world where security and access to children cannot be guaranteed. So I urge the world to finish the job. Eradicate polio while we still have the opportunity."

Fortunately, in the African nations of Angola, Sierra Leone, and Somalia, warring forces have laid down their arms and allowed health workers to reach the children. In 2000, 16 West African countries coordinated immunization activities, and more than 80 million children were immunized. In 2001, conflict-affected countries in central Africa, including Angola and the Democratic Republic of the Congo, synchronized their NIDs for the first time and 16 million children were vaccinated.

To stop the transmission of the wild poliovirus in Africa by the end of 2002, NIDs will target northern Nigeria along the border with Niger. Major progress has been achieved in Angola, Egypt, Ethiopia, Somalia, and Sudan, and the goal is to stop transmission of the wild poliovirus by mid-2002—given continued political commitment in the five nations and no further deterioration of security.

Seventeen nations are considered at particularly high risk of reestablished wild poliovirus transmission, including Central African Republic, Chad, Congo Republic, Democratic Republic of the Congo, Indonesia, Madagascar, and Mozambique. These countries either are recently endemic or border endemic areas. They also have low routine-immunization coverage and inadequate surveillance capability or both. These nations require continuing, supplementary immunization activities and the assurance that no child has been missed.

"If we don't finish this off," warns Piconi, "this virus can creep back rather quickly. We need to stay focused and eliminate a disease that's vaccine preventable."

Anne Stein is a health and fitness writer in Evanston, Ill., USA, who has written frequently about the polio-eradication effort for THE ROTARIAN.

AERIAL WAR AGAINST DISEASE

Satellite tracking of epidemics is soaring

BY JOHN PICKRELL

Imagine this: It's the height of summer a decade from now, and stifling heat blankets New York City. A hidden foe has lurked out of sight for many months, and authorities are on high alert. The streets are deserted, Central Park is sealed, helicopters circle overhead, and residents remain inside with their windows firmly closed. Dousing the city, the surrounding suburbs, and perhaps wider areas with protective chemicals is the only defense.

In this fictional scenario, the city isn't under terrorist attack. The concealed enemy is the deadly virus that causes West Nile fever, which first emerged in New York in 1999. Attempts at vaccines (*SN: 3/16/02, p. 164*) haven't borne fruit, and the heat is creating the ideal conditions for drawing virus-carrying mosquitoes out of their winter refuges.

However, officials of the future are prepared to fight the disease, courtesy of an unlikely medical tool: the National Oceanic and Atmospheric Administration's (NOAA) series of polar-orbiting satellites about 850 kilometers above Earth.

The NOAA satellites aren't sending pictures of crows—the most common avian reservoir of West Nile virus in the United States—but they're producing maps of vegetation and weather affecting the mosquitoes that carry the disease to people.

In 2002, satellites already enable people to forecast weather, spy on enemy armies, broadcast television signals, and phone a friend in the Amazon jungle. A growing group of researchers has also been harnessing satellites in the battle against infectious diseases (*SN: 8/2/97, p. 72; 2/2/02, p. 78*).

The amount of data created by environmental satellites is increasing at an exponential rate, and researchers worldwide are catching on to the value of this information. The February *Photogrammetric Engineering and Remote Sensing* details several research projects that use satellite data to reveal patterns of diseases including schistosomiasis, tuberculosis, Ebola, Rift Valley fever, and West Nile fever.

SATELLITE FEVER—INTREPID's preliminary risk map for West Nile fever. Land-surface temperatures (shading variation) and locations of birds harboring West Nile virus (black dots) help researchers predict where outbreaks are likely to occur.

FOLLOW THE FLORA The rationale behind the use of satellites to fight vector-borne disease lies in the field of landscape epidemiology, which began in Russia in the 1930s. Theory holds that features of the landscape, such as temperature, humidity, rainfall, and plant growth, can help public health officials pinpoint where deadly microbes—or the animals that carry them—dwell. Ideally, that indicates where people are at greatest risk and so where governments should target disease-control measures and medicines.

Researchers first discussed adding satellites to the tools of landscape epidemiology in the early 1970s, with the launch of NASA's Landsat-1 satellite, says Robert Venezia, program manager for public health applications at NASA's headquarters. In recent years, NASA has promoted the use of satellite data for health purposes. The agency is currently working with an international team of universities and U.S.-government bodies under the banner of the International Research Partnership for Infectious Diseases, or INTREPID. The program's researchers are today in the early stages of creating a satellite-borne early-warning system for West Nile virus in the United States, says epidemiologist Simon Hay at the University of Oxford in England. With it, workers will be better prepared, as in the fictional example above.

The scientists are focusing on areas with warm land surfaces, high humidity, and vegetation types likely to harbor mosquitoes. Data on these variables will be combined with other information, such as bird-migration routes, to create risk maps for West Nile fever during each summer.

Ebola—Chlorophyll reflects light in a distinctive way, so satellite images clearly reveal vegetation changes, which act as a surrogate for rainfall patterns.

Two other deadly diseases—Rift Valley fever and Ebola—are also the focus of remote-sensing initiatives. These viral infections, which cause swelling and fever, strike in intermittent outbreaks in Africa. Rift Valley fever was first documented in the 1930s in the area of Kenya from which it takes its name. It kills few people, but it's a major problem in livestock, where it causes spontaneous abortions and many deaths.

Ebola is more deadly to people and more enigmatic. Named after a river in Sudan, it was reported appearing for the first time in 1976. Reappearing periodically, it spreads rapidly and kills up to 88 percent of the people it infects.

Despite the disease's deadliness, scientific understanding of Ebola is limited. Infections first show up in people who have been in tropical forests, but researchers still have no idea which animal species act as vectors or reservoirs for the virus.

Scientists at the NASA-Goddard Space Flight Center in Greenbelt, Md., are approaching the challenges of Ebola and Rift Valley fever in a different way from the strategy employed to study U.S. West Nile virus. They're not only trying to predict future outbreaks of the African diseases but also attempting to better understand their causes.

The only documented Ebola outbreaks have occurred within two limited periods. The disease struck in Congo and Sudan during the 1970s, before satellite data were being collected con-

sistently in Africa, and resurfaced for a further sporadic round of infection in Gabon, Ivory Coast, and Congo during the 1990s.

Compton J. Tucker of Goddard and his colleagues suspect that transitions between wet and dry periods in tropical forests may lie at the heart of Ebola outbreaks. Using images taken by the NOAA and Landsat satellites from 1994 to 1996, the team has studied vegetation changes as a surrogate for rainfall patterns.

That's relatively easy to do from space. Chlorophyll—the green pigment in leaves—reflects light in a distinctive way, so satellite images clearly reveal vegetation. The researchers found that images taken during three separate Ebola outbreaks indicated a transition from below-average to above-average vegetation cover within equatorial Africa's forests. The finding suggests that the disease "may follow when a rare tropical dry period is brought to an abrupt end with a change to very wet conditions," says Tucker.

He notes that it will take additional research to confirm that a wet-to-dry transition is the trigger and to explain why such an event spreads the virus. "It's fortunate for those affected by Ebola that we have so few outbreaks to study," Tucker says, "but it makes our job more difficult."

CREATURE FEATURES The researchers have had more luck nailing down the conditions that promoted Rift Valley fever during the 1980s and 1990s.

From the satellite data, Rift Valley fever appears linked to above-average rainfall in savanna ecosystems. The wet conditions lead to booms in mosquito populations that spread Rift Valley fever to people.

The accuracy with which the researchers can use satellite data to predict the disease's course has already led the team to create a disease-surveillance and early-warning system. Computers analyze daily satellite data for the telltale predictors of mosquito outbreaks in areas prone to the disease, says Tucker. The researchers pass along 2-week summaries of the data to the U.S. Army and the World Health Organization. These groups can warn their personnel and others in threatened areas to take precautions against infection, for instance, by stepping up immunizations. Satellite surveillance programs can now "monitor diseases in both space and time," Hay says.

Schistosomiasis—Researchers turned to data from the Landsat TM satellite to identify areas most likely to harbor snails.

So far, satellite epidemiology has focused mostly on insect-borne diseases because correlations among weather, plant cover, and insect infestations are clearly established. But scientists are starting to use NASA's Landsat satellites—700 km above Earth—to track other disease carriers.

In China, researchers are using satellites to map areas at highest risk from the debilitating disease schistosomiasis. Caused by a parasitic flatworm that spends part of its life in amphibious snails, the illness afflicts 200 million people worldwide. The parasite, *Schistosoma japonica*, burrows into the skin of people who venture into contaminated water. Inside a victim, it migrates to internal organs, such as the lungs or liver, causing serious long-term illness.

In the mid-1990s, Edmund Seto of the University of California, Berkeley was approached by colleagues from China to help find high-risk areas and monitor the spread of the disease as the environment changes during the country's rapid development.

Seto's team of researchers turned to data from the Landsat TM satellite to identify areas most likely to harbor snails. In China, the snail host is divided into two subspecies, one that lives in certain types of mountainous irrigated farmland and the other that favors marshy floodplains of the lower Yangtze River. Both snails are susceptible to environmental change and tightly linked to their favored habitats.

The researchers selected sites in the two habitats and then compared painstakingly collected field data with satellite images of the regions taken with seven visible-light and infrared wavelengths. Different combinations are useful for classifying different types of vegetation and land cover. The team noted whether each 30-meter-by-30-meter pixel in the images matched snail habitat, as determined by the field data.

As the data accumulates, computer programs may be able to recognize snail habitat—and thus, areas of high disease risk—in future images of the same or other areas of China, so health officials can pinpoint risky areas without sending researchers into the field.

As China undergoes major environmental changes, the technology will enable scientists to monitor potential snail habitat. Seto cites the example of the enormous Three Gorges Dam being constructed on the Yangtze River. To be completed in 2003, the dam will create a reservoir covering 395 square miles, alter the course of rivers and streams, and perhaps create large new tracts of snail habitat. "The impacts are huge," says Seto.

POSSUM POSSIBILITIES In New Zealand, veterinary researchers are using remote sensing to spy on a disease culprit that's larger and fluffier than snails or mosquitoes. The brushtail possum carries the bacterium responsible for bovine tuberculosis (TB).

Bovine TB—McKenzie and her colleagues used images from the French SPOT3 satellite to map out vegetation types to predict infected-possum hot spots.

Mycobacterium bovis, a close relative of the germ responsible for tuberculosis in people, preferentially afflicts cattle. It can, however, infect a person who drinks tainted milk, as it commonly did before pasteurization.

Bovine TB still strikes dairy and beef herds, causes major reductions in milk yields, and makes livestock unfit for export to many countries. The disease also continues to pose a threat to farm and slaughterhouse workers.

Eradicating the disease in cattle wouldn't be a problem if possums didn't complicate matters, says Joanna McKenzie, of Massey University in Palmerston North, New Zealand. These marsupials are also susceptible to *M. bovis*.

The brushtail possum is a relentless marauder that invaded New Zealand from Australia. Listed among the world's 100 most-invasive species, it is one of New Zealand's worst pests. It costs the government there $13 million a year in pest-control expenses.

The otherwise nocturnal mammal becomes disorientated during its dying throes from bovine TB, and it then wanders into pastures during daylight, says McKenzie. Inquisitive deer and cattle often come to investigate, sniffing and licking the afflicted animals and, in the process, getting a strong draught of TB bacteria.

The traditional approach to fighting bovine TB in possums is extensive culling of both the healthy and ill animals in wide areas of infection. McKenzie's group hopes to use remote sensing to focus efforts, which would reduce costs and bring hope of eradicating the disease from the country.

The researchers studied a 60-square-kilometer area of New Zealand's North Island. The location has rolling hills, thousands of cattle, and numerous pockets of bovine-TB-infected possums. McKenzie and her colleagues used images from the French SPOT3 satellite to map out vegetation types for each 20-square-meter piece of the overall area. Vegetation type determines food sources and suitable den sites for the possums, says McKenzie.

Because the animals prefer to make dens in large trees or steep slopes, the researchers also took into account topographic data indicating the slope of each pixel. Overlaid with a map of farms within the area, the data enabled McKenzie's team to predict infected possum hot spots.

To test the accuracy of the approach, the researchers and agricultural authorities focused annual possum-eradication campaigns on 30 farms that satellite data had identified as possum hot spots. From 1997 through 1999, the researchers directed intensive trapping and killing programs on those areas, and during the seasons of high risk, farmers kept their grazing cattle out of pastures where they would be likely to encounter sick possums.

After the 3-year trial, the incidence of TB in cattle in the test area was significantly lower than among livestock on a similar group of farms receiving traditional, less-targeted efforts to control possums.

HIGH-TECH INTELLIGENCE Researchers' interest in satellite mapping of infectious diseases has been building steadily, says Hay. "The new generation of satellite sensors is also continually opening up new opportunities," he adds.

One recent advance is the development of radar satellites. Traditionally, satellites rely on passive detection of reflected energy, such as visible light or infrared radiation. Radar satellites provide similar information but generate their own signals and then receive the reflections. So, they can pick up images night and day. The wavelengths that the radar satellites emit also enable them to see through clouds.

Researchers have been finding it tough to keep abreast of technological advances in recent years, with new satellites improving the scope and quality of data. At any given time, NASA has in store about 1 petabyte of data from all its Earth-observing satellites, says Venezia. This breathtaking quantity—one qua-drillion bytes—is roughly equivalent to the amount of data stored in a mile-high stack of CD-ROMS.

Developing the human-health applications of all those data is more important today than at any other time in the last 30 years, says Durland Fish of the Yale University School of Medicine. Many "new vector-borne diseases are emerging... and we have very few people trained in how to handle them," he says.

"Thirty years ago we thought we were on top of infectious disease," says Hay. However, with pests' growing resistance to drugs and insecticides, many of the diseases have resurged. Increasingly mobile populations are escalating the problem, and now there is the additional threat of deliberate introduction of disease by terrorists. The numerous satellites encircling Earth may provide powerful medical intelligence.

From *Science News,* April 6, 2002, pp. 218-220 by John Pickrell. © 2001 by Science Service Inc. Reprinted by permission.

UNIT 7
Values and Visions

Unit Selections

Key Points to Consider

- Comment on the idea that it is naive to speak of global issues in terms of ethics.

- What role can governments, international organizations, and the individual play in making the world a more ethical place?

- How is the political role of women changing, and what impacts are these changes having on conflict resolution and community building?

- The consumption of resources is the foundation of the modern economic system. What are the values underlying this economic system, and how resistant to change are they?

- What are the characteristics of leadership?

- In addition to the ideas presented here, what other new ideas are being expressed, and how likely are they to be widely accepted?

 **Links: www.dushkin.com/online/**
These sites are annotated in the World Wide Web pages.

Human Rights Web
http://www.hrweb.org
InterAction
http://www.interaction.org

The final unit of this book considers how humanity's view of itself is changing. Values, like all other elements discussed in this anthology, are dynamic. Visionary people with new ideas can have a profound impact on how a society deals with problems and adapts to changing circumstances. Therefore, to understand the forces at work in the world today, values, visions, and new ideas in many ways are every bit as important as new technology or changing demographics.

Novelist Herman Wouk, in his book *War and Remembrance*, observed that many institutions have been so embedded in the social fabric of their time that people assumed that they were part of human nature. Slavery and human sacrifice are two examples. However, forward-thinking people opposed these institutions. Many knew that they would never see the abolition of these social systems within their own lifetimes, but they pressed on in the hope that someday these institutions would be eliminated.

Wouk believes the same is true for warfare. He states, "Either we are finished with war or war will finish us." Aspects of society such as warfare, slavery, racism, and the secondary status of women are creations of the human mind; history suggests that they can be changed by the human spirit.

The articles of this unit have been selected with the previous six units in mind. Each explores some aspect of world affairs from the perspective of values and alternative visions of the future.

New ideas are critical to meeting these challenges. The examination of well-known issues from new perspectives can yield new insights into old problems. It was feminist Susan B. Anthony who once remarked that "social change is never made by the masses, only by educated minorities." The redefinition of human values (which, by necessity, will accompany the successful confrontation of important global issues) is a task that few people take on willingly. Nevertheless, in order to deal with the dangers of nuclear war, overpopulation, and environmental degradation, educated people must take a broad view of history. This is going to require considerable effort and much personal sacrifice.

When people first begin to consider the magnitude of contemporary global problems, many often become disheartened and depressed. Some ask: What can I do? What does it matter? Who cares? There are no easy answers to these questions, but people need only look around to see good news as well as bad. How individuals react to the world is not solely a function of so-called objective reality but a reflection of themselves.

As stated at the beginning of the first unit, the study of global issues is the study of people. The study of people, furthermore,

is the study of both values and the level of commitment supporting these values and beliefs.

It is one of the goals of this book to stimulate you, the reader, to react intellectually and emotionally to the discussion and description of various global challenges. In the process of studying these issues, hopefully you have had some new insights into your own values and commitments. In the presentation of the allegory of the balloon, the third color added represented the "meta" component, all of those qualities that make human beings unique. It is these qualities that have brought us to this "special moment in time," and it will be these same qualities that will determine the outcome of our historically unique challenges.

Are Human Rights Universal?

Shashi Tharoor

The growing consensus in the West that human rights are universal has been fiercely opposed by critics in other parts of the world. At the very least, the idea may well pose as many questions as it answers. Beyond the more general, philosophical question of whether anything in our pluri-cultural multipolar world is truly universal, the issue of whether human rights is an essentially Western concept—ignoring the very different cultural, economic, and political realities of the other parts of the world—cannot simply be dismissed. Can the values of the consumer society be applied to societies that have nothing to consume? Isn't talking about universal rights rather like saying that the rich and the poor both have the same right to fly first class and to sleep under bridges? Don't human rights as laid out in the international convenants ignore the traditions, the religions, and the socio-cultural patterns of what used to be called the Third World? And at the risk of sounding frivolous, when you stop a man in traditional dress from beating his wife, are you upholding her human rights or violating his?

This is anything but an abstract debate. To the contrary, ours is an era in which wars have been waged in the name of human rights, and in which many of the major developments in international law have presupposed the universality of the concept. By the same token, the perception that human rights as a universal discourse is increasingly serving as a flag of convenience for other, far more questionable political agendas, accounts for the degree to which the very idea of human rights is being questioned and resisted by both intellectuals and states. These objections need to be taken very seriously.

The philosophical objection asserts essentially that nothing can be universal; that all rights and values are defined and limited by cultural perceptions. If there is no universal culture, there can be no universal human rights. In fact, some philosophers have objected that the concept of human rights is founded on an anthropocentric, that is, a human-centered, view of the world, predicated upon an individualistic view of man as an autonomous being whose greatest need is to be free from interference by the state—free to enjoy what one Western writer summed up as the "right to private property, the right to freedom of contract, and the right to be left alone." But this view would seem to clash with the communitarian one propounded by other ideologies and cultures where society is conceived of as far more than the sum of its individual members.

Who Defines Human Rights?

Implicit in this is a series of broad, culturally grounded objections. Historically, in a number of non-Western cultures, individuals are not accorded rights in the same way as they are in the West. Critics of the universal idea of human rights contend that in the Confucian or Vedic traditions, duties are considered more important than rights, while in Africa it is the community that protects and nurtures the individual. One African writer summed up the African philosophy of existence as: "I am because we are, and because we are therefore I am." Some Africans have argued that they have a complex structure of communal entitlements and obligations grouped around what one might call four "r's": not "rights," but respect, restraint, responsibility, and reciprocity. They argue that in most African societies group rights have always taken precedence over individual rights, and political decisions have been made through group consensus, not through individual assertions of rights.

These cultural differences, to the extent that they are real, have practical implications. Many in developing countries argue that some human rights are simply not relevant to their societies—the right, for instance, to political pluralism, the right to paid vacations (always good for a laugh in the sweatshops of the Third World), and, inevitably, the rights of women. It is not just that some societies claim they are simply unable to provide certain rights to all their citizens, but rather that they see the "universal" conception of human rights as little more than an attempt to impose alien Western values on them.

Rights promoting the equality of the sexes are a contentious case in point. How, critics demand, can women's rights be universal in the face of widespread divergences of cultural practice, when in many societies, for example, marriage is not seen as a contract between two individuals but as an alliance between lineages, and when the permissible behavior of womenfolk is central to the society's perception of its honor?

And, inseparable from the issues of tradition, is the issue of religion. For religious critics of the universalist definition of human rights, nothing can be universal that is not founded on transcendent values, symbolized by God, and sanctioned by the guardians of the various faiths. They point out that the cardinal document of the contemporary human rights movement, the Universal Declaration of Human Rights, can claim no such heritage.

Recently, the fiftieth anniversary of the Universal Declaration was celebrated with much fanfare. But critics from countries that were still colonies in 1948 suggest that its provisions reflect the ethnocentric bias of the time. They go on to argue that the concept of human rights is really a cover for Western interventionism in the affairs of the developing world, and that "human rights" are merely an instrument of Western political neocolonialism. One critic in the 1970s wrote of his fear that "Human Rights might turn out to be a Trojan horse, surreptitiously introduced into other civilizations, which will then be obliged to accept those ways of living, thinking and feeling for which Human Rights is the proper solution in cases of conflict."

In practice, this argument tends to be as much about development as about civilizational integrity. Critics argue that the developing countries often cannot afford human rights, since the tasks of nation building, economic development, and the consolidation of the state structure to these ends are still unfinished. Authoritarianism, they argue, is more efficient in promoting development and economic growth. This is the premise behind the so-called Asian values case, which attributes the economic growth of Southeast Asia to the Confucian virtues of obedience, order, and respect for authority. The argument is even a little more subtle than that, because the suspension or limiting of human rights is also portrayed as the sacrifice of the few for the benefit of the many. The human rights concept is understood, applied, and argued over only, critics say, by a small Westernized minority in developing countries. Universality in these circumstances would be the universality of the privileged. Human rights is for the few who have the concerns of Westerners; it does not extend to the lowest rungs of the ladder.

The Case for the Defense

That is the case for the prosecution—the indictment of the assumption of the universality of human rights. There is, of course, a case for the defense. The philosophical objection is, perhaps surprisingly, the easiest to counter. After all, concepts of justice and law, the legitimacy of government, the dignity of the individual, protection from oppressive or arbitrary rule, and participation in the affairs of the community are found in every society on the face of this earth. Far from being difficult to identify, the number of philosophical common denominators between different cultures and political traditions makes universalism anything but a distortion of reality.

Historically, a number of developing countries—notably India, China, Chile, Cuba, Lebanon, and Panama—played an active and highly influential part in the drafting of the Universal Declaration of Human Rights. In the case of the human rights covenants, in the 1960s the developing world actually made the decisive contribution; it was the "new majority" of the Third World states emerging from colonialism—particularly Ghana and Nigeria—that broke the logjam, ending the East–West stalemate that had held up adoption of the covenants for nearly two decades. The principles of human rights have been widely adopted, imitated, and ratified by developing countries; the fact that therefore they were devised by less than a third of the states now in existence is really irrelevant.

In reality, many of the current objections to the universality of human rights reflect a false opposition between the primacy of the individual and the paramountcy of society. Many of the civil and political rights protect groups, while many of the social and economic rights protect individuals. Thus, crucially, the two sets of rights, and the two covenants that codify them, are like Siamese twins—inseparable and interdependent, sustaining and nourishing each other.

Still, while the conflict between group rights and individual rights may not be inevitable, it would be naïve to pretend that conflict would never occur. But while groups may collectively exercise rights, the individuals within them should also be permitted the exercise of their rights within the group, rights that the group may not infringe upon.

A Hidden Agenda?

Those who champion the view that human rights are not universal frequently insist that their adversaries have hidden agendas. In fairness, the same accusation can be leveled against at least some of those who cite culture as a defense against human rights. Authoritarian regimes who appeal to their own cultural traditions are cheerfully willing to crush culture domestically when it suits them to do so. Also, the "traditional culture" that is sometimes advanced to justify the nonobservance of human rights, including in Africa, in practice no longer exists in a pure form at the national level anywhere. The societies of developing countries have not remained in a pristine, pre-Western state; all have been subject to change and distortion by external influence, both as a result of colonialism in many cases and through participation in modern interstate relations.

You cannot impose the model of a "modern" nation-state cutting across tribal boundaries and conventions on your country, appoint a president and an ambassador to the United Nations, and then argue that tribal traditions should be applied to judge the human rights conduct of the resulting modern state.

In any case, there should be nothing sacrosanct about culture. Culture is constantly evolving in any living society, responding to both internal and external stimuli, and there is much in every culture that societies quite naturally outgrow and reject. Am I, as an Indian, obliged to defend, in the name of my culture, the practice of suttee, which was banned 160 years ago, of obliging widows to immolate themselves on their husbands' funeral pyres? The fact that slavery was acceptable across the world for at least 2,000 years does not make it acceptable to us now; the deep historical roots of anti-Semitism in European culture cannot justify discrimination against Jews today.

The problem with the culture argument is that it subsumes all members of a society under a cultural framework that may in fact be inimical to them. It is one thing to advocate the cultural argument with an escape clause—that is, one that does not seek to coerce the dissenters but permits individuals to opt out and to assert their individual rights. Those who freely choose to live by and to be treated according to their traditional cultures are welcome to do so, provided others who wish to be free are not oppressed in the name of a culture they prefer to disavow.

A controversial but pertinent example of an approach that seeks to strengthen both cultural integrity and individual freedom is India's Muslim Women (Protection of Rights upon Divorce) Act. This piece of legislation was enacted following the famous Shah Banu case, in which the Supreme Court upheld the right of a divorced Muslim woman to alimony, prompting howls of outrage from Muslim traditionalists who claimed this violated their religious beliefs that divorced women were only entitled to the return of the bride price paid upon marriage. The Indian parliament then passed a law to override the court's judgment, under which Muslim women married under Muslim law would be obliged to accept the return of the bride price as the only payment of alimony, but that the official Muslim charity, the Waqf Board, would assist them.

Many Muslim women and feminists were outraged by this. But the interesting point is that if a Muslim woman does not want to be subject to the provisions of the act, she can marry under the civil code; if she marries under Muslim personal law, she will be subject to its provisions. That may be the kind of balance that can be struck between the rights of Muslims as a group to protect their traditional practices and the right of a particular Muslim woman, who may not choose to be subject to that particular law, to exempt herself from it.

It needs to be emphasized that the objections that are voiced to specific (allegedly Western) rights very frequently involve the rights of women, and are usually vociferously argued by men. Even conceding, for argument's sake, that child marriage, widow inheritance, female circumcision, and the like are not found reprehensible by many societies, how do the victims of these practices feel about them? How many teenage girls who have had their genitalia mutilated would have agreed to undergo circumcision if they had the human right to refuse to permit it? For me, the standard is simple: where coercion exists, rights are violated, and these violations must be condemned whatever the traditional justification. So it is not culture that is the test, it is coercion.

Not with Faith, But with the Faithful

Nor can religion be deployed to sanction the status quo. Every religion seeks to embody certain verities that are applicable to all mankind—justice, truth, mercy, compassion—though the details of their interpretation vary according to the historical and geographical context in which the religion originated. As U.N. secretary general Kofi Annan has often said, the problem is usually not with the faith, but with the faithful. In any case, freedom is not a value found only in Western faiths: it is highly prized in Buddhism and in different aspects of Hinduism and Islam.

If religion cannot be fairly used to sanction oppression, it should be equally obvious that authoritarianism promotes repression, not development. Development is about change, but repression prevents change. The Nobel Prize–winning economist Amartya Sen has pointed out in a number of interesting pieces that there is now a generally agreed-upon list of policies that are helpful to economic development— "openness to competition, the use of international markets, a high level of literacy and school education, successful land reforms, and public provision of incentives for investment, export and industrialization"—none of which requires authoritarianism; none is incompatible with human rights. Indeed, it is the availability of political and civil rights that gives people the opportunity to draw attention to their needs and to demand action from the government. Sen's work has established, for example, that no substantial famine has ever occurred in any independent and democratic country with a relatively free press. That is striking; though there may be cases where authoritarian societies have had success in achieving economic growth, a country like Botswana, an exemplar of democracy in Africa, has grown faster than most authoritarian states.

In any case, when one hears of the unsuitability or inapplicability or ethnocentrism of human rights, it is important to ask what the unstated assumptions of this view really are. What exactly are these human rights that it is so unreasonable to promote? If one picks up the more contentious covenant—the one on civil and political rights— and looks through the list, what can one find that someone in a developing country can easily do without? Not the right to life, one trusts. Freedom from torture? The right not to be enslaved, not to be physically assaulted, not to be arbitrarily arrested, imprisoned, executed? No one actually advocates in so many words the abridgement of any of these rights. As Kofi Annan asked at a speech in Tehran University in 1997: "When have you heard a free voice demand an end to freedom? Where have you heard a slave argue for slavery? When have you heard a victim of torture endorse the ways of the torturer? Where have you heard the tolerant cry out for intolerance?"

Tolerance and mercy have always, and in all cultures, been ideals of government rule and human behavior. If we do not unequivocally assert the universality of the rights that oppressive governments abuse, and if we admit that these rights can be diluted and changed, ultimately we risk giving oppressive governments an intellectual justification for the morally indefensible. Objections to the applicability of international human rights standards have all too frequently been voiced by authoritarian rulers and power elites to rationalize their violations of human rights—violations that serve primarily, if not solely, to sustain them in power. Just as the Devil can quote scripture for his purpose, Third World communitarianism can be the slogan of a deracinated tyrant trained, as in the case of Pol Pot, at the Sorbonne. The authentic voices of the Third World know how to cry out in pain. It is time to heed them.

The "Right to Development"

At the same time, particularly in a world in which market capitalism is triumphant, it is important to stress that the right to development is also a universal human right. The very concept of development evolved in tune with the concept of human rights; decolonization and self-determination advanced side by side with a consciousness of the need to improve the standards of living of subject peoples. The idea that human rights could be ensured merely by the state not interfering with individual freedom cannot survive confrontation with a billion hungry, deprived, illiterate, and jobless human beings around the globe. Human rights, in one memorable phrase, start with breakfast.

For the sake of the deprived, the notion of human rights has to be a positive, active one: not just protection from the state but also the protection of the state, to permit these human beings to fulfill the basic aspirations of growth and development that are frustrated by poverty and scarce resources. We have to accept that social deprivation and economic exploitation are just as evil as political oppression or racial persecution. This calls for a more profound approach to both human rights and to development. Without development, human rights could not be

truly universal, since universality must be predicated upon the most underprivileged in developing countries achieving empowerment. We can not exclude the poorest of the poor from the universality of the rich.

After all, do some societies have the right to deny human beings the opportunity to fulfill their aspirations for growth and fulfillment legally and in freedom, while other societies organize themselves in such a way as to permit and encourage human beings freely to fulfill the same needs? On what basis can we accept a double standard that says that an Australian's need to develop his own potential is a right, while an Angolan's or an Albanian's is a luxury?

Universality, Not Uniformity

But it is essential to recognize that universality does not presuppose uniformity. To assert the universality of human rights is not to suggest that our views of human rights transcend all possible philosophical, cultural, or religious differences or represent a magical aggregation of the world's ethical and philosophical systems. Rather, it is enough that they do not fundamentally contradict the ideals and aspirations of any society, and that they reflect our common universal humanity, from which no human being must be excluded.

Most basically, human rights derive from the mere fact of being human; they are not the gift of a particular government or legal code. But the standards being proclaimed internationally can become reality only when applied by countries within their own legal systems. The challenge is to work towards the "indigenization" of human rights, and their assertion within each country's traditions and history. If different approaches are welcomed within the established framework—if, in other words, eclecticism can be encouraged as part of the consensus and not be seen as a threat to it—this flexibility can guarantee universality, enrich the intellectual and philosophical debate, and so complement, rather than undermine, the concept of worldwide human rights. Paradoxical as it may seem, it is a universal idea of human rights that can in fact help make the world safe for diversity.

Note

This article was adapted from the first Mahbub-ul-Haq Memorial Lecture, South Asia Forum, October 1998.

Shashi Tharoor is Director of Communications and Special Projects in the Office of the Secretary General of the United Nations. The views expressed here are the author's own and do not necessarily reflect the positions of the United Nations.

The Grameen Bank

A small experiment begun in Bangladesh has turned into a major new concept in eradicating poverty

by Muhammad Yunus

Over many years, Amena Begum had become resigned to a life of grinding poverty and physical abuse. Her family was among the poorest in Bangladesh—one of thousands that own virtually nothing, surviving as squatters on desolate tracts of land and earning a living as day laborers.

In early 1993 Amena convinced her husband to move to the village of Kholshi, 112 kilometers (70 miles) west of Dhaka. She hoped the presence of a nearby relative would reduce the number and severity of the beatings that her husband inflicted on her. The abuse continued, however—until she joined the Grameen Bank. Oloka Ghosh, a neighbor, told Amena that Grameen was forming a new group in Kholshi and encouraged her to join. Amena doubted that anyone would want her in their group. But Oloka persisted with words of encouragement. "We're all poor—or at least we all were when we joined. I'll stick up for you because I know you'll succeed in business.

Amena's group joined a Grameen Bank Center in April 1993. When she received her first loan of $60, she used it to start her own business raising chickens and ducks. When she repaid her initial loan and began preparing a proposal for a second loan of $110, her friend Oloka gave her some sage advice: "Tell your husband that Grameen does not allow borrowers who are beaten by their spouses to remain members and take loans." From that day on, Amena suffered significantly less physical abuse at the hands of her husband. Today her business continues to grow and provide for the basic needs of her family.

Unlike Amena, the majority of people in Asia, Africa and Latin America have few opportunities to escape from poverty. According to the World Bank, more than 1.3 billion people live on less than a dollar a day. Poverty has not been eradicated in the 50 years since the Universal Declaration on Human Rights asserted that each individual has a right to:

A standard of living adequate for the health and well-being of himself and of his family, including food, clothing, housing and medical care and necessary social services, and the right to security in the event of unemployment, sickness, disability, widowhood, old age or other lack of livelihood in circumstances beyond his control.

Will poverty still be with us 50 years from now? My own experience suggests that it need not.

After completing my Ph.D. at Vanderbilt University, I returned to Bangladesh in 1972 to teach economics at Chittagong University. I was excited about the possibilities for my newly independent country. But in 1974 we were hit with a terrible famine. Faced with death and starvation outside my classroom, I began to question the very economic theories I was teaching. I started feeling there was a great distance between the actual life of poor and hungry people and the abstract world of economic theory.

I wanted to learn the real economics of the poor. Because Chittagong University is located in a rural area, it was easy for me to visit impoverished households in the neighboring village of Jobra. Over the course of many visits, I learned all about the lives of my struggling neighbors and much about economics that is never taught in the classroom. I was dismayed to see how the indigent in Jobra suffered because they could not come up with small amounts of working capital. Frequently they needed less than a dollar a person but could get that money only on extremely unfair terms. In most cases, people were required to sell their goods to moneylenders at prices fixed by the latter.

This daily tragedy moved me to action. With the help of my graduate students, I made a list of those who needed small amounts of money. We came up with 42 people. The total amount they needed was $27.

I was shocked. It was nothing for us to talk about millions of dollars in the classroom, but we were ignoring the minuscule capital needs of 42 hardworking, skilled people next door. From my own pocket, I lent $27 to those on my list.

Still, there were many others who could benefit from access to credit. I decided to approach the university's bank and try to persuade it to lend to the local poor. The branch manager said, however, that the bank could not give loans to the needy: the villagers, he argued, were not creditworthy.

I could not convince him otherwise. I met with higher officials in the banking hierarchy with similar results. Finally, I offered myself as a guarantor to get the loans.

In 1976 I took a loan from the local bank and distributed the money to poverty-stricken individuals in Jobra. Without exception, the villagers paid back their loans. Confronted with this evidence, the bank still refused to grant them loans directly. And so I tried my experiment in another village, and again it was successful. I kept expanding my work, from two to five, to 20, to 50, to 100 villages, all to convince the bankers that they should be lending to the poor. Although each time we expanded to a new village the loans were repaid, the bankers still would not change their view of those who had no collateral.

Because I could not change the banks, I decided to create a separate bank for the impoverished. After a great deal of work and negotiation with the government, the Grameen Bank ("village bank" in Bengali) was established in 1983.

From the outset, Grameen was built on principles that ran counter to the conventional wisdom of banking. We sought out the very poorest borrowers, and we required no collateral. The bank rests on the strength of its borrowers. They are required to join the bank in self-formed groups of five. The group members provide one another with peer support in the form of mutual assistance and advice. In addition, they allow for peer discipline by evaluating business viability and ensuring repayment. If one member fails to repay a loan, all members risk having their line of credit suspended or reduced.

The Power of Peers

Typically a new group submits loan proposals from two members, each requiring between $25 and $100. After these two borrowers successfully repay their first five weekly installments, the next two group members become eligible to apply for their own loans. Once they make five repayments, the final member of the group may apply. After 50 installments have been repaid, a borrower pays her interest, which is slightly above the commercial rate. The borrower is now eligible to apply for a larger loan.

The bank does not wait for borrowers to come to the bank; it brings the bank to the people. Loan payments are made in weekly meetings consisting of six to eight groups, held in the villages where the members live. Grameen staff attend these meetings and often visit individual borrowers' homes to see how the business—whether it be raising goats or growing vegetables or hawking utensils—is faring.

Today Grameen is established in nearly 39,000 villages in Bangladesh. It lends to approximately 2.4 million bor-

rowers, 94 percent of whom are women. Grameen reached its first $1 billion in cumulative loans in March 1995, 18 years after it began in Jobra. It took only two more years to reach the $2-billion mark. After 20 years of work, Grameen's average loan size now stands at $180. The repayment rate hovers between 96 and 100 percent.

A year after joining the bank, a borrower becomes eligible to buy shares in Grameen. At present, 94 percent of the bank is owned by its borrowers. Of the 13 members of the board of directors, nine are elected from among the borrowers; the rest are government representatives, academics, myself and others.

A study carried out by Sydney R. Schuler of John Snow, Inc., a private research group, and her colleagues concluded that a Grameen loan empowers a woman by increasing her economic security and status within the family. In 1998 a study by Shahidur R. Khandker an economist with the World Bank, and others noted that participation in Grameen also has a significant positive effect on the schooling and nutrition of children—as long as women rather than men receive the loans. (Such a tendency was clear from the early days of the bank and is one reason Grameen lends primarily to women: all too often men spend the money on themselves.) In particular, a 10 percent increase in borrowing by women resulted in the arm circumference of girls—a common measure of nutritional status—expanding by 6 percent. And for every 10 percent increase in borrowing by a member the likelihood of her daughter being enrolled in school increased by almost 20 percent.

Not all the benefits derive directly from credit. When joining the bank, each member is required to memorize a list of 16 resolutions. These include commonsense items about hygiene and health—drinking clean water, growing and eating vegetables, digging and using a pit latrine, and so on—as well as social dictums such as refusing dowry and managing family size. The women usually recite the entire list at the weekly branch meetings, but the resolutions are not otherwise enforced.

Even so, Schuler's study revealed that women use contraception more consistently after joining the bank. Curiously, it appears that women who live in villages where Grameen operates, but who are not themselves members, are also more likely to adopt contraception. The population growth rate in Bangladesh has fallen dramatically in the past two decades, and it is possible that Grameen's influence has accelerated the trend.

In a typical year 5 percent of Grameen borrowers—representing 125,000 families—rise above the poverty level. Khandker concluded that among these borrowers extreme poverty (defined by consumption of less than 80 percent of the minimum requirement stipulated by the Food and Agriculture Organization of the United Nations) declined by more than 70 percent within five years of their joining the bank.

To be sure, making a microcredit program work well—so that it meets its social goals and also stays economi-

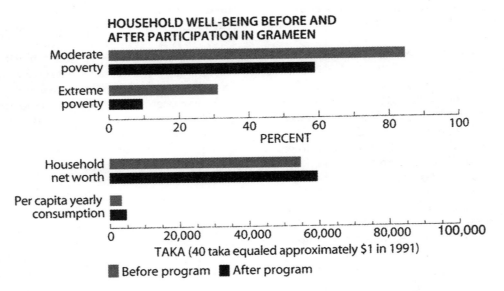

HOUSEHOLD WELL-BEING BEFORE AND AFTER PARTICIPATION IN GRAMEEN

Before program — After program

cally sound—is not easy. We try to ensure that the bank serves the poorest: only those living at less than half the poverty line are eligible for loans. Mixing poor participants with those who are better off would lead to the latter dominating the groups. In practice, however, it can be hard to include the most abjectly poor, who might be excluded by their peers when the borrowing groups are being formed. And despite our best efforts, it does sometimes happen that the money lent to a woman is appropriated by her husband.

Given its size and spread, the Grameen Bank has had to evolve ways to monitor the performance of its branch managers and to guarantee honesty and transparency. A manager is not allowed to remain in the same village for long, for fear that he may develop local connections that impede his performance. Moreover, a manager is never posted near his home. Because of such constraints—and because managers are required to have university degrees—very few of them are women. As a result, Grameen has been accused of adhering to a paternalistic pattern. We are sensitive to this argument and are trying to change the situation by finding new ways to recruit women.

Grameen has also often been criticized for being not a charity but a profit-making institution. Yet that status, I am convinced, is essential to its viability. Last year a disastrous flood washed away the homes, cattle and most other belongings of hundreds of thousands of Grameen borrowers. We did not forgive the loans, although we did issue new ones, and give borrowers more time to repay. Writing off loans would banish accountability, a key factor in the bank's success.

Liberating Their Potential

The Grameen model has now been applied in 40 countries. The first replication, begun in Malaysia in 1986, currently serves 40,000 poor families; their repayment rate has consistently stayed near 100 percent. In Bolivia, microcredit has allowed women to make the transition from "food for work" programs to managing their own businesses. Within two years the majority of women in the program acquire enough credit history and financial skills to qualify for loans from mainstream banks. Similar success stories are coming in from programs in poor countries everywhere. These banks all target the most impoverished, lend to groups and usually lend primarily to women.

The Grameen Bank in Bangladesh has been economically self-sufficient since 1995. Similar institutions in other countries are slowly making their way toward self-reliance. A few small programs are also running in the U.S., such as in innercity Chicago. Unfortunately, because labor costs are much higher in the U.S. than in developing countries—which often have a large pool of educated unemployed who can serve as managers or accountants—the operations are more expensive there. As a result, the U.S. programs have had to be heavily subsidized.

In all, about 22 million poor people around the world now have access to small loans. Microcredit Summit, an institution based in Washington, D.C., serves as a resource center for the various regional microcredit institutions and organizes yearly conferences. Last year the attendees pledged to provide 100 million of the world's poorest families, especially their women, with credit by the year 2005. The campaign has grown to include more than 2,000 organizations, ranging from banks to religious institutions to nongovernmental organizations to United Nations agencies.

The standard scenario for economic development in a poor country calls for industrialization via investment. In this "topdown" view, creating opportunities for employment is the only way to end poverty. But for much of the developing world, increased employment exacerbates migration from the countryside to the cities and creates low-paying jobs in miserable conditions. I firmly believe that, instead, the eradication of poverty starts with people

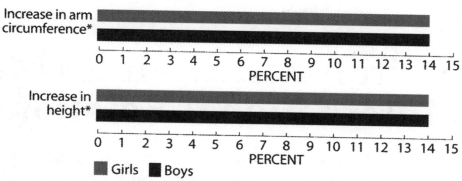

IMPACT OF GRAMEEN ON NUTRITIONAL MEASURES OF CHILDREN

Increase in arm circumference*

0 1 2 3 4 5 6 7 8 9 10 11 12 13 14 15
PERCENT

Increase in height*

0 1 2 3 4 5 6 7 8 9 10 11 12 13 14 15
PERCENT

■ Girls ■ Boys

*Bars reflect changes accompanying a 10 percent increase in credit to women.

being able to control their own fates. It is not by creating jobs that we will save the poor but rather by providing them with the opportunity to realize their potential. Time and time again I have seen that the poor are poor not because they are lazy or untrained or illiterate but because they cannot keep the genuine returns on their labor.

Self-employment may be the only solution for such people, whom our economies refuse to hire and our taxpayers will not support. Microcredit views each person as a potential entrepreneur and turns on the tiny economic engines of a rejected portion of society. Once a large number of these engines start working, the stage can be set for enormous socioeconomic change.

Applying this philosophy, Grameen has established more than a dozen enterprises, often in partnership with other entrepreneurs. By assisting microborrowers and microsavers to take ownership of large enterprises and even infrastructure companies, we are trying to speed the process of overcoming poverty. Grameen Phone, for instance, is a cellular telephone company that aims to serve urban and rural Bangladesh. After a pilot study in 65 villages, Grameen Phone has taken a loan to extend its activities to all villages in which the bank is active. Some 50,000 women, many of whom have never seen a telephone or even an electric light, will become the providers of telephone service in their villages. Ultimately, they will become the owners of the company itself by buying its shares. Our latest innovation, Grameen Investments, allows U.S. individuals to support companies such as Grameen Phone while receiving interest on their investment. This is a significant step toward putting commercial funds to work to end poverty.

I believe it is the responsibility of any civilized society to ensure human dignity to all members and to offer each in-

dividual the best opportunity to reveal his or her creativity. Let us remember that poverty is not created by the poor but by the institutions and policies that we, the better off, have established. We can solve the problem not by means of the old concepts but by adopting radically new ones.

The Author

MUHAMMAD YUNUS, the founder and managing director of the Grameen Bank, was born in Bangladesh. He obtained a Ph.D. in economics from Vanderbilt University in 1970 and soon after returned to his home country to teach at Chittagong University. In 1976 he started the Grameen project, to which he has devoted all his time for the past decade. He has served on many advisory committees: for the government of Bangladesh, the United Nations, and other bodies concerned with poverty, women and health. He has received the World Food Prize, the Ramon Magsaysay Award, the Humanitarian Award, the Man for Peace Award and numerous other distinctions as well as six honorary degrees.

Further Reading

GRAMEEN BANK: PERFORMANCE AND SUSTAINABILITY. Shahidur R. Khandker, Baqui Khalily and Zahed Khan. World Bank Discussion Papers, No. 306. ISBN 0-8213-3463-8. World Bank, 1995.

GIVE US CREDIT. Alex Counts. Times Books (Random House), 1996.

FIGHTING POVERTY WITH MICROCREDIT: EXPERIENCE IN BANGLADESH. Shahidur R. Khandker. Oxford University Press, 1998.

Grameen Bank site is available at www.grameenfoundation.org on the World Wide Web.

Why Environmental Ethics to International Relations

"Environmental ethics [should] not be seen as an add-on to be approached after the important issues of security and economics have been settled. Instead, we [should] recognize that all our important social choices are inherently about the 'natural' world we create."

JOHN BARKDULL

What challenge does environmental ethics pose for international relations? International relations is usually understood as the realm of power politics, a world in which military might and the quest to survive dominate. In this world, moral concern for other human beings, much less nature, is limited or entirely lacking. Environmental ethics—a set of principles to guide human interaction with the earth—calls on us to extend moral consideration beyond humans to other living things and to natural "wholes" such as bioregions and ecosystems. Is it possible to introduce environmental ethics' far-reaching moral claims into the competitive, militarized, economically unequal world political system?

Although explorations in environmental ethics now have a long resumé, the dialogue over the human debt to the natural environment has proceeded largely without reference to international politics, to international relations theory, or even to the literature on international ethics. Practical politics is thus often removed from consideration. And scholars of international relations have barely considered the relationship between their studies and environmental ethics.

Bringing the two fields into the same conversation is possible. International political theory has profound implications for understanding how humans ought to relate to the environment. Realism and liberal institutionalism (the mainstream of international relations theory), by suggesting what political, economic, and social goals are desirable, also imply what environmental values should prevail. They indicate what kind of world humans should or can create and thus tell us how we should relate to the environment. The question then is not whether environmental ethics should matter in world politics, but in which way: which environmental ethic does in fact matter, which should, and what obstacles prevent needed changes in political practices from being made?

WHICH ENVIRONMENTAL ETHICS?

Environmental ethics can be anthropocentric, biocentric, or ecocentric. Anthropocentric ethics is about what humans owe each other. It evaluates environmental policies with regard to how they affect human well-being. For example, exploitation of natural resources such as minerals can destroy forests on which indigenous peoples depend. Moral evaluation of the environmental destruction proceeds in terms of the rights, happiness, or just treatment of all human parties, including the displaced tribes and the consumers who benefit from the minerals. Anthropocentric environmental ethics generally calls for more environmental protection than we now undertake; current unsustainable resource-use patterns and conversion of land to agricultural or urban uses mean that existing practices do more harm to humans than good, especially when future generations are considered. Still, many observers find anthropocentric environmental ethics unsatisfactory because it appears not to recognize other creatures' inherent right to share the planet and considers only their value to human beings.

Biocentric environmental ethics seeks to correct this deficiency by according moral standing to non-human creatures. Humans have moral worth but only as one species among many living things that also have moral standing. The grizzly bear's right to sufficient domain for sustaining life and reproduction has as much moral weight (if not more) as a logging company's desire to make a profit in that domain. Even if maintaining the grizzly bear's habitat means some humans must live in somewhat less spacious homes, the loss of human utility by no means cancels the animal's moral claim to the forest. In short, animals have rights. Which animals have moral standing and whether plants do as well remain matters of dispute among biocentric theorists. Nonetheless, biocentric theory expands the moral realm beyond humans and hence implies greater moral obligations than anthropocentric ethics.[1]

Ecocentric theory tackles a problem at the heart of biocentric theory. In reality, ecosystems work on the principle of eat and be eaten. We may accord the grizzly "rights" but the bear survives by consuming salmon, rodents, and so forth, thus violating other living creatures' right to life. Humans are simply part of a complex food chain or web of life. Given this, ecocentric theory asserts that moral status should attach to ecological wholes, from bioregions to the planetary ecosystem (sometimes called Gaia). Ecocentric theorists are not concerned about particular animals or even species, but with the entire evolutionary process. Evolution involves the "land" broadly understood to include all its organic and nonorganic components. To disrupt or destroy the evolutionary process, reducing the diversity of life and the stability and beauty of the natural system, is unethical. As Aldo Leopold, the environmental philosopher who first developed the land ethic, put it in his 1949 book, *A Sand County Almanac and Sketches Here and There*, "A thing is right when it tends to preserve the integrity, stability, and beauty of the biotic community. It is wrong when it tends otherwise." The emphasis here is on the word "community."

Each of these approaches suggests the need for change in the practice of international politics. Anthropocentric environmental ethics implies the least extensive reform, although these still could be far-reaching, especially with regard to current economic arrangements. Developed industrial economies rely heavily on the global commons for "free" natural services, such as areas to dispose of pollutants. For example, reliance on fossil fuels leads to increases in CO_2 in the atmosphere, and in turn to global warming. Developing countries undergoing industrialization will draw on the atmosphere's capacity to absorb greenhouse gases. The added load, along with already high levels of emissions from developed countries, could push the environment beyond a critical threshold, setting off catastrophic climate changes because of global warming. These climatic upheavals could lead to crop failures and destructive storms battering coastal cities. What is fair under these circumstances? Should developed countries make radical changes—such as decentralizing and deindustrializing—in their economic arrangements? Should they refrain from adding the potentially disastrous increment of greenhouse gases that will push the climate over the threshold of climatic catastrophe? If yes, then anthropocentric environmental ethics calls for far-reaching social and economic reform.

Biocentric environmental ethics also implies considerable economic reform. If animals have moral standing, then killing them or destroying their habitat for human benefit is unacceptable. In particular, the massive species loss resulting from deforestation is a moral failure even if humans profit. Likewise, agricultural practices that rely on pesticides and fertilizers that harm nonhuman species should be curtailed. Warfare's effects on nonhuman living things would also need to be evaluated. Just-war theory generally evaluates collateral damage's significance in the context of civilians killed or injured due to military operations. Yet collateral damage also kills and injures animals that have even less stake and less say in the conflict than civilians. Should their right to life be considered? Biocentric ethics would say yes. If so, virtually the entire practice of modern war might be held as inherently immoral.

Ecocentric ethics implies the strongest critique of current practices. Disrupting the ecological cycle or the evolutionary process is morally unacceptable. Most current economic or military practices would not pass muster. Indeed, in its strong form, ecocentric ethics would require a major reduction in the human population, since the 6 billion people now on earth are already disrupting the evolutionary process and will continue to do so as world population grows to 10 billion or more. Political institutions must be replaced, either with one-world government capable of implementing ecocentric environmental policy, or with ecologically based bioregional political units (ecocentric theorists hold differing views on whether authoritarian government or more democracy is needed to make ecocentrism effective in practice). If bioregionalism were adopted, world trade would come to a halt since each bioregion would be self-sustaining. Wasteful resource use would be curtailed. Long-term sustainability in harmony with the needs of other living things would be the desired end. For some ecocentric thinkers, the model is a hunter-gatherer society or a peasant agriculture society.

The gap between what environmental ethics calls for and what international political theory postulates may find its bridge in the land ethic.

Environmental ethics in each form carries important implications for the practice of international politics. Yet the environmental ethics literature usually pays little attention to obvious features of the international system. This is not to say that environmental ethics bears no relationship to political realities. If realism (the theory of power politics) and liberal institutionalism (the theory emphasizing interdependence and the possibilities for cooperation) both contain implicit environmental ethics, then environmental ethics contains implicit political theory. Yet without explicit attention to international political theory, environmental ethics lacks the basis to determine which of its recommendations is feasible, and which utopian.

BRIDGING THE GAP

The gap between what environmental ethics calls for and what international political theory postulates may find its bridge in the land ethic. The land ethic, as formulated by J. Baird Callicott, recognizes that environmental obligations are only part of our moral world.[2] Although the land ethic implies significant change in existing practices, it does not necessarily call for abandoning the sovereign state, relinquishing national identity or authority to a world government, or even abolishing capitalism. Rather, it asks for balance between human needs and the requisites of preserving the diversity of life flowing from the evolutionary process. The land ethic simply states that which enhances the integrity, stability, and beauty of the land is good, and that which does not is bad. Human intervention can serve good purposes by this standard. (Indeed, Aldo Leopold was himself a hunter, and found no contradiction between that pursuit and his commitment to the land ethic.) Presumably, human-induced changes to the landscape must be evaluated in context, assessing positive and negative effects.

Yet if the land ethic is to provide guidance for international politics, it needs to identify the other values that must be balanced with its requirements. In international politics, these can be determined in terms of mainstream international political theory, realism and liberal institutionalism. Realism and liberal insitutionalism capture much about how the international system works and what values shape international political practices. Thus we can observe the world for clues as to what international theory entails for the kind of world we should create. Although the verdict is not positive for existing practices, this does not foreclose the possibility of change within either paradigm. But how specifically does realism and liberal institutionalism see the relationship of humans to nature?

REALISM AND THE ENVIRONMENT

Realism is generally understood to be amoral. States do as they must to survive. Survival can justify breaking agreements, lying, deception, violence, and theft. Those who fail to play the game disappear. Those who are best at the game dominate the others. Morality, when invoked, is usually a cover for state interests. Certainly, some prominent realists have said otherwise. Hans Morgenthau recognized the moral content of foreign policy, as did E. H. Carr and Reinhold Niebuhr. Nonetheless, realists usually observe the human capacity for "evil" when the stakes are high.

But this negative perspective on morality obscures realism's highly moral claims. Realism asserts that humans naturally form groups, which experience conflicts of in-

terest because resources are scarce. Maintaining the group's autonomy and freedom is the highest good. On it depends the ability of a people to work out their destiny within the borders of the state. Implicitly, this moral project justifies the extreme measures states undertake. Environmental ethics must recognize this as an extremely powerful moral claim. At the same time, international relations theory must recognize that staking this claim, which superficially appears to be a social question, implies a view of how humans should relate to the natural world.

Realism assumes that the state system, or at least some form of power politics involving contending groups, will characterize human relations as long as humans inhabit the planet. The possibilities for environmental (or any other) ethics are limited by this evidently permanent institutional arrangement. Virtually every state action must be evaluated in terms of the relative gains it offers with other states. The struggle for survival and dominance is an endless game in which any minor advantage today could have profound consequences tomorrow. Moreover, as Machiavelli observed, chance plays such a large role in human affairs that immediate advantage is all the prudent policymaker can consider. Thus, to think about long-term environmental trends, for example, is impractical, because an actor that sacrifices present advantage for future gains may not be around to enjoy the fruits.

Perhaps the most significant implication of realism lies in its emphasis on military security. Military imperatives dictate that states develop and deploy the most effective military technology available. The effects on the land of the particular choices made are little considered. No military technology could be more environmentally damaging than nuclear weapons, but these weapons confer maximum national power. Thus the environmental effects of producing, storing, deploying, and dismantling them (not to mention the effects they would have on the environment if ever used) are considered secondary. Here we see how realism as an international political theory is at the same time an implicit environmental ethic: land has little or no moral worth. This is a choice about how humans are to relate to the natural world, not only a choice about how states (or humans) are to relate to one another. A similar argument could be made about the entire range of military technologies, from cluster bombs to napalm to defoliants to biological weapons.

According to realism, the economic institutions of a society must support the most effective military establishment. Societies that attempt to structure economic relations along other lines, such as long-term sustainability, will soon find themselves overwhelmed by other states that make choices geared toward military dominance. States that wish to survive will emulate the most successful economic systems of other states and constantly seek economic innovations that will give them the edge. Capitalism as practiced in the United States and other major Western nations seems to be most compatible with military preponderance.

Realism thus implicitly endorses capitalism, albeit only because it is the most successful economic system at present for enhancing national power (as the former Soviet Union discovered). Capitalism has put the United States at the top of the international order. Others fail to emulate the United States at their peril. Moreover, realism would suggest that because economic growth facilitates military preparedness, autonomous economic growth should override other goals, including environmental protection. Saving a wetland will not contribute as much to national security as producing goods for export. Consequently, realism's emphasis on security leads to embracing the market in its most environmentally heedless form.

Realism's attitude toward the land is that it is territory, an asset of the state, a form of property. The land's status is as a mere resource with no moral standing apart from human uses. It has no life of its own. Prudent management is the most that is morally required. Hence realism shares the modern notion of nature as a spiritless "other" that humans can rightfully manipulate to serve their own ends.

We see that realism's strong moral claim—that a people's right to determine their own destiny, to define and develop their own idea of freedom and the good society, without interference from others—contradicts the institutional arrangements that realism produces. In practice, few alternatives are available. A people who decide that their destiny is to live in harmony with the land, to follow the land ethic's central precept, would quickly lose the freedom to do so through conquest and domination by other states. Aside from the evident fact that many environmental problems require cooperation across political jurisdictions, the competitiveness of the state system ensures that environmental consciousness will not long guide state policy.[3] Realism thus implies an environmental ethic; unfortunately, it is a most pernicious one. Equally unfortunate for environmental ethics is that realism is an undoubtedly incomplete, but not inaccurate, description of how the international system works.

Yet realism's historical and social argument rests on the moral claim that the group is the highest value: that it is within the group that some conception of the good society can be pursued. But surely a good society is one that fosters environmental values, a goal that suffers when nations pursue national security at all costs. As environmental crises mount, the contradiction at the heart of realist ethics becomes more obvious. Perhaps this can lead to changed conceptions of morality.

Ethical standards change. Nationalism, which underpins today's state system, has not driven human behavior for all history (nationalism, for example, had little influence in feudal Europe). Humans can change their view of how best to pursue a vision of the good society. To the extent that the land ethic becomes part of the moral dialogue, institutional change to bring about a healthier human-environment relationship is possible. Nonetheless, realism reminds us that the road to ethical change toward a land ethic likely will be long and hard.

LIBERAL INSTITUTIONALISM AND THE ENVIRONMENT

Liberal institutionalism is far more ready to accept that universal values such as respect for the land exist. Unlike realism, the liberal perspective considers human rights standards to apply across boundaries and cultures. Individuals are the moral agents and moral objects of liberal thought. Individuals have rights that exist regardless of their cultural heritage.

Furthermore, liberalism asserts that these individuals have a particular character. Partly self-interested and partly altruistic, individuals are aware of their dependence on collective action to obtain the good life. The liberal individual also acts, or should, through enlightened self-interest, which is the best way to secure the means of life and protection against bodily harm. Liberal individuals are predisposed to make certain choices. But would they choose a different way of life, namely, one more in harmony with the land?

The question becomes pertinent because it is not at all clear that liberal society is sustainable. Liberalism as manifested in practice is strongly committed to the market system. Indeed, the entire point of liberal institutional international political economy is to find the means to open the world economy to free trade and investment. From this perspective, environmental problems become unintended side-effects of otherwise desirable industrialization and economic growth. The problem for liberalism is simply managing these unfortunate consequences in ways that maintain the open economy. But as the modern market system encompasses more of the globe and penetrates deeper into social life, profound social choices occur, the result of the incremental effects of countless discrete, uncoordinated individual actions. Liberals are comfortable with this way of making social choices due to their faith in progress; the mounting ecological catastrophe might speak against this optimistic view.

Liberal institutionalism cannot escape its entanglement with and commitment to the capitalist market system. In effect, this means that liberal institutionalism can only with difficulty critique that system as it has developed in history. Hence liberal institutionalism will continue to see normal diplomacy and statecraft, the operations of multinational corporations, the growth of free trade and investment, and rising interdependence as progress toward a better world. This in turn exhibits liberal institutionalism's environmental ethic: managerial, limited to mitigation of the market's worst effects, and committed to economic growth and development. The world we should build is on display. It is embodied in the more enlightened liberal states, those that combine commitment to individual liberty, representative democracy, and free enterprise with some degree of environmental awareness. It is industrialized, or postindustrial. It is technologically advanced. It provides a wide range of goods and services to consumers. Environmental concerns enter by way of interest groups devoted to the "issue" rather than as fundamental values that determine which practices to retain and which to abandon.

Like realism, liberal institutionalism captures a large part of the truth about how contemporary international politics operates. It suggests emphasizing certain trends in the hope of dampening others; strengthening the forces for globalization to reduce the impact of military competition. But its commitment to the predominant global economic institutions leaves little room for a land ethic. Liberal institutionalism's anthropocentrism and consequent emphasis on economic growth leads to a relative lack of concern for the stability, integrity, and beauty of the land. Nonetheless, liberal institutionalism is far more open to the possibility of value change and political transformation than is realism. Liberal theory's faith in progress can imply that liberalism itself eventually will be transcended in favor of more earth-centered ethics. Yet current liberal international theory does not recognize or embrace this possibility. To the extent that theory is practice, liberal international theory contributes to the worsening environmental crisis rather than offering a way out.

A NEW DIALOGUE

Both realism and liberal institutionalism are implicitly environmental ethics. They tell us the relationship humans should have with nature, even if they largely base their claims not on ethical choice but on what we must do under existing circumstances. But humans can make conscious choices about what kind of international order to create and maintain. The realist imperative to play the game of power politics or be eliminated from the system depends on a prior choice about ethics and practice. It precludes the possibility of collaboratively engaging the "other" in democratic dialogue aimed at discovering different social practices that do not, for example, lead to environmentally heedless arms races. The "other" must always remain other in realist thought, an assumption that is far from proven. Likewise, liberalism's imperative to rely on the market if we are to achieve individual liberty and social progress is open to question. If the individual is constituted in community—that is, by social practices—then the self-definition of the community can

change. Acquisitive individualism and consumerism need not define the individuals.

How is change to come about? More authentic democracy, based on unforced, open discourse, affords the opportunity to choose consciously the kind of world we are to build. The choice need not come about indirectly, as the result of more immediate decisions on how to achieve national security, nor need it occur unintentionally as individuals make the best of circumstances not of their choosing. Engaging in this dialogue will require abandoning the notion that nature and the social are distinct. The social and the natural are inextricable. We constitute nature through our practices at the same time that we constitute the social world. Thus an ecologically informed political discourse will be one that recognizes the environmental ethics embedded in all political worldviews.

Environmental ethics will not be seen as an add-on to be approached after the important issues of security and economics have been settled. Instead, we will recognize that all our important social choices are inherently about the "natural" world we create. We will consciously raise the question of what this particular action means for that world, and we will recognize that it is our responsibility, not something external to us.

NOTES

1. For more on anthropocentric and biocentric ethics, see J. Baird Callicott, *In Defense of the Land Ethic: Essays for Environmental Philosophy* (Albany: State University of New York Press, 1989).
2. See Callicott, *In Defense of the Land Ethic.*
3. This competition also influences the abilities of states to cooperate to deal with transnational environmental problems. States are expected to attempt to free ride or otherwise exploit the global "commons." If environmental cooperation occurs, it is likely due to a hegemonic power or small group of large powers imposing an international regime. Of course, the Hobbesian use of power to make and enforce law is the antithesis of democratic decision making (which could well be a major element of the society's vision of the good life). But because states are and must be short-sighted and self-interested, no alternative to coercive imposition of regimes exists. Whether such regimes would conform to the requisites of long-term environmental sustainability—much less to the integrity, beauty, and stability of the land—is doubtful.

JOHN BARKDULL *is an associate professor of political science at Texas Tech University. His research interests include international political theory, international ethics, and environmental policy.*

Women Waging Peace

You can't end wars simply by declaring peace. "Inclusive security" rests on the principle that fundamental social changes are necessary to prevent renewed hostilities. Women have proven time and again their unique ability to bridge seemingly insurmountable divides. So why aren't they at the negotiating table?

By Swanee Hunt and Cristina Posa

Allowing men who plan wars to plan peace is a bad habit. But international negotiators and policymakers can break that habit by including peace promoters, not just warriors, at the negotiating table. More often than not, those peace promoters are women. Certainly, some extraordinary men have changed the course of history with their peacemaking; likewise, a few belligerent women have made it to the top of the political ladder or, at the grass-roots level, have taken the roles of suicide bombers or soldiers. Exceptions aside, however, women are often the most powerful voices for moderation in times of conflict. While most men come to the negotiating table directly from the war room and battlefield, women usually arrive straight out of civil activism and—take a deep breath—family care.

Yet, traditional thinking about war and peace either ignores women or regards them as victims. This oversight costs the world dearly. The wars of the last decade have gripped the public conscience largely because civilians were not merely caught in the crossfire; they were targeted, deliberately and brutally, by military strategists. Just as warfare has become "inclusive"—with civilian deaths more common than soldiers'—so too must our approach toward ending conflict. Today, the goal is not simply the absence of war, but the creation of sustainable peace by fostering fundamental societal changes. In this respect, the United States and other countries could take a lesson from Canada, whose innovative "human security" initiative—by making human beings and their communities, rather than states, its point of reference—focuses on safety and protection, particularly of the most vulnerable segments of a population.

The concept of "inclusive security," a diverse, citizen-driven approach to global stability, emphasizes women's agency, not their vulnerability. Rather than motivated by gender fairness, this concept is driven by efficiency: Women are crucial to inclusive security since they are often at the center of nongovernmental organizations (NGOs), popular protests, electoral referendums, and other citizen-empowering movements whose influence has grown with the global spread of democracy. An inclusive security approach expands the array of tools available to police, military, and diplomatic structures by adding collaboration with local efforts to achieve peace. Every effort to bridge divides, even if unsuccessful, has value, both in lessons learned and links to be built on later. Local actors with crucial experience resolving conflicts, organizing political movements, managing relief efforts, or working with military forces bring that experience into ongoing peace processes.

International organizations are slowly recognizing the indispensable role that women play in preventing war and sustaining peace. On October 31, 2000, the United Nations Security Council issued Resolution 1325 urging the secretary-general to expand the role of women in U.N. field-based operations, especially among military observers, civilian police, human rights workers, and humanitarian personnel. The Organization for Security and Co-operation in Europe (OSCE) is working to move women off the gender sidelines and into the everyday activities of the organization—particularly in the Office for Democratic Institutions and Human Rights, which has been useful in monitoring elections and human rights throughout Europe and the former Soviet Union. Last November, the European Parliament passed a hard-hitting resolution calling on European Union members (and the European Commission and Council) to promote the equal participation of women in diplomatic conflict resolution; to ensure that women fill at least 40 percent of all reconciliation, peacekeeping, peace-enforcement, peace-building, and conflict-prevention posts; and to support the creation and strengthening of NGOs (including women's organizations) that focus on conflict prevention, peace building, and post-conflict reconstruction.

Ironically, women's status as second-class citizens is a source of empowerment, since it has made women adept at finding innovative ways to cope with problems.

But such strides by international organizations have done little to correct the deplorable extent to which local women have been relegated to the margins of police, military, and diplomatic efforts. Consider that Bosnian women were not invited to participate in the Dayton talks, which ended the war in Bosnia, even though during the conflict 40 women's associations remained organized and active across ethnic lines. Not surprisingly, this exclusion has subsequently characterized—and undermined—the implementation of the Dayton accord. During a 1997 trip to Bosnia, U.S. President Bill Clinton, Secretary of State Madeleine Albright, and National Security Advisor Samuel Berger had a miserable meeting with intransigent politicians elected under the ethnic-based requirements of Dayton. During the same period, First Lady Hillary Rodham Clinton engaged a dozen women from across the country who shared story after story of their courageous and remarkably effective work to restore their communities. At the end of the day, a grim Berger faced the press, offering no encouraging word from the meetings with the political dinosaurs. The first lady's meeting with the energetic women activists was never mentioned.

We can ignore women's work as peacemakers, or we can harness its full force across a wide range of activities relevant to the security sphere: bridging the divide between groups in conflict, influencing local security forces, collaborating with international organizations, and seeking political office.

BRIDGING THE DIVIDE

The idea of women as peacemakers is not political correctness run amok. Social science research supports the stereotype of women as generally more collaborative than men and thus more inclined toward consensus and compromise. Ironically, women's status as second-class citizens is a source of empowerment, since it has made women adept at finding innovative ways to cope with problems. Because women are not ensconced within the mainstream, those in power consider them less threatening, allowing women to work unimpeded and "below the radar screen." Since they usually have not been behind a rifle, women, in contrast to men, have less psychological distance to reach across a conflict line. (They are also more accepted on the "other side," because it is assumed that they did not do any of the actual killing.) Women often choose an identity, notably that of mothers, that cuts across international borders and ethnic enclaves. Given their roles as family nurturers,

women have a huge investment in the stability of their communities. And since women know their communities, they can predict the acceptance of peace initiatives, as well as broker agreements in their own neighborhoods.

As U.N. Secretary-General Kofi Annan remarked in October 2000 to the Security Council, "For generations, women have served as peace educators, both in their families and in their societies. They have proved instrumental in building bridges rather than walls." Women have been able to bridge the divide even in situations where leaders have deemed conflict resolution futile in the face of so-called intractable ethnic hatreds. Striking examples of women making the impossible possible come from Sudan, a country splintered by decades of civil war. In the south, women working together in the New Sudan Council of Churches conducted their own version of shuttle diplomacy—perhaps without the panache of jetting between capitals—and organized the Wunlit tribal summit in February 1999 to bring an end to bloody hostilities between the Dinka and Nuer peoples. As a result, the Wunlit Covenant guaranteed peace between the Dinka and the Nuer, who agreed to share rights to water, fishing, and grazing land, which had been key points of disagreement. The covenant also returned prisoners and guaranteed freedom of movement for members of both tribes.

On another continent, women have bridged the seemingly insurmountable differences between India and Pakistan by organizing huge rallies to unite citizens from both countries. Since 1994, the Pakistan-India People's Forum for Peace and Democracy has worked to overcome the hysterics of the nationalist media and jingoistic governing elites by holding annual conventions where Indians and Pakistanis can affirm their shared histories, forge networks, and act together on specific initiatives. In 1995, for instance, activists joined forces on behalf of fishers and their children who were languishing in each side's jails because they had strayed across maritime boundaries. As a result, the adversarial governments released the prisoners and their boats.

In addition to laying the foundation for broader accords by tackling the smaller, everyday problems that keep people apart, women have also taken the initiative in drafting principles for comprehensive settlements. The platform of Jerusalem Link, a federation of Palestinian and Israeli women's groups, served as a blueprint for negotiations over the final status of Jerusalem during the Oslo process. Former President Clinton, the week of the failed Camp David talks in July 2000, remarked simply, "If we'd had women at Camp David, we'd have an agreement."

Sometimes conflict resolution requires unshackling the media. Journalists can nourish a fair and tolerant vision of society or feed the public poisonous, one-sided, and untruthful accounts of the "news" that stimulate violent conflict. Supreme Allied Commander of Europe Wesley Clark understood as much when he ordered NATO to bomb transmitters in Kosovo to prevent the Milosevic media machine from spewing ever more inflammatory rhetoric. One of the founders of the independent Kosovo radio station RTV-21 realized that there were "many instances of male colleagues reporting with anger, which served to raise the tensions rather than lower them." As a result, RTV-

21 now runs workshops in radio, print, and TV journalism to cultivate a core of female journalists with a noninflammatory style. The OSCE and the BBC, which train promising local journalists in Kosovo and Bosnia, would do well to seek out women, who generally bring with them a reputation for moderation in unstable situations.

Nelson Mandela suggested at last summer's Arusha peace talks that if Burundian men began fighting again, their women should withhold "conjugal rights" (like cooking, he added).

INFLUENCING SECURITY FORCES

The influence of women on warriors dates back to the ancient Greek play *Lysistrata*. Borrowing from that play's story, former South African President Nelson Mandela suggested at last summer's Arusha peace talks on the conflict in Burundi that if Burundian men began fighting again, their women should withhold "conjugal rights" (like cooking, he added).

Women can also act as a valuable interface between their countries' security forces (police and military) and the public, especially in cases when rapid response is necessary to head off violence. Women in Northern Ireland, for example, have helped calm the often deadly "marching season" by facilitating mediations between Protestant unionists and Catholic nationalists. The women bring together key members of each community, many of whom are released prisoners, as mediators to calm tensions. This circle of mediators works with local police throughout the marching season, meeting quietly and maintaining contacts on a 24-hour basis. This intervention provides a powerful extension of the limited tools of the local police and security forces.

Likewise, an early goal of the Sudanese Women's Voice for Peace was to meet and talk with the military leaders of the various rebel armies. These contacts secured women's access to areas controlled by the revolutionary movements, a critical variable in the success or failure of humanitarian efforts in war zones. Women have also worked with the military to search for missing people, a common element in the cycle of violence. In Colombia, for example, women were so persistent in their demands for information regarding 150 people abducted from a church in 1999 that the army eventually gave them space on a military base for an information and strategy center. The military worked alongside the women and their families trying to track down the missing people. In short, through moral suasion, local women often have influence where outsiders, such as international human rights agencies, do not.

That influence may have allowed a female investigative reporter like Maria Cristina Caballero to go where a man could not go, venturing on horseback alone, eight hours into the jungle to tape a four-hour interview with the head of the paramilitary forces in Colombia. She also interviewed another guerilla leader and published an award-winning comparison of the transcripts, showing where the two mortal enemies shared the same vision. "This [was] bigger than a story," she later said, "this [was] hope for peace." Risking their lives to move back and forth across the divide, women like Caballero perform work that is just as important for regional stabilization as the grandest Plan Colombia.

INTERNATIONAL COLLABORATION

Given the nature of "inclusive" war, security forces are increasingly called upon to ensure the safe passage of humanitarian relief across conflict zones. Women serve as indispensable contacts between civilians, warring parties, and relief organizations. Without women's knowledge of the local scene, the mandate of the military to support NGOs would often be severely hindered, if not impossible.

In rebel-controlled areas of Sudan, women have worked closely with humanitarian organizations to prevent food from being diverted from those who need it most. According to Catherine Loria Duku Jeremano of Oxfam: "The normal pattern was to hand out relief to the men, who were then expected to take it home to be distributed to their family. However, many of the men did what they pleased with the food they received: either selling it directly, often in exchange for alcohol, or giving food to the wives they favored." Sudanese women worked closely with tribal chiefs and relief organizations to establish a system allowing women to pick up the food for their families, despite contrary cultural norms.

In Pristina, Kosovo, Vjosa Dobruna, a pediatric neurologist and human rights leader, is now the joint administrator for civil society for the U.N. Interim Administration Mission in Kosovo (UNMIK). In September 2000, at the request of NATO, she organized a multiethnic strategic planning session to integrate women throughout UNMIK. Before that gathering, women who had played very significant roles in their communities felt shunned by the international organizations that descended on Kosovo following the bombing campaign. Vjosa's conference pulled them back into the mainstream, bringing international players into the conference to hear from local women what stabilizing measures they were planning, rather than the other way around. There, as in Bosnia, the OSCE has created a quota system for elected office, mandating that women comprise one third of each party's candidate list; leaders like Vjosa helped turn that policy into reality.

In addition to helping aid organizations find better ways to distribute relief or helping the U.N. and OSCE implement their ambitious mandates, women also work closely with them to locate and exchange prisoners of war. As the peace processes in Northern Ireland, Bosnia, and the Middle East illustrate, a deadlock on the exchange and release of prisoners can be a major obstacle to achieving a final settlement. Women activists in Armenia and Azerbaijan have worked closely with the International Helsinki Citizens Assembly and the OSCE for the release

The Black and the Green

Grass-roots women's organizations in Israel come in two colors: black and green. The Women in Black, founded in 1988, and the Women in Green, founded in 1993, could not be further apart on the political spectrum, but both claim the mantle of "womanhood" and "motherhood" in the ongoing struggle to end the Israeli-Palestinian conflict.

One month after the Palestinian intifada broke out in December 1988, a small group of women decided to meet every Friday afternoon at a busy Jerusalem intersection wearing all black and holding hand-shaped signs that read: "Stop the Occupation." The weekly gatherings continued and soon spread across Israel to Europe, the United States, and then to Asia.

While the movement was originally dedicated to achieving peace in the Middle East, other groups soon protested against repression in the Balkans and India. For these activists, their status as women lends them a special authority when it comes to demanding peace. In the words of the Asian Women's Human Rights Council: "We are the Women in Black... women, unmasking the many horrific faces of more public 'legitimate' forms of violence—state repression, communalism, ethnic cleansing, nationalism, and wars...."

Today, the Women in Black in Israel continue their nonviolent opposition to the occupation in cooperation with the umbrella group Coalition of Women for a Just Peace. They have been demonstrating against the closures of various Palestinian cities, arguing that the blockades prevent pregnant women from accessing healthcare services and keep students from attending school. The group also calls for the full participation of women in peace negotiations.

While the Women in Black stood in silent protest worldwide, a group of "grandmothers, mothers, wives, and daughters; housewives and professionals; secular and religious" formed the far-right Women in Green in 1993 out of "a shared love, devotion and concern for Israel." Known for the signature green hats they wear at rallies, the Women in Green emerged as a protest to the Oslo accords on the grounds that Israel made too many concessions to Yasir Arafat's Palestinian Liberation Organization. The group opposes returning the Golan Heights to Syria, sharing sovereignty over Jerusalem with the Palestinians, and insists that "Israel remain a Jewish state."

The Women in Green boast some 15,000 members in Israel, and while they have not garnered the global support of the Women in Black, 15,000 Americans have joined their cause. An ardent supporter of Israeli Prime Minister Ariel Sharon, the group seeks to educate the Israeli electorate through weekly street theater and public demonstrations, as well as articles, posters, and newspaper advertisements.

White the groups' messages and methods diverge, their existence and influence demonstrate that women can mobilize support for political change—no matter what color they wear.

—FP

of hostages in the disputed region of Nagorno-Karabakh, where tens of thousands of people have been killed. In fact, these women's knowledge of the local players and the situation on the ground would make them indispensable in peace negotiations to end this 13-year-old conflict.

REACHING FOR POLITICAL OFFICE

In 1977, women organizers in Northern Ireland won the Nobel Peace Prize for their nonsectarian public demonstrations. Two decades later, Northern Irish women are showing how diligently women must still work not only to ensure a place at the negotiating table but also to sustain peace by reaching critical mass in political office. In 1996, peace activists Monica McWilliams (now a member of the Northern Ireland Assembly) and May Blood (now a member of the House of Lords) were told that only leaders of the top 10 political parties—all men—would be included in the peace talks. With only six weeks to organize, McWilliams and Blood gathered 10,000 signatures to create a new political party (the Northern Ireland Women's Coalition, or NIWC) and got themselves on the ballot. They were voted into the top 10 and earned a place at the table.

The grass-roots, get-out-the-vote work of Vox Femina convinced hesitant Yugoslav women to vote for change; those votes contributed to the margin that ousted President Slobodan Milosevic.

The NIWC's efforts paid off. The women drafted key clauses of the Good Friday Agreement regarding the importance of mixed housing, the particular difficulties of young people, and the need for resources to address these problems. The NIWC also lobbied for the early release and reintegration of political prisoners in order to combat social exclusion and pushed for a comprehensive review of the police service so that all members of society would accept it. Clearly, the women's prior work with individuals and families affected by "the Troubles" enabled them to formulate such salient contributions to the agreement. In the subsequent public referendum on the Good Friday Agreement, Mo Mowlam, then British secretary of state for Northern Ireland, attributed the overwhelming success of the YES Campaign to the NIWC's persistent canvassing and lobbying.

Women in the former Yugoslavia are also stepping forward to wrest the reins of political control from extremists (including women, such as ultranationalist Bosnian Serb President Biljana Plavsic) who destroyed their country. Last December, Zorica Trifunovic, founding member of the local Women in Black (an antiwar group formed in Belgrade in October 1991), led a meeting that united 90 women leaders of pro-democracy political campaigns across the former Yugoslavia. According to polling by the National Democratic Institute, the grass-roots, get-out-the-vote work of groups such as Vox Femina (a local NGO that participated in the December meeting) convinced hesitant women to vote for change; those votes contributed to the margin that ousted President Slobodan Milosevic.

International security forces and diplomats will find no better allies than these mobilized mothers, who are tackling the toughest, most hardened hostilities.

Argentina provides another example of women making the transition from protesters to politicians: Several leaders of the Madres de la Plaza de Mayo movement, formed in the 1970s to protest the "disappearances" of their children at the hands of the military regime, have now been elected to political office. And in Russia, the Committee of Soldiers' Mothers—a protest group founded in 1989 demanding their sons' rights amidst cruel conditions in the Russian military—has grown into a powerful organization with 300 chapters and official political status. In January, U.S. Ambassador to Moscow Jim Collins described the committee as a significant factor in countering the most aggressive voices promoting military force in Chechnya. Similar mothers' groups have sprung up across the former Soviet Union and beyond—including the Mothers of Tiananmen Square. International security forces and diplomats will find no better allies than these mobilized mothers, who are tackling the toughest, most hardened hostilities.

YOU'VE COME A LONG WAY, MAYBE

Common sense dictates that women should be central to peacemaking, where they can bring their experience in conflict resolution to bear. Yet, despite all of the instances where women have been able to play a role in peace negotiations, women remain relegated to the sidelines. Part of the problem is structural: Even though more and more women are legislators and soldiers, underrepresentation persists in the highest levels of political and military hierarchies. The presidents, prime ministers, party leaders, cabinet secretaries, and generals who typically negotiate peace settlements are overwhelmingly men. There is also a psychological barrier that precludes women from sitting in on negotiations: Waging war is still thought of as a "man's job," and as such, the task of stopping war often is delegated to men

(although if we could begin to think about the process not in terms of stopping war but promoting peace, women would emerge as the more logical choice). But the key reason behind women's marginalization may be that everyone recognizes just how good women are at forging peace. A U.N. official once stated that, in Africa, women are often excluded from negotiating teams because the war leaders "are afraid the women will compromise" and give away too much.

Some encouraging signs of change, however, are emerging. Rwandan President Paul Kagame, dismayed at his difficulty in attracting international aid to his genocide-ravaged country, recently distinguished Rwanda from the prevailing image of brutality in central Africa by appointing three women to his negotiating team for the conflict in the Democratic Republic of the Congo. In an unusually healthy tit for tat, the Ugandans responded by immediately appointing a woman to their team.

Will those women make a difference? Negotiators sometimes worry that having women participate in the discussion may change the tone of the meeting. They're right: A British participant in the Northern Ireland peace talks insightfully noted that when the parties became bogged down by abstract issues and past offenses, "the women would come and talk about their loved ones, their bereavement, their children and their hopes for the future." These deeply personal comments, rather than being a diversion, helped keep the talks focused. The women's experiences reminded the parties that security for all citizens was what really mattered.

The role of women as peacemakers can be expanded in many ways. Mediators can and should insist on gender balance among negotiators to ensure a peace plan that is workable at the community level. Cultural barriers can be overcome if high-level visitors require that a critical mass (usually one third) of the local interlocutors be women (and not simply present as wives). When drafting principles for negotiation, diplomats should determine whether women's groups have already agreed upon key conflict-bridging principles, and whether their approach can serve as a basis for general negotiations.

Moreover, to foster a larger pool of potential peacemakers, embassies in conflict areas should broaden their regular contact with local women leaders and sponsor women in training programs, both at home and abroad. Governments can also do their part by providing information technology and training to women activists through private and public partnerships. Internet communication allows women peace builders to network among themselves, as well as exchange tactics and strategies with their global counterparts.

"Women understood the cost of the war and were genuinely interested in peace," recalls retired Admiral Jonathan Howe, reflecting on his experience leading the U.N. mission in Somalia in the early 1990s. "They'd had it with their warrior husbands. They were a force willing to say enough is enough. The men were sitting around talking and chewing qat, while the women were working away. They were such a positive force.... You have to look at all elements in society and be ready to tap into those that will be constructive."

Want to Know More?

The Internet is invaluable in enabling the inclusive security approach advocated in this article. The Web offers not only a wealth of information but, just as important, relatively cheap and easy access for citizens worldwide. Most of the women's peace-building activities and strategies explored in this article can be found on the Web site of **Women Waging Peace**—a collaborative venture of Harvard University's John F. Kennedy School of Government and the nonprofit organization Hunt Alternatives, which recognize the essential role and contribution of women in preventing violent conflict, stopping war, reconstructing ravaged societies, and sustaining peace in fragile areas around the world. On the site, women active in conflict areas can communicate with each other without fear of retribution via a secure server. The women submit narratives detailing their strategies, which can then be read on the public Web site. The site also features a video archive of interviews with each of these women. You need a password to view these interviews, so contact Women Waging Peace online or call (617) 868-3910.

The Organization for Security and Co-operation in Europe (OSCE) is an outstanding resource for qualitative and quantitative studies of women's involvement in conflict prevention. Start with the final report of the *OSCE Supplementary Implementation Meeting: Gender Issues* (Vienna: UNIFEM, 1999), posted on the group's Web site. **The United Nations Development Fund for Women** (UNIFEM) also publishes reports on its colorful and easy-to-navigate site. The fund's informative book, *Women at the Peace Table: Making a Difference* (New York: UNIFEM, 2000), available online, features interviews with some of today's most prominent women peacemakers, including Hanan Ashrawi and Mo Mowlam.

For a look at how globalization is changing women's roles in governments, companies, and militaries, read Cynthia Enloe's *Bananas, Beaches and Bases: Making Feminist Sense of International Politics* (Berkeley: University of California Press, 2001). In *Maneuvers: The International Politics of Militarizing Women's Lives* (Berkeley: University of California Press, 2000), Enloe examines the military's effects on women, whether they are soldiers or soldiers' spouses. For a more general discussion of where feminism fits into academia and policymaking, see **"Searching for the Princess? Feminist Perspectives in International Relations"** (*The Harvard International Review*, Fall 1999) by J. Ann Tickner, associate professor of international relations at the University of Southern California.

The Fall 1997 issue of FOREIGN POLICY magazine features two articles that highlight how women worldwide are simultaneously gaining political clout but also bearing the brunt of poverty: **"Women in Power: From Tokenism to Critical Mass"** by Jane S. Jaquette and **"Women in Poverty: A New Global Underclass"** by Mayra Buvinic.

• For links to relevant Web sites, as well as a comprehensive index of related FOREIGN POLICY articles, access **www.foreignpolicy.com**.

Lasting peace must be homegrown. Inclusive security helps police forces, military leaders, and diplomats do their jobs more effectively by creating coalitions with the people most invested in stability and most adept at building peace. Women working on the ground are eager to join forces. Just let them in.

Swanee Hunt is director of the Women in Public Policy Program at Harvard University's John F. Kennedy School of Government. As the United States' ambassador to Austria (1993–97), she founded the "Vital Voices: Women in Democracy" initiative. Cristina Posa, a former judicial clerk at the United Nations International Criminal Tribunal for the former Yugoslavia, is an attorney at Cleary, Gottlieb, Steen & Hamilton in New York.

THE NEXT CHRISTIANITY

We stand at a historical turning point, the author argues—one that is as epochal for the Christian world as the original Reformation. Around the globe Christianity is growing and mutating in ways that observers in the West tend not to see. Tumultuous conflicts within Christianity will leave a mark deeper than Islam's on the century ahead

BY PHILIP JENKINS

Ever since the sexual-abuse crisis erupted in the U.S. Roman Catholic Church in the mid-1980s, with allegations of child molestation by priests, commentators have regularly compared the problems faced by the Church to those it faced in Europe at the start of the sixteenth century, on the eve of the Protestant Reformation—problems that included sexual laxity and financial malfeasance among the clergy, and clerical contempt for the interests of the laity. Calls for change have become increasingly urgent since January, when revelations of widespread sexual misconduct and grossly negligent responses to it emerged prominently in the Boston archdiocese. Similar, if less dramatic, problems have been brought to light in New Orleans, Providence, Palm Beach, Omaha, and many other dioceses. The reform agendas now under discussion within the U.S. hierarchy involve ideas about increased lay participation in governance—ideas of the sort heard when Martin Luther confronted the Roman Catholic orthodoxy of his day. They also include such ideas as admitting women to the priesthood and permitting priests to marry.

Explicit analogies to the Reformation have become commonplace not only among commentators but also among anticlerical activists, among victims' groups, and, significantly, among ordinary lay believers. One representative expert on sexual misconduct, much quoted, is Richard Sipe, a former monk who worked at the sexual-disorders clinic at Johns Hopkins University and is now a psychotherapist based in California. Over the years Sipe has spoken regularly of "a new Reformation." "We are at 1515," he has written, "between when Martin Luther went to Rome in 1510 and 1517 when he nailed his 95 theses on the door in Wittenberg." That act can reasonably be seen as the symbolic starting point of the Reformation, when a united Christendom was rent asunder.

Historians continue to debate the causes and consequences of the Reformation, and of the forces that it unleashed. Among other things, the Reformation broke the fetters that constrained certain aspects of intellectual life during the Middle Ages. Protestants, of course, honor the event as the source of their distinctive religious traditions; many Protestant denominations celebrate Reformation Day, at the end of October, commemorating the posting of the theses at Wittenberg. And liberal Catholics invoke the word these days to emphasize the urgency of reform—changes both broad and specific that they demand from the

Church. Their view is that the crisis, which exposes fault lines of both sexuality and power, is the most serious the Church has faced in 500 years—as serious as the one it faced in Luther's time.

The first Reformation was an epochal moment in the history of the Western world—and eventually, by extension, of the rest of the world. The status quo in religious affairs was brought to an end. Relations between religions and governments, not to mention among different denominations, took a variety of forms—sometimes symbiotic, often chaotic and violent. The transformations wrought in the human psyche by the Reformation, and by the Counter-Reformation it helped to provoke, continue to play themselves out. This complex historical episode, which is now often referred to simply as "the Reformation," touched everything. It altered not just the practice of religion but also the nature of society, economics, politics, education, and the law.

Commentators today, when speaking of the changes needed in the Catholic Church, generally do not have in mind the sweeping historical aftermath of the first Reformation—but they should. The Church has developed a fissure whose size most people do not fully appreciate. The steps that liberal Catholics would take to resolve some of the Church's urgent issues, steps that might quell unease or revolt in some places, would prove incendiary in others. The problem with reform, 500 years ago or today, is that people disagree—sometimes violently—on the direction it should take.

The fact is, we are at a moment as epochal as the Reformation itself—a Reformation moment not only for Catholics but for the entire Christian world. Christianity as a whole is both growing and mutating in ways that observers in the West tend not to see. For obvious reasons, news reports today are filled with material about the influence of a resurgent and sometimes angry Islam. But in its variety and vitality, in its global reach, in its association with the world's fastest-growing societies, in its shifting centers of gravity, in the way its values and practices vary from place to place—in these and other ways it is Christianity that will leave the deepest mark on the twenty-first century. The process will not necessarily be a peaceful one, and only the foolish would venture anything beyond the broadest predictions about the religious picture a century or two ahead. But the twenty-first century will almost certainly be regarded by future historians as a century in which religion replaced ideology as the prime animating and destructive force in human affairs, guiding attitudes to political liberty and obligation, concepts of nationhood, and, of course, conflicts and wars.

The original Reformation was far more than the rising up of irate lay people against corrupt and exploitative priests, and it was much more than a mere theological row. It was a far-reaching social movement that sought to return to the original sources of Christianity. It challenged the idea that divine authority should be mediated through institutions or hierarchies, and it denied the value of tradition. Instead it offered radical new notions of the supremacy of written texts (that is, the books of the Bible), interpreted by individual consciences. The Reformation made possible a religion that could be practiced privately, rather than mainly in a vast institutionalized community.

This move toward individualism, toward the privatization of religious belief, makes the spirit of the Reformation very attractive to educated people in the West. It stirs many liberal Catholic activists, who regard the aloof and arrogant hierarchy of the Church as not only an affront but something inherently corrupt. New concepts of governance sound exciting, even intoxicating, to reformers, and seem to mesh with likely social and technological trends. The invention of movable type and the printing press, in the fifteenth century, was a technological development that spurred mass literacy in the vernacular languages—and accelerated the forces of religious change. In the near future, many believe, the electronic media will have a comparably powerful impact on our ways of being religious. An ever greater reliance on individual choice, the argument goes, will help Catholicism to become much more inclusive and tolerant, less judgmental, and more willing to accept secular attitudes toward sexuality and gender roles. In the view of liberal Catholics, much of the current crisis derives directly from archaic if not primitive doctrines, including mandatory celibacy among the clergy, intolerance of homosexuality, and the prohibition of women from the priesthood, not to mention a more generalized fear of sexuality. In their view, anyone should be able to see that the idea that God, the creator and lord of the universe, is concerned about human sexuality is on its way out.

If we look beyond the liberal West, we see that another Christian revolution, quite different from the one being called for in affluent American suburbs and upscale urban parishes, is already in progress. Worldwide, Christianity is actually moving toward supernaturalism and neo-orthodoxy, and in many ways toward the ancient world view expressed in the New Testament: a vision of Jesus as the embodiment of divine power, who overcomes the evil forces that inflict calamity and sickness upon the human race. In the global South (the areas that we often think of primarily as the Third World) huge and growing Christian populations—currently 480 million in Latin America, 360 million in Africa, and 313 million in Asia, compared with 260 million in North America—now make up what the Catholic scholar Walbert Buhlmann has called the Third Church, a form of Christianity as distinct as Protestantism or Orthodoxy, and one that is likely to become dominant in the faith. The revolution taking place in Africa, Asia, and Latin America is far more sweeping in its implications than any current shifts in North American religion, whether Catholic or Protestant. There is increasing tension between what one might call a liberal Northern Reformation and the surging Southern religious revolution, which one might equate with the Counter-Reformation, the internal Catholic reforms that took place at the same time as the Reformation—although in references to the past and the present the term "Counter-Reformation" misleadingly implies a simple reaction instead of a social and spiritual explosion. No matter what the terminology, however, an enormous rift seems inevitable.

Although Northern governments are still struggling to come to terms with the notion that Islam might provide a powerful and threatening supranational ideology, few seem to realize the potential political role of ascendant Southern Christianity. The religious rift between Northern and Southern Europe in the sixteenth century suggests just how dramatic the political consequences of a North-South divide in the contemporary Christian world might be. The Reformation led to nothing less than the creation of the modern European states and the international order we recognize today. For more than a century Europe was rent by sectarian wars between Protestants and Catholics, which by the 1680s had ended in stalemate. Out of this impasse, this failure to impose a monolithic religious order across the Continent, there arose such fundamental ideas of modern society as the state's obligation to tolerate minorities and the need to justify political authority without constantly invoking God and religion. The Enlightenment—and, indeed, Western modernity—could have occurred only as a consequence of the clash, military and ideological, between Protestants and Catholics.

Today across the global South a rising religious fervor is coinciding with declining autonomy for nation-states, making useful an analogy with the medieval concept of Christendom—the Res Publica Christiana—as an overarching source of unity and a focus of loyalty transcending mere kingdoms or empires. Kingdoms might last for only a century or two before being supplanted by new states or dynasties, but rational people knew that Christendom simply endured. The laws of individual nations lasted only as long as the nations themselves; Christendom offered a higher set of standards and mores that could claim to be universal. Christendom was a primary cultural reference, and it may well re-emerge as such in the Christian South—as a new transnational order in which political, social, and personal identities are defined chiefly by religious loyalties.

The first Reformation was a lot less straightforward than some histories suggest. The sixteenth-century Catholic Church, after all, did not collapse after Luther kicked in the door. The Counter-Reformation was moving in a diametrically opposite direction, reasserting older forms of devotion and tradition, and reformulating the Church's controversial claims for hierarchy and spiritual authority. The Counter-Reformation was not just survivalist and defensive, as is commonly assumed; it was also innovative and dynamic. For at least a century after Luther's Reformation, in fact, the true political, cultural, and social centers of Europe were as much in the Catholic South as in the Protestant North. The Catholic states—Spain, Portugal, and France—were launching missionary ventures into Africa, Asia, North and South America. By the 1570s Catholic missionaries were creating a transoceanic Church structure: the see of Manila was an offshoot of the archdiocese of Mexico City.

By about 1600 the Catholic Church had become the first religious body—indeed, the first institution of any sort—to operate on a global scale. Even in the Protestant heartlands of Northern and Western Europe—England, Sweden, and the German lands—the heirs of the Reformation had to spend many years discouraging their people from succumbing to the attractions of Catholicism. Conversions to Catholicism were steady throughout the century or so after 1580. It looked as if the Reformation had effectively cut Protestant Europe off from the mainstream of the Christian world. Only in the eighteenth century would Protestantism find a secure and then strategically preponderant place on the global stage, through the success of booming commercial states such as England and the Netherlands, whose political triumphs ultimately contained and in some cases pushed back the earlier empires.

The changes that Catholic and other reformers today are trying to inspire in North America and Europe (and that seem essential if Christianity is to be preserved as a modern, relevant force on those continents) run utterly contrary to the dominant cultural movements in the rest of the Christian world, which look very much like the Counter-Reformation. But this century is unlike the sixteenth in that we are not facing a roughly equal division of Christendom between two competing groups. Rather, Christians are facing a shrinking population in the liberal West and a growing majority of the traditional Rest. During the past half century the critical centers of the Christian world have moved decisively to Africa, to Latin America, and to Asia. The balance will never shift back.

The growth in Africa has been relentless. In 1900 Africa had just 10 million Christians out of a continental population of 107 million—about nine percent. Today the Christian total stands at 360 million out of 784 million, or 46 percent. And that percentage is likely to continue rising, because Christian African countries have some of the world's most dramatic rates of population growth. Meanwhile, the advanced industrial countries are experiencing a dramatic birth dearth. Within the next twenty-five years the population of the world's Christians is expected to grow to 2.6 billion (making Christianity by far the world's largest faith). By 2025, 50 percent of the Christian population will be in Africa and Latin America, and another 17 percent will be in Asia. Those proportions will grow steadily. By about 2050 the United States will still have the largest single contingent of Christians, but all the other leading nations will be Southern: Mexico, Brazil, Nigeria, the Democratic Republic of the Congo, Ethiopia, and the Philippines. By then the proportion of non-Latino whites among the world's Christians will have fallen to perhaps one in five.

The population shift is even more marked in the specifically Catholic world, where Euro-Americans are already in the minority. Africa had about 16 million Catholics in the early 1950s; it has 120 million today, and is expected to have 228 million by 2025. The *World Christian Encyclopedia* suggests that by 2025 almost three quarters of all Catholics will be found in Africa, Asia, and Latin America. The likely map of twenty-first-century Catholicism represents an unmistakable legacy of the Counter-Reformation and its global missionary ventures.

These figures actually understate the Southern predominance within Catholicism, and within world Christianity more generally, because they fail to take account of Southern emigrants to Europe and North America. Even as this migration continues, established white communities in Europe are declining demographically, and their religious beliefs and practices are moving further away from traditional Christian roots. The result is that skins of other hues are increasingly evident in European churches; half of all London churchgoers are now black. African and West Indian churches in Britain are reaching out to whites, though members complain that their religion is often seen as "a black thing" rather than "a God thing."

In the United States a growing proportion of Roman Catholics are Latinos, who should represent a quarter of the nation by 2050 or so. Asian communities in the United States have sizable Catholic populations. Current trends suggest that the religious values of Catholics with a Southern ethnic and cultural heritage will long remain quite distinct from those of other U.S. populations. In terms of liturgy and worship Latino Catholics are strikingly different from Anglo believers, not least in maintaining a fervent devotion to the Virgin Mary and the saints.

European and Euro-American Catholics will within a few decades be a smaller and smaller fragment of a worldwide Church. Of the 18 million Catholic baptisms recorded in 1998, eight million took place in Central and South America, three million in Africa, and just under three million in Asia. (In other words, these three regions already account for more than three quarters of all Catholic baptisms.) The annual baptism total for the Philippines is higher than the totals for Italy, France, Spain, and Poland combined. The number of Filipino Catholics could grow to 90 million by 2025, and perhaps to 130 million by 2050.

The demographic changes within Christianity have many implications for theology and religious practice, and for global society and politics. The most significant point is that in terms of both theology and moral teaching, Southern Christianity is more conservative than the Northern—especially the American—version. Northern reformers, even if otherwise sympathetic to the indigenous cultures of non-Northern peoples, obviously do not like this fact. The liberal Catholic writer James Carroll has complained that "world Christianity [is falling] increasingly under the sway of anti-intellectual fundamentalism." But the cultural pressures may be hard to resist.

The denominations that are triumphing across the global South—radical Protestant sects, either evangelical or Pentecostal, and Roman Catholicism of an orthodox kind—are stalwartly traditional or even reactionary by the standards of the economically advanced nations. The Catholic faith that is rising rapidly in Africa and Asia looks very much like a pre-Vatican II faith, being more traditional in its respect for the power of bishops and priests and in its preference for older devotions. African Catholicism in particular is far more comfortable with notions of authority and spiritual charisma than with newer ideas of consultation and democracy.

This kind of faith is personified by Nigeria's Francis Cardinal Arinze, who is sometimes touted as a future Pope. He is sharp and articulate, with an attractively self-deprecating style, and he has served as the president of the Pontifical Council for Inter-Religious Dialogue, which has given him invaluable experience in talking with Muslims, Hindus, Jews, and members of other faiths. By liberal Northern standards, however, Arinze is rigidly conservative, and even repressive on matters such as academic freedom and the need for strict orthodoxy. In his theology as much as his social views he is a loyal follower of Pope John Paul II. Anyone less promising for Northern notions of reform is difficult to imagine.

Meanwhile, a full-scale Reformation is taking place among Pentecostal Christians—whose ideas are shared by many Catholics. Pentecostal believers reject tradition and hierarchy, but they also rely on direct spiritual revelation to supplement or replace biblical authority. And it is Pentecostals who stand in the vanguard of the Southern Counter-Reformation. Though Pentecostalism emerged as a movement only at the start of the twentieth century, chiefly in North America, Pentecostals today are at least 400 million strong, and heavily concentrated in the global South. By 2040 or so there could be as many as a billion, at which point Pentecostal Christians alone will far outnumber the world's Buddhists and will enjoy rough numerical parity with the world's Hindus.

The booming Pentecostal churches of Africa, Asia, and Latin America are thoroughly committed to re-creating their version of an idealized early Christianity (often described as the restoration of "primitive" Christianity). The most successful Southern churches preach a deep personal faith, communal orthodoxy, mysticism, and puritanism, all founded on obedience to spiritual authority, from whatever source it is believed to stem. Pentecostals—and their Catholic counterparts—preach messages that may appear simplistically charismatic, visionary, and apocalyptic to a Northern liberal. For them prophecy is an everyday reality, and many independent denominations trace their foundation to direct prophetic authority. Scholars of religion customarily speak of these proliferating congregations simply as the "prophetic churches."

Of course, American reformers also dream of a restored early Church; but whereas Americans imagine a Church freed from hierarchy, superstition, and dogma, Southerners look back to one filled with spiritual power and able to exorcise the demonic forces that cause sickness and poverty. And yes, "demonic" is the word. The most successful Southern churches today speak openly of spiritual healing and exorcism. One controversial sect in the process of developing an international following is the Brazilian-based Universal Church of the Kingdom of God, which claims to offer "strong prayer to destroy witchcraft, demon possession, bad luck, bad dreams, all spiritual problems," and promises that members will gain "prosperity and financial breakthrough." The Cherubim and Seraphim movement of West Africa claims to have "conscious knowledge of the evil spirits which sow the seeds of discomfort, set afloat ill-luck, diseases, induce barrenness, sterility and the like."

Americans and Europeans usually associate such religious ideas with primitive and rural conditions, and assume that the older world view will disappear with the coming of modernization and urbanization. In the contemporary South, however, the success of highly supernatural churches should rather be seen as a direct by-product of urbanization. (This should come as no surprise to Americans; look at the Pentecostal storefronts in America's inner cities.) As predominantly rural societies have become more urban over the past thirty or forty years, millions of migrants have been attracted to ever larger urban areas, which lack the resources and the infrastructure to meet the needs of these wanderers. Sometimes people travel to cities within the same nation, but often they find themselves in different countries and cultures, suffering a still greater sense of estrangement. In such settings religious communities emerge to provide health, welfare, and education.

This sort of alternative social system, which played an enormous role in the earliest days of Christianity, has been a potent means of winning mass support for the most committed religious groups and is likely to grow in importance as the gap between people's needs and government's capacities to fill them becomes wider. Looking at the success of Christianity in the Roman Empire, the historian Peter Brown has written, "The Christian community suddenly came to appeal to men who felt deserted... Plainly, to be a Christian in 250 brought more protection from one's fellows than to be a *civis Romanus*." Being a member of an active Christian church today may well bring more tangible benefits than being a mere citizen of Nigeria or Peru.

Often the new churches gain support because of the way they deal with the demons of oppression and want: they interpret the horrors of everyday urban life in supernatural terms. In many cases these churches seek to prove their spiritual powers in struggles against witchcraft. The intensity of belief in witchcraft across much of Africa can be startling. As recently as last year at least 1,000 alleged witches were hacked to death in a single "purge" in the Democratic Republic of the Congo. Far from declining with urbanization, fear of witches has intensified. Since the collapse of South Africa's apartheid regime, in 1994, witchcraft has emerged as a primary social fear in Soweto, with its three million impoverished residents.

The desperate public-health situation in the booming mega-cities of the South goes far toward explaining the emphasis of the new churches on healing mind and body. In Africa in the early twentieth century an explosion of Christian healing movements and new prophets coincided with a dreadful series of epidemics, and the religious upsurge of those years was in part a quest for bodily health. Today African churches stand or fall by their success in healing, and elaborate rituals have formed around healing practices (though church members disagree on whether believers should rely entirely on spiritual assistance). The same interest in spiritual healing is found in what were once the mission churches—bodies such as the Anglicans and the Lutherans. Nowhere in the global South do the various spiritual healers find serious competition from modern scientific medicine: it is simply beyond the reach of most of the poor.

Disease, exploitation, pollution, drink, drugs, and violence, taken together, can account for why people might easily accept that they are under siege from demonic forces, and that only divine intervention can save them. Even radical liberation theologians use apocalyptic language on occasion. When a Northerner asks, in effect, where the Southern churches are getting such ideas, the answer is not hard to find: they're getting them from the Bible. Southern Christians are reading the New Testament and taking it very seriously; in it they see the power of Jesus fundamentally expressed through his confrontations with demonic powers, particularly those causing sickness and insanity. "Go back and report to John what you hear and see," Jesus says in the Gospel according to Matthew (11: 4–5). "The blind receive sight, the lame walk, those who have leprosy are cured, the deaf hear, the dead are raised, and the good news is preached to the poor." For the past two hundred years Northern liberals have employed various nonliteral interpretations of these healing passages—perhaps Jesus had a good sense of the causes and treatment of psychosomatic ailments? But that is not, of course, how such scenes are understood within the Third Church.

Today, as in the early sixteenth century, a literal interpretation of the Bible can be tremendously appealing. To quote a modern-day follower of the African prophet Johane Masowe, cited in Elizabeth Isichei's *A History of Christianity in Africa*, "When we were in these synagogues [the European churches], we used to read about the works of Jesus Christ... cripples were made to walk and the dead were brought to life... evil spirits driven out... That was what was being done in Jerusalem. We Africans, however, who were being instructed by white people, never did anything like that... We were taught to read the Bible, but we ourselves never did what the people of the Bible used to do."

Alongside the fast-growing churches have emerged apocalyptic and messianic movements that try to bring in the kingdom of God through armed violence. Some try to establish the thousand-year reign of Jesus Christ on earth, as prophesied in the Book of Revelation. This phenomenon would have been instantly familiar to Europeans 500 years ago, when the Anabaptists and other millenarian groups flourished. Perhaps the most traumatic event of the Reformation occurred in the German city of Münster in 1534–1535, when Anabaptist rebels established a radical social order that abolished property and monogamy; a homicidal king-messiah held dictatorial power until the forces of state authority conquered and annihilated the fanatics. Then as now, it was difficult to set bounds to religious enthusiasm.

Extremist Christian movements have appeared regularly across parts of Africa where the mechanisms of the state are weak. They include groups such as the Lumpa Church, in Zambia, and the terrifying Lord's Resistance Army (LRA), in Uganda. In 2000 more than a thousand people in another Ugandan sect, the Movement for the Restoration of the Ten Commandments of God, perished in an apparent mass suicide. In each case a group emerged from orthodox roots and then gravitated toward apocalyptic fanaticism. The Ten Commandments sect grew out of orthodox Catholicism. The Lumpa Church began, in the 1950s, with Alice Lenshina, a Presbyterian convert who claimed to receive divine visions urging her to fight witchcraft. She became the *lenshina*, or queen, of her new church, whose name, Lumpa, means "better than all others." The group attracted a hundred thousand followers, who formed a utopian community in order to await the Second Coming of Jesus Christ. Since it rejected worldly regimes to the point of refusing to pay taxes, the Lumpa became increasingly engaged in confrontations with the Zambian government, leading to open rebellion in the 1960s.

Another prophetic Alice appeared in Uganda during the chaotic civil wars that swept that country in the 1980s. Alice Lakwena was a former Catholic whose visions led her to establish the Holy Spirit Mobile Force, also pledged to fight witches. She refused to accept the national peace settlement established under President Yoweri Museveni, and engaged in a holy war against his regime. Holy Spirit soldiers, many of them children and young teenagers, were ritually anointed with butter on the understanding that it would make them bulletproof. When Lakwena's army was crushed, in 1991, most of her followers merged with the LRA, which is notorious for filling its ranks by abducting children. Atrocities committed by the group include mass murder, rape, and forced cannibalism. Today as in the sixteenth century, an absolute conviction that one is fighting for God's cause makes moot the laws of war.

The changing demographic balance between North and South helps to explain the current shape of world Catholicism, including the fact that the Church has been headed by Pope John Paul II. In the papal election of 1978 the Polish candidate won the support of Latin American cardinals, who were not prepared to accept yet another Western European. In turn, John Paul has recognized the growing Southern presence in the Church. Last year he elevated forty-four new cardinals, of whom eleven were Latin American, two Indian, and three African. The next time a papal election takes place, fifty-seven of the 135 cardinals eligible to vote, or more than 40 percent, will be from Southern nations. Early this century they will constitute a majority.

It may be true that from the liberal Northern perspective, pressure for a Reformation-style solution to critical problems in the Church—the crisis in clerical celibacy, the shortage of priests, the sense that the laity's concerns are ignored—seems overwhelming. Poll after poll in the United States and Europe indicates significant distrust of clerical authority and support for greater lay participation and women's equality. The obvious question in the parishes of the developed world seems to be how long the aloof hierarchy can stave off the forces of history.

From Rome, however, the picture looks different, as do the "natural" directions that history is going to take. The Roman church operates on a global scale and has done so for centuries. Long before the French and British governments had become aware of global politics—and well before their empires came into being—papal diplomats were thinking through their approaches to China, their policies in Peru, their views on African affairs, their stances on the issues facing Japan and Mexico. To adapt a popular activist slogan, the Catholic Church not only thinks globally, it acts globally. That approach is going to have weighty consequences. On present evidence, a Southern-dominated Catholic Church is likely to react traditionally to the issues that most concern American and European reformers: matters of theology and devotion, sexual ethics and gender roles, and, most fundamentally, issues of authority within the Church.

Neatly illustrating the cultural gulf that separates Northern and Southern churches is an incident involving Moses Tay, the Anglican archbishop of Southeast Asia, whose see is based in Singapore. In the early 1990s Tay traveled to Vancouver, where he encountered the totem poles that are a local tourist attraction. To him, they were idols possessed by evil spirits, and he concluded that they required handling by prayer and exorcism. This horrified the local Anglican Church, which was committed to building good relationships with local Native American communities, and which regarded exorcism as absurd superstition. The Canadians, like other good liberal Christians throughout the North, were long past dismissing alien religions as diabolically inspired. It's difficult not to feel some sympathy with the archbishop, however. He was quite correct to see the totems as authentic religious symbols, and considering the long history of Christian writing on exorcism and possession, he could also summon many precedents to support his position. On that occasion Tay personified the global Christian confrontation.

The cultural gap between Christians of the North and the South will increase rather than diminish in the coming decades, for reasons that recall Luther's time. During the early modern period Northern and Southern Europe were divided between the Protestantism of the word and the Catholicism of the senses—between a religious culture of preaching, hymns, and Bible reading, and one of statues, rituals, and processions. Today we might see as a parallel the impact of electronic technologies, which is being felt at very different rates in the Northern and Southern worlds. The new-media revolution is occurring in Europe, North America, and the Pacific Rim while other parts of the globe are focusing on—indeed, still catching up with—the traditional world of book learning. Northern communities will move to ever more decentralized and privatized forms of faith as Southerners maintain older ideals of community and traditional authority.

On moral issues, too, Southern churches are far out of step with liberal Northern churches. African and Latin American churches tend to be very conservative on issues such as homosexuality and abortion. Such disagreement can pose real political difficulties for churches that aspire to a global identity and that try to balance diverse opinions. At present this is scarcely an issue for the Roman Catholic Church, which at least officially preaches the same conservatism for all regions. If, however, Church officials in North America or Europe proclaimed a moral stance more in keeping with progressive secular values, they would be divided from the growing Catholic churches of the South by a de facto schism, if not a formal breach.

For thirty years Northern liberals have dreamed of a Third Vatican Council to complete the revolution launched by Pope John XXIII—one that would usher in a new age of ecclesiastical democracy and lay empowerment. It would be a bitter irony for the liberals if the council were convened but turned out to be a conservative, Southern-dominated affair that imposed moral and theological litmus tests intolerable to North Americans and Europeans—if, in other words, it tried to implement not a new Reformation but a new Counter-Reformation. (In that sense we would be witnessing not a new Wittenberg but, rather, a new Council of Trent—that is, a strongly traditional gathering that would restate the Church's older ideology and attempt to set it in stone for all future ages.) If a future Southern Pope struggled to impose a new vision of orthodoxy on America's Catholic bishops, universities, and seminaries, the result could well be an actual rather than a de facto schism.

The experience of the world's Anglicans and Episcopalians may foretell the direction of conflicts within the Roman Catholic Church. In the Anglican Communion, which is also torn by a global cultural conflict over issues of gender and sexuality, orthodox Southerners seek to re-evangelize a Euro-American world that they view as coming close to open heresy. This uncannily recalls the situation in sixteenth-century Europe, in which Counter-Reformation Catholics sent Jesuits and missionary priests to reconvert those regions that had fallen into Protestantism.

Anglicans in the North tend to be very liberal on homosexuality and the ordination of women. In recent years, however, liberal clerics have been appalled to find themselves outnumbered and regularly outvoted. In these votes the bishops of Africa and Asia have emerged as a rock-solid conservative bloc. The most ferocious battle to date occurred at the Lambeth World Conference in 1998, which adopted, over the objections of the liberal bishops, a forthright traditional statement proclaiming the impossibility of reconciling homosexual conduct with Christian ministry. As in the Roman Catholic Church, the predominance of Southerners at future events of this kind will only increase. Nigeria already has more practicing Anglicans than any other country, far more than Britain itself, and Uganda is not far behind. By mid-century the global total of Anglicans could approach 150 million, of whom only a small minority will be white Europeans or North Americans. The shifting balance with-in the church could become a critical issue very shortly, since the new Archbishop of Canterbury, Rowan Williams, is notably gay-friendly and has already ordained a practicing homosexual as a priest.

The Lambeth debate also initiated a series of events that Catholic reformers should study carefully. Briefly, American conservatives who were disenchanted with the liberal establishment in the U.S. Episcopal Church realized that they had powerful friends overseas, and transferred their religious allegiance to more-conservative authorities in the global South. Since 2000 some conservative American Episcopalians have traveled to Moses Tay's cathedral in Singapore, where they were consecrated as bishops by Asian and African Anglican prelates, including the Rwandan archbishop Emmanuel Kolini. By tradition an Anglican archbishop is free to ordain whomever he pleases within his province, so although the Americans live and work in South Carolina, Pennsylvania, and other states, they are now technically bishops within the province of Rwanda. They have become missionary bishops, charged with ministering to conservative congregations in the United States, where they support a dissident "virtual province" within the church. They and their conservative colleagues are now part of the Anglican Mission in America, which is intended officially to "lead the Episcopal Church back to its biblical foundations." The mission aims to restore traditional teachings and combat what it sees as the "manifest heresy" and even open apostasy of the U.S. Church leadership. Just this past summer Archbishop Kolini offered his protection to dissident Anglicans in the Vancouver area, who were rebelling against liberal proposals to allow same-sex couples to receive a formal Church blessing.

Ultimately, the first Christendom—the politico–religious order that dominated Europe from the sixth century through the sixteenth—collapsed in the face of secular nationalism, under the overwhelming force of what Thomas Carlyle described as "the three great elements of modern civilization, gunpowder, printing, and the Protestant religion." Nation-states have dominated the world ever since. Today, however, the whole concept of national autonomy is under challenge, partly as a result of new technologies. In the coming decades, according to a recent CIA report, "Governments will have less and less control over flows of information, technology, diseases, migrants, arms, and financial transactions, whether licit or illicit, across their borders. The very concept of 'belonging' to a particular state will probably erode." If a once unquestionable construct like Great Britain is under threat, it is not surprising that people are questioning the existence of newer and more artificial entities in Africa and Asia.

For a quarter of a century social scientists analyzing the decline of the nation-state have drawn parallels between the world today and the politically fragmented yet cosmopolitan world of the Middle Ages. Some scholars have even predicted the emergence of some secular movement or ideology that would command loyalty across nations like the Christendom of old. Yet the more we look at the Southern Hemisphere, the more we see that although supranational ideas are flourishing, they are not in the least secular. The parallels to the Middle Ages may be closer than anyone has guessed.

Across the global South cardinals and bishops have become national moral leaders in a way essentially unseen in the West since the seventeenth century. The struggles of South African churches under apartheid spring to mind, but just as impressive were the pro-democracy campaigns of many churches and denominations elsewhere in Africa during the 1980s and 1990s. Prelates know that they are expected to speak for their people, even though if they speak boldly, they may well pay with their lives. Important and widely revered modern martyrs include Archbishop Luwum, of Uganda; Archbishop Munzihirwa, of Zaire; and Cardinal Biayenda, of Congo-Brazzaville.

As this sense of moral leadership grows, we might reasonably ask whether Christianity will also provide a guiding political ideology for much of the world. We might even imagine a new wave of Christian states, in which political life is inextricably bound up with religious belief. Zambia declared itself a Christian nation in 1991, and similar ideas have been bruited in Zimbabwe, Kenya, and Liberia. If this ideal does gain popularity, the Christian South will soon be dealing with some debates, of long standing in the North, over the proper relationship between Church and State and between rival churches under the law. Other inevitable questions involve tolerance and diversity, the relationship between majority and minority communities, and the extent to which religiously inspired laws can (or should) regulate private morality and behavior. These issues were all at the core of the Reformation.

Across the regions of the world that will be the most populous in the twenty-first century, vast religious contests are already in progress, though so far they have impinged little on Western opinion. The most significant conflict is in Nigeria, a nation that by rights should be a major regional power in this century and perhaps even a global power; but recent violence between Muslims and Christians raises the danger that Nigerian society might be brought to ruin by the clash of *jihad* and crusade. Muslims and Christians are at each other's throats in Indonesia, the Philippines, Sudan, and a growing number of other African nations; Hindu extremists persecute Christians in India. Demographic projections suggest that these feuds will simply worsen. Present-day battles in Africa and Asia may anticipate the political outlines to come, and the roots of future great-power alliances. These battles are analogous to the ideological conflicts of the twentieth century, the alternating hot and cold wars between advocates of fascism and of democracy, of socialism and of capitalism. This time, however, the competing ideologies are explicitly religious, promising their followers a literal rather than merely a metaphorical kingdom of God on earth.

Let us imagine Africa in the throes of fiery religious revivals, as Muslim and Christian states jostle for political influence. Demographic change alone could provoke more-aggressive international policies, as countries with swollen populations tried to appropriate living space or natural resources. But religious tensions could make the situation far worse. If mega-cities are not to implode through social unrest and riot, governments have to find some way to mobilize the teeming masses of unemployed teenagers and young adults. Persuading them to fight for God is a proven way of siphoning off internal tension, especially if the religion in question already has a powerful ideal of martyrdom. Liberia, Uganda, and Sierra Leone have given rise to ruthless militias ready to kill or die for whatever warlord directs them, often following some notionally religious imperative. In the 1980s the hard-line Shiite mullahs of Iran secured their authority by sending hundreds of thousands of young men to martyr themselves in human-wave assaults against the Iraqi front lines. In contemporary Indonesia, Islamist militias can readily find thousands of poor recruits to fight against the nation's Christian minorities.

Some of the likely winners in the religious economy of the new century are precisely those groups with a strongly apocalyptic mindset, in which the triumph of righteousness is associated with the vision of a world devastated by fire and plague. This could be a perilously convenient ideology for certain countries with weapons of mass destruction. (The candidates that come to mind include not only Iraq and Iran but also future regional powers such as Indonesia, Nigeria, the Congo, Uganda, and South Africa.) All this means that our political leaders and diplomats should pay at least as much attention to religions and sectarian frontiers as they ever have to the location of oil fields.

Perhaps the most remarkable point about these potential conflicts is that the trends pointing toward them have registered so little on the consciousness of even well-informed Northern observers. What, after all, do most Americans know about the distribution of Christians worldwide? I suspect that most see Christianity very much as it was a century ago—a predominantly European and North American faith. In discussions of the recent sexual-abuse crisis "the Catholic Church" and "the American Church" have been used more or less synonymously.

As the media have striven in recent years to present Islam in a more sympathetic light, they have tended to suggest that Islam, not Christianity, is the rising faith of Africa and Asia, the authentic or default religion of the world's huddled masses. But Christianity is not only surviving in the global South, it is enjoying a radical revival, a return to scriptural roots. We are living in revolutionary times.

But we aren't participating in them. By any reasonable assessment of numbers, the most significant transformation of Christianity in the world today is not the liberal Reformation that is so much desired in the North. It is the Counter-Reformation coming from the global South. And it's very likely that in a decade or two neither component of global Christianity will recognize its counterpart as fully or authentically Christian.

Philip Jenkins is a Distinguished Professor of History and Religious Studies at Pennsylvania State University. His most recent book is The Next Christendom *(2002).*

From *The Atlantic Monthly,* October 2002, pp. 53-55, 58-62, 64, 66-68. © 2002 by Philip Jenkins. Reprinted by permission.

Modernization's Challenge to Traditional Values: Who's Afraid of Ronald McDonald?

"Modernization" means "Americanization" to many who fear a coming McWorld. But a study by two social researchers indicates that traditional values will keep most countries from becoming clones of the United States.

By Ronald Inglehart and Wayne E. Baker

The World Values Survey—a two-decade-long examination of the values of 65 societies coordinated by the University of Michigan's Institute for Social Research—is the largest investigation ever conducted of attitudes, values, and beliefs around the world. This study has carried out three waves of representative national surveys: the first in 1981–1982, the second in 1990–1991, and the third in 1995–1998. The fourth wave is being completed in 1999–2001. The study now represents some 80% of the world's population. These societies have per capita GNPs ranging from $300 to more than $30,000. Their political systems range from long-established stable democracies to authoritarian states.

The World Values Survey data have been used by researchers around the world for hundreds of publications in more than a dozen languages. Studies that have been based on the data cover a wide range of topics, including volunteerism in Europe, political partisanship and social class in Ireland, democratization in Korea, liberalization in Mexico, future values in Japan, and the religious vote in Western Europe.

This article examines the relationship between cultural values and economic globalization and modernization: What impact does economic development have on the values of a culture, and vice versa? Is a future "McWorld" inevitable?

Rich Values, Poor Values

The World Values Survey data show us that the world views of the people of rich societies differ systematically from those of low-income societies across a wide range of political, social, and religious norms and beliefs. The two most significant dimensions that emerged reflected, first, a polarization between *traditional* and *secular-rational* orientations toward authority and, second, a polarization between *survival* and *self-expres-*

sion values. By *traditional* we mean those societies that are relatively authoritarian, place strong emphasis on religion, and exhibit a mainstream version of preindustrial values such as an emphasis on male dominance in economic and political life, respect for authority, and relatively low levels of tolerance for abortion and divorce. Advanced societies, or *secular-rational*, tend to have the opposite characteristics.

A central component of the survival vs. self-expression dimension involves the polarization between materialist and postmaterialist values. Massive evidence indicates that a cultural shift throughout advanced industrial society is emerging among generations who have grown up taking survival for granted. Values among this group emphasize environmental protection, the women's movement, and rising demand for participation in decision making in economic and political life. During the past 25 years, these values have become increasingly widespread in almost all advanced industrial societies for which extensive time-series evidence is available.

Economic development brings with it sweeping cultural change, some modernization theorists tell us. Others argue that cultural values are enduring and exert more influence on society than does economic change. Who's right?

One goal of the World Values Survey is to study links between economic development and changes in values. A key question that we ask is whether the globalization of the economy will necessarily produce a homogenization (or, more specifically, an Americanization) of culture—a so-called "McWorld."

In the nineteenth century, modernization theorists such as Karl Marx and Friedrich Nietzsche made bold predictions about the future of industrial society, such as the rise of labor and the decline of religion. In the twentieth century, non-Western societies were expected to abandon their traditional cultures and assimilate the technologically and morally "superior" ways of the West.

Clearly now, at the start of the twenty-first century, we need to rethink "modernization." Few people today anticipate a proletarian revolution, and non-western societies such as East Asia have surpassed their Western role models in key aspects of modernization, such as rates of economic growth. And few observers today attribute moral superiority to the West.

Two Dimensions of Cross-Cultural Variation

1. Traditional vs. Secular-Rational Values
   Traditional values emphasize the following:
   - God is very important in respondent's life.
   - Respondent believes it is more important for a child to learn obedience and religious faith than independence and determination.
   - Respondent believes abortion is never justifiable.
   - Respondent has strong sense of national pride.
   - Respondent favors more respect for authority.
   Secular-Rational values emphasize the opposite.
2. Survival vs. Self-Expression Values
   Survival values emphasize the following:
   - Respondent gives priority to economic and physical security over self-expression and quality of life.
   - Respondent describes self as not very happy.
   - Respondent has not signed and would not sign a petition.
   - Respondent believes homosexuality is never justifiable.
   - Respondent believes you have to be very careful about trusting people.
   Self-Expression values emphasize the opposite.

Source: World Values Survey (http://wvs.isr.umich.edu)

On the other hand, one core concept of modernization theory still seems valid: Industrialization produces pervasive social and cultural consequences, such as rising educational levels, shifting attitudes toward authority, broader political participation, declining fertility rates, and changing gender roles. On the basis of the World Values Surveys, we believe that economic development has systematic and, to some extent, predictable cultural and political consequences. Once a society has embarked on industrialization—the central element of the modernization process—certain changes are highly likely to occur. But economic development is not the *only* force at work.

In the past few decades, modernization has become associated with *post*-industrialization: the rise of the knowledge and service-oriented economy. These changes in the nature of work had major political and cultural consequences, too. Rather than growing more materialistic with increased prosperity, postindustrial societies are experiencing an increasing emphasis on quality-of-life issues, environmental protection, and self-expression.

While industrialization increased human dominance over the environment—and consequently created a dwindling role for religious belief—the emergence of postindustrial society is stimulating further evolution of prevailing world views in a different direction. Life in postindustrial societies centers on services rather than material objects, and more effort is focused on communicating and processing information. Most people spend their productive hours dealing with other people and symbols.

Thus, the rise of postindustrial society leads to a growing emphasis on self-expression. Today's unprecedented wealth in advanced societies means an increasing share of the population grows up taking survival for granted. Their value priorities shift from an overwhelming emphasis on economic and physical security toward an increasing emphasis on subjective well-being and quality of life. "Modernization," thus, is not linear—it moves in new directions.

How Values Shape Culture

Different societies follow different trajectories even when they are subjected to the same forces of economic development, in part because situation-specific factors, such as a society's cultural heritage, also shape how a particular society develops. Recently, Samuel Huntington, author of *The Clash of Civilizations* (Simon & Schuster, 1996), has focused on the role of religion in shaping the world's eight major civilizations or "cultural zones": Western Christianity, Orthodox, Islam, Confucian, Japanese, Hindu, African, and Latin American. These zones were shaped by religious traditions that are still powerful today, despite the forces of modernization.

Distinctive cultural zones persist two centuries after the industrial revolution began.

Other scholars observe other distinctive cultural traits that endure over long periods of time and continue to shape a society's political and economic performance. For example, the regions of Italy in which democratic institutions function most successfully today are those in which civil society was relatively well developed in the nineteenth century and even earlier, as Robert Putnam notes in *Making Democracy Work* (Princeton University Press, 1993). And a cultural heritage of "low trust" puts a society at a competitive disadvantage in global markets because it is less able to develop large and complex social institutions, Francis Fukuyama argues in *Trust: The Social Virtues and the Creation of Prosperity* (Free Press, 1995).

The impression that we are moving toward a uniform "McWorld" is partly an illusion. The seemingly identical McDonald's restaurants that have spread throughout the world actually have different social meanings and fulfill dif-

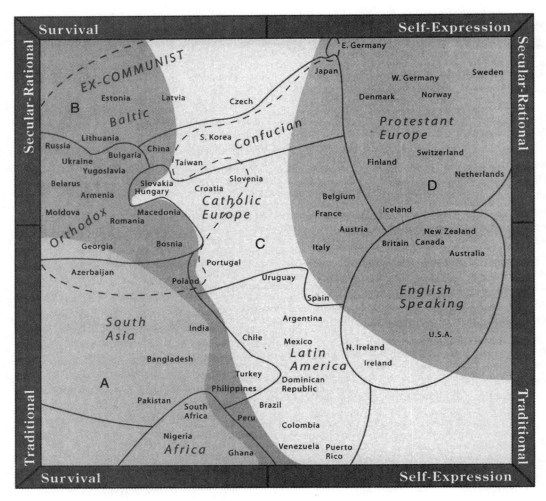

Less than $2,000 GNP per capita A ▨ ▨ C $5,000 to $15,000 GNP per capita

$2,000 to $5,000 GNP per capita B ▨ ▨ D More than $15,000 GNP per capita

ferent social functions in different cultural zones. Eating in a McDonald's restaurant in Japan is a different social experience from eating in one in the United States, Europe, or China.

Likewise, the globalization of communication is unmistakable, but its effects may be overestimated. It is certainly apparent that young people around the world are wearing jeans and listening to U.S. pop music; what is less apparent is the persistence of underlying value differences.

Mapping and Predicting Values

Using the 1995–1998 wave of the World Values Survey, we produced a map of the world's values, showing the locations of 65 societies on the two cross-cultural dimensions—traditional vs. secular-rational values and survival vs. self-expression values.

What the map shows us is that cross-cultural variation is highly constrained. That is, if the people of a given society place a strong emphasis on religion, that society's relative position on many other variables can be predicted—such as attitudes toward abortion, national pride, respect for authority, and child-drearing. Similarly, survival vs. self-expression values reflect wide-ranging but tightly correlated clusters of values: Materialistic (survival-oriented) societies can be predicted to value maintaining order and fighting inflation, while postmaterialistic (self-expression-oriented) societies can be predicted to value freedom, interpersonal trust, and tolerance of outgroups.

Economic development seems to have a powerful impact on cultural values: The value systems of rich countries differ systematically from those of poor countries. If we superimpose an income "map" over the values map, we see that all 19 societies with an annual per capita GNP of over $15,000 rank relatively high on both dimensions, placing them in the upper right-hand corner. This economic zone cuts across the boundaries of the Protestant, ex-Communist, Confucian, Catholic, and English-speaking cultural zones.

On the other hand, all societies with per capita GNPs below $2,000 fall into a cluster at the lower left of the map, in an economic zone that cuts across the African, South Asian, ex-Communist, and Orthodox cultural zones. The remaining societies fall into two intermediate cultural-economic zones. Economic development seems to move societies in a common direction, regardless of their cultural heritage. Nevertheless, distinctive cultural zones persist two centuries after the industrial revolution began.

Of course, per capita GNP is only one indicator of a society's level of economic development. Another might be the percentage of the labor force engaged in the agricultural sector, the industrial sector, or the service sector. The shift from an agrarian mode of production to industrial production seems to bring with it a shift from traditional values toward increasing rationalization and secularization.

But a society's cultural heritage also plays a role: All four of the Confucian-influenced societies (China, Taiwan, South Korea, and Japan) have relatively secular values, regardless of the proportion of their labor forces in the industrial sector. Conversely, the historically Roman Catholic societies (e.g., Italy, Portugal, and Spain) display relatively traditional values when compared with Confucian or ex-Communist societies with the same proportion of industrial workers. And virtually all of the historically Protestant societies (e.g., West Germany, Denmark, Norway, and Sweden) rank higher on the survival/self-expression dimension than do all of the historically Roman Catholic societies, regardless of the extent to which their labor forces are engaged in the service sector.

We can conclude from this that changes in GNP and occupational structure have important influences on prevailing world views, but traditional cultural influences persist.

Religious traditions appear to have had an enduring impact on the contemporary value systems of the 65 societies. But a society's culture reflects its entire historical heritage. A central historical event of the twentieth century was the rise and fall of a Communist empire that once ruled one-third of the world's population. Communism left a clear imprint on the value systems of those who lived under it. East Germany remains culturally close to West Germany despite four decades of Communist

rule, but its value system has been drawn toward the Communist zone. And although China is a member of the Confucian zone, it also falls within a broad Communist-influenced zone. Similarly, Azerbaijan, though part of the Islamic cluster, also falls within the Communist superzone that dominated it for decades.

The Deviant U.S.

The World Value Map clearly shows that the United States is a deviant case. We do not believe it is a prototype of cultural modernization for other societies to follow, as some postwar modernization theorists have naively assumed. The United States has a much more traditional value system than any other advanced industrial society.

On the traditional/secular-rational dimension, the United States ranks far below other rich societies, with levels of religiosity and national pride comparable to those found in developing societies. The United States does rank among the most advanced societies along the survival/self-expression dimension, but even here it does not lead the world. The Swedes and the Dutch seem closer to the cutting edge of cultural change than do the Americans.

Modernization theory implies that as societies develop economically their cultures tend to shift in a predictable direction. Our data supports this prediction. Economic differences are linked with large and pervasive cultural differences. But we find clear evidence of the influence of long-established cultural zones.

Do these cultural clusters simply reflect economic differences? For example, do the societies of Protestant Europe have similar values simply because they are rich? No. The impact of a society's historical-cultural heritage persists when we control for GDP per capita and the structure of the labor force. On a value such as *interpersonal trust* (a variable on the surival / self-expression dimension), even rich Catholic societies rank lower than rich Protestant ones.

Within a given society, however, Catholics rank about as high on *interpersonal trust* as do Protestants. The shared historical experience of given nations, not individual personality, is crucial. Once established, the cross-cultural differences linked with religion have become part of a national culture that is transmitted by the educational institutions and mass media of given societies to the people of that nation. Despite globalization, the nation remains a key unit of shared experience, and its educational and cultural institutions shape the values of almost everyone in that society.

The Persistence of Religious and Spiritual Beliefs

As a society shifts from an agrarian to an industrial economy and survival comes to be taken for granted, traditional religious beliefs tend to decline. Nevertheless, as the twenty-first century opens, cleavages along religious lines remain strong. Why has religion been so slow to disappear?

History has taken an ironic turn: Communist-style industrialization was especially favorable to secularization, but the collapse of Communism has given rise to pervasive insecurity—

and a return to religious beliefs. Five of the seven ex-Communist societies for which we have time-series data show rising church attendance.

Throughout advanced industrial societies we see two contrasting trends: the decline of attendance at religious services on the one hand, and on the other the persistence of religious beliefs and the rise of spirituality. The need for answers to spiritual questions such as why we are here and where we are going does not die out in postindustrial society. Spiritual concerns will probably always be part of the human outlook. In fact, in the three successive waves of the World Values Survey, concern for the meaning and purpose of life became *stronger* in most advanced industrial societies.

Conclusion: Whither Modernization?

Economic development is associated with pervasive, and to an extent predictable, cultural changes. Industrialization promotes a shift from traditional to secular-rational values; postindustrialization promotes a shift toward more trust, tolerance, and emphasis on well-being. Economic collapse propels societies in the opposite direction.

Economic development tends to push societies in a common direction, but rather than converging they seem to move along paths shaped by their cultural heritages. Therefore, we doubt that the forces of modernization will produce a homogenized world culture in the foreseeable future.

Certainly it is misleading to view cultural change as "Americanization." Industrializing societies in general are not becoming like the United States. In fact, the United States seems to be a deviant case: Its people hold much more traditional values and beliefs than do those in any other equally prosperous society. If any societies exemplify the cutting edge of cultural change, it would be the Nordic countries.

Finally, modernization is probabilistic, not deterministic. Economic development tends to transform a given society in a predictable direction, but the process and path are not inevitable. Many factors are involved, so any prediction must be contingent on the historical and cultural context of the society in question.

Nevertheless, the central prediction of modernization theory finds broad support: Economic development is associated with major changes in prevailing values and beliefs. The world views of rich societies differ markedly from those of poor societies. This does not necessarily imply cultural convergence, but it does predict the general direction of cultural change and (insofar as the process is based on intergenerational population replacement) even gives some idea of the rate at which such change is likely to occur.

In short, economic development will cause shifts in the values of people in developing nations, but it will not produce a uniform global culture. The future may *look* like McWorld, but it won't feel like one.

Modernization and McDonald's

McDonald's restaurants have become a dominant symbol of the globalization of the economy and target of the wrath of globalization's many opponents. But local values still wield great influence on culture, so don't look for McWorld to emerge anytime soon, say social researchers Ronald Inglehart and Wayne E. Baker.

About the Authors

Ronald Inglehart is professor of political science and program director at the Institute for Social Research, University of Michigan, Ann Arbor, Michigan 48106. E-mail RFI@umich.edu. The World Values Survey Web site is http://wvs.isr.umich.edu/.

Wayne E. Baker is professor of organizational behavior and director of the Center for Society and Economy, University of Michigan Business School, and faculty associate at the Institute for Social Research. He may be reached by e-mail at wayneb@umich.edu; his Web site is www.bus.umich.edu/cse.

This article draws on their paper "Modernization, Cultural Change, and the Persistence of Traditional Values" in the *American Sociological Review* (February 2000).

Originally published in the March/April 2001 issue of *The Futurist*, pp. 16-21. Used with permission from the World Future Society, 7910 Woodmont Avenue, Suite 450, Bethesda, Maryland 20814. Telephone: 310/656-8274; Fax: 301/951-0394; (http://www.wfs.org).

Index

Index

Test Your Knowledge Form

We encourage you to photocopy and use this page as a tool to assess how the articles in *Annual Editions* expand on the information in your textbook. By reflecting on the articles you will gain enhanced text information. You can also access this useful form on a product's book support Web site at *http://www.dushkin.com/online/*.

NAME: DATE:

TITLE AND NUMBER OF ARTICLE:

BRIEFLY STATE THE MAIN IDEA OF THIS ARTICLE:

LIST THREE IMPORTANT FACTS THAT THE AUTHOR USES TO SUPPORT THE MAIN IDEA:

WHAT INFORMATION OR IDEAS DISCUSSED IN THIS ARTICLE ARE ALSO DISCUSSED IN YOUR TEXTBOOK OR OTHER READINGS THAT YOU HAVE DONE? LIST THE TEXTBOOK CHAPTERS AND PAGE NUMBERS:

LIST ANY EXAMPLES OF BIAS OR FAULTY REASONING THAT YOU FOUND IN THE ARTICLE:

LIST ANY NEW TERMS/CONCEPTS THAT WERE DISCUSSED IN THE ARTICLE, AND WRITE A SHORT DEFINITION:

We Want Your Advice

ANNUAL EDITIONS revisions depend on two major opinion sources: one is our Advisory Board, listed in the front of this volume, which works with us in scanning the thousands of articles published in the public press each year; the other is you—the person actually using the book. Please help us and the users of the next edition by completing the prepaid article rating form on this page and returning it to us. Thank you for your help!

ANNUAL EDITIONS: Global Issues 03/04

ARTICLE RATING FORM

Here is an opportunity for you to have direct input into the next revision of this volume.
We would like you to rate each of the articles listed below, using the following scale:

1. **Excellent: should definitely be retained**
2. **Above average: should probably be retained**
3. **Below average: should probably be deleted**
4. **Poor: should definitely be deleted**

Your ratings will play a vital part in the next revision.
Please mail this prepaid form to us as soon as possible.
Thanks for your help!

RATING	ARTICLE
	1. A Special Moment in History
	2. Clash of Globalizations
	3. A New Grand Strategy
	4. Mr. Order Meets Mr. Chaos
	5. The Big Crunch
	6. Breaking *Out* or Breaking *Down*
	7. Bittersweet Harvest: The Debate Over Genetically Modified Crops
	8. The Challenges We Face
	9. The Heat Is On
	10. We *Can* Build a Sustainable Economy
	11. The Complexities and Contradictions of Globalization
	12. Dueling Globalizations: A Debate Between Thomas L. Friedman and Ignacio Ramonet
	13. Will Globalization Go Bankrupt?
	14. America's Two-Front Economic Conflict
	15. What's Wrong With This Picture?
	16. Overcoming Japan's China Syndrome
	17. Leasing the Rain
	18. Going Cheap
	19. The Reluctant Imperialist: Terrorism, Failed States, and the Case for American Empire
	20. Nasty, Brutish, and Long: America's War on Terrorism
	21. Nuclear Nightmares
	22. "Why Do They Hate Us?"
	23. India, Pakistan, and the Bomb
	24. China as Number One
	25. Battlefield: Space
	26. Strategies for World Peace: The View of the UN Secretary-General
	27. Justice Goes Global
	28. Meet the World's Top Cop
	29. The New Containment: An Alliance Against Nuclear Terrorism
	30. Countdown to Eradication
	31. Aerial War Against Disease: Satellite Tracking of Epidemics Is Soaring

RATING	ARTICLE
	32. Are Human Rights Universal?
	33. The Grameen Bank
	34. Why Environmental Ethics Matters to International Relations
	35. Women Waging Peace
	36. The Next Christianity
	37. Modernization's Challenge to Traditional Values: Who's Afraid of Ronald McDonald?

(Continued on next page)

221

ABOUT YOU

Name _____   Date _____

Are you a teacher? ☐   A student? ☐
Your school's name _____

Department _____

Address _____   City _____   State _____   Zip _____

School telephone # _____

YOUR COMMENTS ARE IMPORTANT TO US!

Please fill in the following information:
For which course did you use this book?
_____

Did you use a text with this ANNUAL EDITION?   ☐ yes ☐ no
What was the title of the text?
_____

What are your general reactions to the *Annual Editions* concept?
_____

Have you read any pertinent articles recently that you think should be included in the next edition? Explain.
_____

Are there any articles that you feel should be replaced in the next edition? Why?
_____

Are there any World Wide Web sites that you feel should be included in the next edition? Please annotate.
_____

May we contact you for editorial input?   ☐ yes ☐ no
May we quote your comments? ☐ yes ☐ no